Chloroplasts

CELLULAR ORGANELLES

Series Editor: PHILIP SIEKEVITZ
Rockefeller University, New York, New York

MITOCHONDRIA
Alexander Tzagoloff

CHLOROPLASTS
J. Kenneth Hoober

Chloroplasts

J. Kenneth Hoober
Temple University School of Medicine
Philadelphia, Pennsylvania

Plenum Press · New York and London

Library of Congress Cataloging in Publication Data

Hoober, J. Kenneth, 1938–
    Chloroplasts.

    (Cellular organelles)
    Includes bibliographies and index.
    1. Chloroplasts. I. Title. II. Series.
QK725.H597   1984   581.87′33   84-9934
ISBN 0-306-41643-3
ISBN 0-306-41686-7 (pbk.)

©1984 Plenum Press, New York
A Division of Plenum Publishing Corporation
233 Spring Street, New York, N.Y. 10013

Printed in the United States of America

Preface

This monograph is intended to provide an overview of the structure, function, and development of the chloroplast. It should be viewed as a beginning of the study of chloroplasts and not as an end. In keeping with an introductory approach, abbreviations generally have not been used, so that substance is not replaced by symbol.

The principal aim has been to provide a teaching tool to introduce students to the major characteristics of the chloroplast, with as much emphasis on mechanisms as possible at this level. It was written for students with an advanced college level education in biology and chemistry who also have some knowledge of biochemistry. The fundamentals of these subjects cannot be included in a book of this type. However, to provide a meaningful description of how the chloroplast works, i.e., what the mechanisms of photosynthetic reactions are, the subject must be dealt with at the molecular level. Living systems are chemical systems, and the importance of understanding these systems at the molecular level cannot be overstated. Therefore, although attempts were made to keep the chemistry at a relatively simple level, occasionally statements are made that can be understood only with a sufficient background knowledge of chemistry.

It is important for students to realize in broad outline form the functions of the chloroplast and where its functions fit into the scheme of life. It is important to appreciate the evolutionary as well as the biochemical development of the structure. Also important is an awareness of the history of research on photosynthesis. Given the vast amount of information available today, there is often too little time or, more unfortunately, interest in knowing where the information came from. Although a detailed history of research on photosynthesis cannot be covered in a small book such as this, a brief account of its first hundred years is included with the hope that the factors that gave this research its direction, and the pioneers who ventured into unexplored areas, will be appreciated.

This monograph has several features that should be pointed out to the reader. Most books of this type present the areas that are most interesting and important to the author, i.e., how he perceives the subject. This one is no exception. Emphasis is placed on those aspects that are particularly appealing to me. Therefore, the discussion is not complete, for which I apologize. Although I have tried to include the essentials of the subjects, there undoubtedly are errors and omissions because of my own lack of knowledge or recognition of importance. Advanced students may find this monograph disappointing in its depth of coverage. For such students the current literature has no substitute. This monograph cannot provide the detail found in recent in-depth reviews on specific subjects.

I wish to express my gratitude to the many investigators who allowed me to reprint their results and to those who generously provided micrographs. Special appreciation is expressed to Dr. Philip Siekevitz for providing the opportunity to undertake this task and for his expert editorial guidance.

J. Kenneth Hoober

Philadelphia

Contents

Chapter 4

The Process of Photosynthesis: The Light Reactions

Chapter 5

The Process of Photosynthesis: The Dark Reactions

Chapter 6
The Chloroplast Genome and Its Expression

Chapter 7
Development of Chloroplasts: Structure and Function

Chapter 8

Development of Chloroplasts: Biosynthetic Pathways and Regulation

Chapter 9

Evolutionary Aspects of Chloroplast Development

Chloroplasts

Introduction

The astrophysicists tell us that the universe was created about 16 billion years ago. The visible universe contains slightly more than 10^{50} metric tons (10^{56}g) of material. Since matter was created from energy ($M = E/c^2$), according to the current concept of the origin of the universe, rough calculations indicate that creation was initiated by an outburst of energy in excess of 10^{70} J. Our minds are unable to comprehend the magnitude of this quantity, yet every thoughtful person must ponder its source.

The primordial conversion of energy to matter produced only the smallest atoms, with hydrogen accounting for about 75% of the matter and the remainder nearly all helium. Hydrogen still constitutes most of the total mass of the universe. The heavier elements (e.g., carbon, oxygen, nitrogen, and iron) apparently are made by nuclear fusion reactions in stars and presently account for only about 0.1% of the total mass of the universe.

A few billion years after the initial event of creation, our own galaxy, the Milky Way, was formed. However, the earth was not formed until much later, about 4.5 billion years ago. It is not known with certainty how soon after it was formed the planet became hospitable for life. Perhaps only a few hundred million years were required for this, since it is becoming quite clear that events leading to the beginning of life happened relatively rapidly thereafter. Paleobiologists have found possible remnants of early life that are 3.5–3.8 billion years old. Considering the complexity required for a living organism, the development of life on earth seems to have occurred quite abruptly. Again, one must ponder over the veritable eruption of highly ordered living things out of a chaotic environment. Calculations based on information theory suggest that the appearance of life could not have been simply a random process, an inevitable consequence of chemical evolution. Even a simple, small protein could not have been formed by a random process, over the span of time during which life developed. Indeed, much of evolutionary history, with a flow to greater states of complexity against an apparent countercurrent of entropy, has a non-random character.

Photosynthesis, the most fundamental and necessary process in biology, developed over three billion years ago, only shortly after the appearance of the first living organisms. Without this development, i.e., the ability to synthesize organic material from carbon dioxide and water, the early forms of life may have expired shortly after their arrival as they consumed the supply of food in the early biosphere. The first photosynthetic process probably had characteristics similar to photosystem 1. The ability of photosynthetic organisms to produce oxygen was gained a little more than two billion years ago, when photosystem 2 developed. (See Chapter 4 for descriptions of the photosystems.) Although there is some evidence that the earliest appearance of oxygen was the result of photodissociation of water in the atmosphere, production of the oxygen-rich environment we currently enjoy undoubtedly was the work of these early photosynthetic organisms. Nevertheless, millions of years were required to oxidize ferrous salts and other reducing agents in the early environment before the concentration of free molecular oxygen reached significant concentrations.

Photosynthetic generation of oxygen had a tremendous impact on the subsequent course of development of living organisms. Because oxygen is toxic, as illustrated by the extreme sensitivity of anaerobic bacteria to molecular oxygen, its arrival provided a strong selective pressure to survival. If an organism was to continue living in the new aerobic environment, defense mechanisms against metabolic products of molecular oxygen were required. Yet, on the other hand, the availability of molecular oxygen also offered the opportunity for the development of oxidative metabolism, by which the energy yield from organic nutrients could be tremendously increased. This remarkably complex adaption to oxygen seems to have occurred without delay.

The earliest photosynthetic organisms were prokaryotes, analogous to present-day photosynthetic bacteria and cyanobacteria. A question that continues to be the subject of lively interest is where did the chloroplast, the photosynthetic organelle in eukaryotic cells, come from? The evolutionary lineage of living things may be more complicated than was thought only a few years ago. Instead of their ancestors being among the prokaryotes, eukaryotes may have had their own separate ancestral lineage, which was established more than three billion years ago. According to the fossil record, eukaryotic cells definitely were in existence about 1.5 billion years ago. Their arrival heralded the start of a lineage from which an amazing degree of complexity and variation has resulted. The primitive eukaryotes were apparently little more than cells with a defined, membrane-bound nucleus into which the chromosomes were collected.

Current evidence supports the hypothesis that the organelles involved in the metabolism of oxygen originated by multiple cellular fusions. Mitochondria may have arisen first, apparently as the result of symbiotic entrapment of aerobic bacteria in the cytoplasm of some protoeukaryotic cells. Between 1 and 2 billion years ago, development of the chloroplast was initiated by the fusion of these early eukaryotic cells with oxygen-evolving photosynthetic prokaryotes. From these combined cells have descended the modern eukaryotic algae and the higher plants. This scenario for the origin of chloroplasts was first proposed about a hundred years ago by A.F.W. Schimper, a German botanist, and is still the most favored conception of how this event occurred.

Photosynthesis not only produces oxygen, but also is the process by which the reserves of organic carbon are replenished (Fig. I.1). The chloroplast

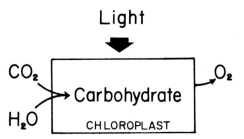

FIG. I.1. The overview of photosynthesis. Carbon dioxide and water are the chemical raw materials, and carbohydrates and oxygen are the main products. Since this process is thermodynamically highly unfavorable, a source of energy must be available to drive the system. The required energy is provided by light.

enzyme ribulose 1,5-bisphosphate carboxylase (see p. 113) is the principal catalyst for the fixation of carbon from CO_2 into carbohydrate material. This enzyme apparently appeared early during evolution, since it is found in the photosynthetic prokaryotes as well as within the chloroplasts of algae and higher plants. The principal products of organic photosynthesis are glucose and its polymers, starch and cellulose.

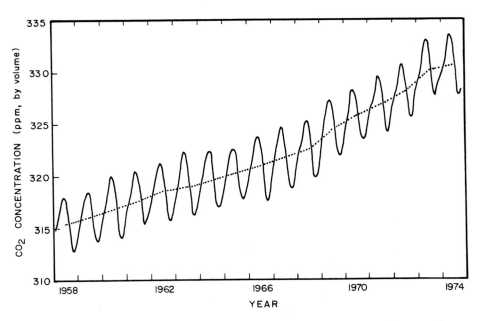

FIG. I.2. Seasonal variations and annual trends in the concentration of CO_2 in the atmosphere at Mauna Loa Observatory, Hawaii, for the period 1958–1974. The seasonal variations are produced by yearly photosynthetic cycles and other effects. Mean annual concentrations are well above the preindustrial levels of 290–300 ppm. (Data from Eckdahl and Keeling, 1973. Also given in Woodwell *et al.*, 1978.)

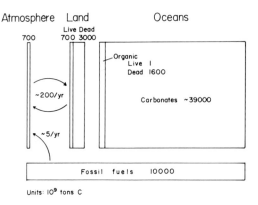

FIG. I.3. Distribution of carbon on the earth and in the atmosphere. (From Woodwell *et al.*, 1978, and Stuiver 1978).

The importance of photosynthesis can be illustrated by considering the magnitude of the process. The flux of carbon from atmospheric CO_2 into organic compounds by green plants on land and in the surface waters of the oceans is about 200×10^9 metric tons/year. Currently, the rate of flux of CO_2 in the opposite direction from respiration, fires, and decay is slightly higher (Fig. I.2). But it is significant to relate these rates to the total stores of exchangeable carbon. The present concentration of CO_2 in the atmosphere (330 ppm by volume or 0.033%) provides a reservoir of 700×10^9 tons (Fig. I.3). There is slightly more carbon in living things (primarily plant life), which is estimated at $700-800 \times 10^9$ tons. Therefore, a significant portion of the readily exchangeable carbon in the biosphere is cycled each year through photosynthetic organisms. [The stores of carbon as humus in soil (3×10^{12} tons), as fossil fuels on land (10×10^{12} tons), and, in particular, as carbonates in the deep oceans (39×10^{12} tons) are much greater. But the exchange of carbon between these stores and the biosphere is relatively slow.]

The driving force for the process of photosynthesis is the gradual release, by our sun, of some of the energy that was initially converted to hydrogen during the early stages of the universe. The thermonuclear fusion reaction in stars, in which hydrogen is converted to helium, consumes only 0.8% of the original mass of the hydrogen. Yet each day 36×10^{10} tons of mass are converted to energy within our own star, the sun. The energy from the sun striking the earth annually is 3×10^{24} J, but less than 1% of this energy is actually captured and converted to a useful form. Although up to 7% of the impinging solar energy can be captured by photosynthesis, under optimal conditions, not all the earth's surface is covered by a thick growth of plant life.

We exist in an energy-rich environment, yet have the increasingly difficult problem of maintaining a readily available energy supply. Only photosynthetic organisms and in particular the organelle within plant cells—the chloroplast— efficiently harvest the abundant energy coming from the sun. It is important for us to understand this organelle, for it is the provider of our food, our warmth, and our prosperity and indeed is the sustainer of life itself.

Additional Reading

Baes, C. F., Jr., Goeller, H. E., Olson, J. S., and Rotty, R. M. (1977) Carbon dioxide and climate: The uncontrolled experiment, *Am. Sci.* **65**:310–320.

Barrow, J. D., and Silk, J. (1980) The structure of the early universe, *Sci. Am.* **242**(4):118–128.

Bolin, B. (1977) Changes of land biota and their importance for the carbon cycle, *Science* **196**:613–615.

Dickerson, R. E. (1978) Chemical evolution and the origin of life, *Sci. Am.* **239**(3):70–86.

Doolittle, W. F. (1980) Revolutionary concepts in evolutionary cell biology, *Trends Biochem Sci.* **5**:146–149.

Eckdahl, C. A., and Keeling, C. D. (1973) in *Carbon and the Biosphere* (G. M. Woodwell and E. V. Pecan, eds.) AEC Technical Information Center, Washington, D. C.

Fredrick, J. F., ed. (1981) Conference on origins and evolution of eukaryotic intracellular organelles, *Ann. N. Y. Acad. Sci.* **361**.

Hall, D. O. (1976) Photobiological energy conversion, *FEBS Lett.* **64**:6–16.

Javoy, M., Pineau, F., and Allègre, C. J. (1982) Carbon geodynamic cycle, *Nature (London)* **300**:171–173 (appended *Nature* **303**:730–731).

Penzias, A. A. (1980) Nuclear processing and isotopes in the galaxy, *Science* **208**:663–669.

Schimper, A. F. W. (1883) Uber die Entwickelung der Chlorophyllkörner und Farbkörper, *Botanische Zeitung* **41**:105–114.

Schopf, J. W. (1978) The evolution of the earliest cells, *Sci. Am.* **239**(3):111–138.

Schwarz, R. M., and Dayhoff, M. O. (1978) Origins of prokaryotes, eukaroytes, mitochondria and chloroplasts, *Science* **199**:395–403.

Stuiver, M. (1978) Atmospheric carbon dioxide and carbon reservoir changes, *Science* **199**:253–258.

Woodwell, G. M., Whittaker, R. H., Reiners, W. A., Likens, G. E., Delwiche, C. C., and Botkin, D. B. (1978) The biota and the world carbon budget, *Science* **199**:141–146.

Yockey, H. P. (1977) A calculation of the probability of spontaneous biogenesis by information theory, *J. Theoret. Biol.* **76**:377–398.

Historical Perspectives
The Beginnings of Research on Photosynthesis

I. The Discovery of Chloroplasts

The human eye cannot discriminate two points that are less than about 100 μm (0.1 mm) apart. Most chloroplasts measure in their long dimension 4–6 μm. Therefore, the discovery of chloroplasts had to await the development of a microscope capable of magnifying an object greater than 20 diameters.

The earliest compound microscope apparently was built late in the 16th century by Hans and Zaccharius Janssen of Holland. By the latter half of the 17th century, remarkable advances had occurred in the technology of making lenses and building microscopes. Robert Hooke (1635–1703), an English physicist, developed a compound microscope capable of magnifying an object about 30 diameters. In 1665, while studying the microscopic appearance of thinly sliced cork, Hooke observed the arrangement of the vestigial cell walls of the plant tissue. In describing his observations, he was the first to put on record the term **cell.** Shortly thereafter, Nehemiah Grew (1641–1712), an English physician and contemporary of Hooke, carried out a detailed investigation of the microanatomy of plants. Grew, using the microscopes constructed by Hooke, noticed during his studies that the green matter in plant leaves had a particulate quality. Thus Grew was the first to observe chloroplasts.

The most remarkable accomplishments in microscopy during the 17th century, however, were achieved by Anton van Leeuwenhoek (1632–1723), a linen draper and town official in Delft, Holland. Leeuwenhoek constructed a number of simple, one-lens microscopes, some of which were capable of magnifications as high as several hundred times. With these microscopes, and possibly the discovery of dark-field illumination,* Leeuwenhoek was the first to see single-celled organisms. Among his remarkable achievements during the period 1675–1685 were the discoveries of protozoa, algae, and bacteria. Leeuwenhoek's remarkably productive study of microscopic life spanned a period of 50 years.

---

*L. E. Casida, Jr., (1976) suggests that Leeuwenhoek may have discovered dark-field illumination, which would have enhanced his ability to see moving organisms in samples.

His technical prowess in lens-making was unmatched in his time. More detailed observations of single-celled organisms were not made until nearly 200 years later.

Louis Joblot (1645–1723), a French physicist and contemporary of Leeuwenhoek, studied microscope design extensively. He constructed improved versions of both simple and compound microscopes. His observations of microscopic life resulted in the first published treatise specifically on microorganisms. In 1718 Joblet wrote:

> ... I have furnished several Microscopes to renowned Physicians, famous Anatomists, and to several other illustrious amateurs in these novelties, and it is almost impossible that each one of them, while pleasantly occupied with the examination of the surprising effects of nature, will not discover every day new and singular things, which they will be able to communicate to the Public. . . . (See Lechevalier, 1976)

However, development of microscopy as a research tool lagged because of technical problems. Most problematic were imperfections in the lens that behaved as prisms and gave colored fringes to an image. The first achromatic telescope lens was made by Chester Moor Hall, an Englishman, in 1733. But not until 1830 had progress in technology advanced sufficiently to make possible the construction of achromatic compound microscopes. The superior lenses as in the instruments developed in Germany by E. Abbe and O. Schott, eliminated much of the image distortion found with earlier models and permitted much higher magnifications. These microscopes were much easier to use than Leeuwenhoek's single-lens microscope but had comparable resolving power. Commercial production of these compound microscopes, which followed shortly, opened the microscopic world to all who cared to look. As a result of the available technology, significant developments in microscopic biology occurred in the years following 1830. In 1837 the German botanist Hugo von Mohl (1805–1872) provided the first definitive description of the chloroplast (*chlorophyllkörnen*) as discrete bodies within the green plant cell. Later, the name **plastid** was given to these bodies by another German botanist, A. F. W. Schimper, in 1883. By the end of the century the name **chloroplast** had become the commonly accepted term in reference to the green organelle within cells of plant tissue.

In 1883 Schimper also deduced from his observations of living cells that chloroplasts are capable of division. He proposed that plastids develop not *de novo* but by fission of existing plastids. He also extended his proposal to suggest that green plants possibly owe their origin to a symbiotic relationship between chlorophyll-containing and colorless organisms. This idea is still very much alive today.

One more notable cytological observation of the 19th century was the description and naming of **grana** (see Chapter 2) within chloroplasts by A. Meyer in 1883. With this accomplishment, the structural features of the chloroplast, resolvable with the light microscope, were described.

Yet by 1900 many of the finer details of chloroplast structure were still unknown. The next significant advances in the structural aspects were not made until nearly 80 years after Meyer reported his findings, when another major technological development was achieved. In the mid-20th century, the electron microscope was developed. This instrument permitted magnifications of up to and exceeding 100,000 diameters. With such powerful microscopes individual membranes and other structures, only several nanometers in thickness, were revealed. It was also learned just how extensive is the array of membranes within a chloroplast and how these membranes are organized to form the grana. Furthermore, it was discovered that the chloroplast is surrounded not by a single membrane but by a double-membrane envelope. Numerous small granules, the chloroplast ribosomes, as well as localized fibrous material, identified as DNA, were observed in the stroma.

The detailed knowledge of chloroplast structure is still expanding. Yet by 1900 the general features were known, and along with biochemical studies that had been in progress during the 19th century, the overall function of the chloroplast in photosynthesis was established.

II. Chemical Studies—The Development of Gas Chemistry

The discovery of photosynthesis emerged as an application of the growing knowledge of the chemistry of gases. From the days of Aristotle until the end of the 18th century, it was thought that plants obtain all their nutrients from the soil. This was perhaps understandable, since not until the 17th century did it become clear that "air" has substance. It is therefore obvious that the discovery of photosynthesis, which involves an exchange of gases, could not occur until knowledge of the properties of these invisible materials had been obtained and methods developed to measure them.

Of the three states of matter, the solid and liquid states were recognized already in ancient times. A concept of the third state of matter—the gaseous state—emerged slowly during the Middle Ages and was formalized only early in the 17th century by the Flemish chemist Jan Baptista von Helmont (1577–1644), a contempory of the English natural philosopher Francis Bacon (1561–1626). Van Helmont collected several gases and was the first to employ the term **gas.** Then, in 1654, the Prussian Otto von Guericke (1602–1686) demonstrated that air possesses weight and that a vacuum could be produced with an air pump he had invented. Shortly thereafter, in 1661 Robert Boyle (1627–1691), a British physicist, developed his law of gases ($PV = k$ at constant temperature). Boyle commented in 1673 that generally "men are so accustomed to judge of things by their senses that because air is invisible they ascribe but little to it, and think of it as but one remove from nothing." Boyle is perhaps the founder of the science of chemistry, since he introduced important ideas concerning chemical reactions and analysis and, in particular, advocated a rigorous experimental method for the testing of theories and for the establishment of the reliability of facts. Boyle also for the first time distinguished among elements, compounds, and mixtures.

Still, another century passed until the chemistry of gases came of age. A necessary condition for this new stage was the realization that "air" is more than a simple elementary substance. Van Helmont in the early 17th century had prepared carbon dioxide, which he called *gas sylvestre*, by the action of acids on chalk and by the combustion of charcoal. He distinguished this gas from common air, but had not realized that it was different from other gases that also would not support combustion.

In the early 18th century, Stephen Hales (1677–1761), an English theologian and pioneer in the study of animal and plant physiology, introduced a technique for collecting gases by displacement of water from a filled, inverted jar. Hales collected several gases, including carbon dioxide, but like van Helmont did not recognize them as separate substances. In 1754 Joseph Black (1728–1799), a physician and professor of chemistry at the University of Edinburgh, was the first to realize that the gas released by the action of acids on limestone ($CaCO_3$) was a distinct substance. He also prepared the gas by heating limestone and magnesia alba ($MgCO_3$), which he noticed lost weight as a result. Since Black found that carbon dioxide was "fixed" in the solid state as alkaline carbonates, he called it "fixed air." Lavoisier showed in 1781 that the gas was a compound of carbon and oxygen.

It had been known for a long time that the action of sulfuric acid on iron produced a gas. Paracelsus (Aureolus Philippus Theophrastus Bombastus von Hohenheim, (1493–1541), a Swiss physician, had studied this reaction. Later, van Helmont discovered that the gas given off by this reaction was combustible, yet it would not support combustion. But he did not distinguish it from other flammable gases. Chemists for the next 150 years included this gas with other combustible gases in the name "inflammable air." Henry Cavendish (1731–1810), the famous English physicist and chemist, took up the study and showed in 1766 that the gas was a distinct and definite substance. Cavendish therefore is given credit for the discovery of hydrogen, the name given to the gas in 1783 by Lavoisier, from the Greek words meaning "water-forming." Cavendish and Lavoisier independently discovered in 1783 that the combustion of hydrogen produces water. Thus, the chemical composition of water was established.

The discovery of oxygen, one of the most important advances in the history of chemistry, was made independently by the remarkable Swedish chemist Karl Wilhelm Scheele (1742–1786), between 1771 and 1773 and by the English theologian–historian and chemist Joseph Priestley (1733–1804) in 1774. Scheele produced oxygen, which he called "fire-air," by heating sulfates and phosphates of manganese, magnesium nitrate, and mercuric oxide. He also determined that "common air" was about one-fifth "fire-air" while the remainder was an inert gas. Scheele's work was not published until 1777, one year after Priestley reported his work, and thus Scheele was deprived the priority for the discovery of oxygen.

Priestley's interest in gases grew out of his studies on carbon dioxide produced during the fermentation process at a local brewery. His major early interest was how to make seltzer water. In 1773 Priestley was awarded the Copely Medal by the Royal Society for the supposed health benefits (such as the prevention of sea scurvey) of his artifically carbonated water. This was undoubtedly one of his lesser contributions.

Priestley's experiments on oxygen began by heating the red powder mercuric oxide with sunlight focused by a large lens. On August 1, 1774, after collecting the gas that was expelled, Priestley noted, " . . . what surprised me more than I can well express, was, that a candle burned in this air with a remarkably vigorous flame." This gas also permitted a mouse to live longer than when placed into an equal volume of air. Priestley called the gas "dephlogisticated air," but this gas was renamed oxygen by Lavoisier in 1777. The name *oxygène*, from Greek words meaning "acid-producer," was somewhat of a misnomer, since Lavoisier initially thought that *oxygène* was a necessary ingredient of acids. In fact, what he had observed was that the combustion of sulfur and phosphorus in this gas yielded acidic products.

Actually, oxygen had been prepared by similar means long before Scheele and Priestley's work was done. Eck de Sultzbach carried out the same reaction with mercuric oxide in 1489, but he was probably more impressed with the nature of the elemental mercury ("quicksilver") produced than with the gas given off. Had he collected the "spirit" that he observed also was produced and named it, he may have been credited with its discovery. Other investigators, including Stephen Hales in 1727, collected the gas produced by heating saltpeter (KNO_3). P. Payen also obtained it in the same year as Priestley. But in these cases, no attempts were made to characterize the specific chemical properties of the gas.

Priestley developed an ingenious assay for oxygen based on its reaction with "nitrous air" (nitric oxide), which he also discovered in 1772. When added to other "airs" in inverted jars over water, nitric oxide would cause first a decrease in the volume of the gaseous mixture until the oxygen was consumed ($2NO + O_2 \rightarrow 2NO_2$; NO_2 dissolves into the water, forming nitric and nitrous acids). The subsequent addition of NO to the oxygen-free mixture would then increase the total volume of the gas phase. By this titration assay, he observed that the "goodness" of pure oxygen was about five times greater than that of "common air." In other words, Priestley also had found that air contains about 20% oxygen.

In the year 1774–1794 a revolution occurred in chemical theory, largely as the result of the work of one man, the remarkable French chemist Antoine Laurent Lavoisier (1743–1794). He placed chemical reactions on a quantitative basis. Shortly after the properties of oxygen were communicated to him in 1774 by Priestley, Lavoisier was the first to recognize the chemical basis of oxidation, and he quantitated the oxidations of a number of elements known at the time, particularly those of phosphorus, sulfur, and mercury. As an example of his experiments, Lavoisier placed a fixed weight of mercury into an enclosure filled with air. When the apparatus was warmed over a period of 12 days the oxygen was removed from the air. Concomitantly, the volume of the gaseous phase was reduced by one-fifth and the mercury was converted to a red powder, mercuric oxide. The air at this stage would not support a candle's flame. Lavoisier named the remaining gas *azote* (from Greek words meaning "not life-supporting"), still the French word for nitrogen. But when the red powder was heated to a higher temperature, as in Priestley's experiment, the mercuric oxide decomposed and liberated a volume of gas (oxygen) equal to the initially observed decrease. By measuring the weight of mercury at each stage, Lavoisier was able to measure first an increase in weight as the mercury combined with oxygen and later a decrease as the mercuric oxide decomposed. He also observed that phosphorus and sulfur gained weight when burned. Lavoisier was able thereby to explain chemically and quantitatively the phenomenon of oxidation and proposed his theory on this subject in 1777. Unfortunately, his work was cut short when he was beheaded on May 8, 1794, a casualty of the political vicissitudes of the French Revolution.

III. The Phlogiston Theory

Before the discovery of oxygen, chemists were puzzled by the process of combustion. Geber (Abu Musa Jabir ibn Hayyan al Sufi), an Arabian alchemist, had taught in 776 that combustible substances burn because they contain the "principle of inflammability." Geber equated this principle with sulfur, but other investigators showed later that substances lacking sulfur also burned. Georg Ernst Stahl (1660–1734), a German physician, at the beginning of the 18th century combined these ideas, along with those formulated by Johann Joachin Becher (1635–1682), into his Phlogiston Theory. Stahl proposed that "phlogiston" (from the Greek, meaning "to set on fire") was the principle of combustibility, which was known only by its effects. Only those materials containing phlogiston were thought capable of burning. And those that burned readily, such as carbon and sulfur, contained more of the principle than substances that burned poorly. According to the theory, as combustion proceeded, phlogiston was released into the air to yield "phlogisticated air." The conversion of a metal into a calx (now called an oxide) also involved the escape of phlogiston. The function given for air, which was known to be necessary, was that of a vehicle or a space to accommodate the escaping phlogiston. Air saturated with the principle, or "phlogisticated air," would no longer support combustion. In this context, when Priestley discovered gaseous oxygen it was logical for him to call it "dephlogisticated air."

The phlogiston theory eventually was disproven when it was quantitatively tested. It had been known since 1630, from the observations of Jean Rey (c.1582–c.1645), that a calx is heavier than the metal from which it was derived. The staunch phlogistonists tried to explain this by suggesting that phlogiston possessed negative mass. The final demise of the theory was the work of Lavoisier. His quantitative studies of the process of oxidation established that during combustion, substances gained oxygen rather than lost phlogiston. As Lavoisier in 1783 rather caustically put it, "Chemists have made phlogiston a vague principle, which is not strictly defined and which consequently fits all the explanation demands of it. . . . It is a veritable Proteus that changes its form every instant!" But the early chemists accepted, in the absence of specific knowledge, the existence of "principles." And the phlogiston theory served a valuable purpose by making it possible to organize chemical phenomena in a logical manner. Both Priestley and Cavendish remained believers in the theory even after the properties of oxygen were known.

IV. The Discovery of Photosynthesis

Upon the background of the development of experimental chemistry and physics, particularly that of gases, the discovery of photosynthesis emerged. This discovery required the recognition not only of gas exchange but also of the fact that plants obtain *food* from the air. This concept was particularly difficult to accept. However, already in the early 1600s an experiment had been done that clearly discredited the soil as the major source of plant food. Jan Baptista van Helmont grew a willow tree over a period of five years from a weight of 2 kg to 75 kg (not counting the weight of the leaves that fell over four autumns) in a tub of soil, to which was added only water. At the end of the experiment, the tub of soil had lost less than 0.1 kg in weight. Van Helmont mistakenly concluded that growth of the plant was produced entirely by the water he added to the tub. Although a reasonable conclusion, considering the knowledge of the times, it was still wrong. The significance of this experiment was missed until nearly 200 years later, when the chemical composition of water was determined by Cavendish and Lavoisier in 1783.

In the meantime, Stephen Hales in 1727 reported that gases were both absorbed and evolved by animal and plant tissues. What these gases were he did not know. Hales surmised that "plants very probably draw through their leaves some part of their nourishment from the air." He continued, " . . . may not light also . . . contribute much to ennobling the principles of vegetables?" Hales had observed that plants take up water by their roots and transpire it from their leaves, and he thought that perhaps the reverse was also possible. But such statements concerning plants taking food from the air were clearly ahead of their time, for the evidence to support these statements would not be forthcoming for another 50 years.

The initial discovery related to photosynthesis was made in 1771, assayed with the flame of a candle. In that year, as part of his studies on carbon dioxide, Joseph Priestley performed a simple but remarkable experiment. Since he knew that a mouse died quickly when placed into a jar filled with "fixed air," Priestley was prompted to test the effects of "fixed air" on plant life. Unexpectedly, a sprig of mint flourished in the jar! Furthermore, he found that within the jar of "fixed air," depleted of oxygen by burning a candle, the mint plant would restore the air's ability to support a candle's flame. Priestley wrote (in 1772):

> I have been so happy as by accident to hit upon a method of restoring air which has been injured by the burning of candles and to have discovered at least one of the restoratives which Nature employs for this purpose. It is *vegetation*. One might have imagined that since common air is necessary to vegetable as well as to animal life, both plants and animals had affected it in the same manner; and I own that I had that expectation when I first put a sprig of mint into a glass jar standing inverted in a vessel of water; but when it had continued growing there for some months, I found that the air would neither extinguish a candle, nor was it at all inconvenient to a mouse which I put into it.
>
> Finding that candles would burn very well in air in which plants had grown a long time . . . I thought it was possible that plants might also restore the air which had been injured by the burning of candles. Accordingly, on the 17th of August 1771 I put a sprig of mint into a quantity of air in which a wax candle had burned out and found that on the 27th of the same month another candle burnt perfectly well in it.

Priestley concluded that plants had the ability to absorb phlogiston from the "fixed air," yielding purified or "dephlogisticated air." This was a classic example of the role of serendipity in scientific discovery. But as time went on, Priestley apparently did not realize the relationship between this ability of plants to restore air and his work on mercuric oxide, or the fact that both processes *added* something to the air. Nor did he fully realize the role of light in the process, a problem that later led to some confusion when he could not always repeat the experiment.

The situation was clarified in 1779 by approximately 500 experiments done during the summer months of that year by the Dutchman Jan Ingen-Housz (1730–1799), the physician to the court of Empress Maria Theresa of Austria, while he was on "vacation" in England. Ingen-Housz established the role of light in the restoration of air by plants. He also established, by observing the formation of bubbles on submerged, illuminated leaves, that plants produce oxygen. Although this observation had been made by others, its significance had been overlooked. Ingen-Housz showed that only the green parts of plants have the ability to restore air, an observation that was confirmed by Priestley in 1781. Later, the overall process of photosynthesis was first recognized by Ingen-Housz, who proposed in 1796 that plants acquire carbon from "carbonic acid" in the air and at the same time produce oxygen. Furthermore, he attributed both these capacities directly to the leaves of plants.

Jean Senebier (1742–1809) of Geneva initially showed in 1782 and 1783 that evolution of oxygen is facilitated by, or in some manner is dependent upon, the presence of carbon dioxide. Senebier discovered that plants absorb carbon dioxide and that the absorption of carbon dioxide and the evolution of oxygen were interrelated processes. But Senebier thought the carbon dioxide first dissolved into the soil water and reached the leaves via the roots.

It was not apparent that plants could derive all their carbon from carbon dioxide until the exchange of gases during photosynthesis was placed on a quantitative basis. In 1804, Nicolas Theodore de Saussure (1767–1845), another chemist from Geneva, found that the weight of a plant increased as it absorbed carbon dioxide. But when he carefully balanced the weight of carbon dioxide assimilated with that of the oxygen evolved, the plant had gained more weight than could be accounted for. He then recognized that the additional weight was obtained from water. Importantly, de Saussure set a precedent of arriving at conclusions only after carefully collecting a sufficient amount of data.

So, the knowledge of photosynthesis at this stage of its study can be summarized by the reaction

$$CO_2 + H_2O + light \rightarrow [H_2CO] + O_2 \qquad [1.1]$$

V. Association of Photosynthesis with Chloroplasts

Although Ingen-Housz had shown that only the green parts of plants were involved in photosynthesis, de Saussure doubted that the green color was necessary, being misled by his observations that red leaves also performed photosynthesis. In 1818 Pelletier and Caventou separated the green matter from plants and called it **chlorophyll**. R. J. H. Dutrochet (1776–1847) in 1837 convincingly emphasized that the green chlorophyll was necessary for photosynthesis, and by this time the cytologists had demonstrated that chlorophyll resided in the chloroplasts. Von Mohl observed in 1837 that starch particles existed within chloroplasts. But it was another German, Julius Sachs (1832–

1897), who clearly perceived in 1865 that chloroplasts synthesize sugars and, subsequently, starch as the result of the assimilation of carbon dioxide. Sachs also observed that a plant assimilated no carbon dioxide as long as it contained no chlorophyll, but that this ability appeared as the plant developed chloroplasts. Moreover, he concluded that since starch grains appeared only within the chloroplasts, the assimilation of carbon dioxide must also occur there.

With the biochemical information available, and in conjunction with observations made with the microscope, botanists over 100 years ago had focused attention on the chloroplast as the intracellular site of the process they called **carbon dioxide assimilation.** Yet proof that photosynthesis occurred within chloroplasts had to await studies with isolated chloroplasts. T. W. Englemann began such studies and in 1881 succeeded in showing by direct measurements that chloroplast preparations, when illuminated, had the ability to produce oxygen outside the living cell. During the 1940s important studies were carried out by Hill, Kandler, Strehler, and others on the properties of isolated chloroplasts, in particular their ability to evolve oxygen and incorporate inorganic phosphate into organic form as sugar phosphates. However, only when procedures for isolating *intact* organelles had been adequately developed did the proof emerge that the chloroplast is the *complete* photosynthetic organelle. In 1954 Daniel Arnon and his colleagues showed that preparations of purified, intact chloroplasts would evolve oxygen, fix carbon dioxide into organic materials, and synthesize chemical energy in the form of adenosine triphosphate (ATP) (see Chapter 4).

VI. Development of a Major Research Tool—Isotopic Compounds

A factor that had a tremendous impact on biochemical research in general and on studies of photosynthesis in particular was the introduction of isotopically labeled compounds. Radioactivity was discovered in 1896 by Antoine Henri Becquerel (1852–1908), who worked with phosphorescent uranium salts. In 1898 Marie (1867–1934) and Pierre (1859–1906) Curie announced in Paris the discovery of polonium and radium. Four years later they isolated radium, the first radioactive element to be purified. The earliest biological experiments with isotopes were studies of the exchange of lead ions in roots of plants that were carried out by George Hevesy in 1923, with thorium B as the tracer isotope.

The process of photosynthesis was literally illuminated by the application of isotopes of carbon. Studies of the incorporation of CO_2 labeled with radioactive ^{11}C, which was produced in a cyclotron and has a half-life of only 20 min, were initiated by Ruben, Hassid, and Kamen in 1939. The stable isotope ^{13}C was used extensively during the following years, but its use required the availability of a mass spectrograph. Ruben and Kamen in 1940 discovered ^{14}C, an isotope of carbon that emits beta rays of convenient energy and has a half-life of 5730 years. Several years later production of ^{14}C by the uranium pile was initiated, making it readily available. Along with an extremely sensitive technique for quantitatively detecting the emitted beta rays, the Geiger-Müller discharge tube, this isotope revolutionized studies of photosynthesis. By careful analysis of the incorporation of $^{14}CO_2$ into organic compounds during illumination of cells of the green algae *Chlorella* and *Scenedesmus*, Melvin Calvin and his colleagues in 1954 were able to deduce the cyclic metabolic pathway of carbohydrate synthesis. The early description of isotopes as **tracers** was quite appropriate.

VII. Summary

Details of the mechanism of photosynthesis remain the subject of intensive present-day research activities. Discoveries such as those related throughout this chapter, will continue to be made as new windows to nature are opened by advancing technology. Throughout its history, the study of photosynthesis is replete with exceptional examples of the dependency of the process of discovery on prior developments in chemistry and methodology. As in all areas of science, before a process can be discovered, the technology must be available for detecting and measuring it.

We should look no more critically on the concepts held by the pioneers, who often interpreted their data incorrectly, than we look at our own modern concepts, which also may be based on incomplete data. The major advances in science will continue to be made by those careful investigators who obtain their results in an analytical, quantitative fashion. Out of a simple but carefully performed experiment can come great ideas. Another important aspect, which was exemplified by the early discoverers, is the ability to recognize the significance of the results. The words that Galileo (1564–1642) wrote in the early 1600s (translated by Drake, 1980) about Gilbert's discovery of geomagnetism apply also to photosynthesis:

> I do not doubt that in the course of time this new science will be improved with still further observations, and even more by true and conclusive demonstrations. But that need not diminish the glory of the first observer. I do not have smaller regard for the original inventor of the harp because of the certainty that his instrument was very crudely constructed and even more crudely played; rather, I admire him much more than I do a hundred artists who in ensuing centuries have brought this profession to the highest perfection . . .. To apply oneself to great inventions, starting from the smallest beginnings, and to judge that wonderful arts lie hidden behind trivial and childish things, is not for ordinary minds; these are concepts and ideas for superhuman souls.

Literature Cited

Arnon, D. I. (1955) The chloroplast as a complete photosynthetic unit, *Science* **122**:9–16.

Black, J. (1755) Experiments upon magnesia alba, quicklime, and some other alcaline substances, in *Essays and Observations, Physical and Literary*. Read before a Society in Edinburgh, and Published by them, Vol. II, Edinburgh, pp. 157–225. (Reprinted by the Alembic Club, Edinburgh, No. 1, 1910).

Casida, L. E, Jr. (1976) Leeuwenhoek's observation of bacteria, *Science* **192**:1348–1349.

Cavendish, H. (1784) Experiments on air, *Phil. Trans.* **74**:119–153, **75**:372–384. (Reprinted by The Alembic Club, Edinburgh, No. 3, 1926).

Drake, S. (1980) Newton's apple and Galileo's dialogue, *Sci. Am.* **243** (2):151–156.

Hales, S. (1727) *Statical Essays, Containing Vegetable Staticks, or, an Account of Some Statical Experiments on the Sap of Vegetation*, W. Innys, London. [See also Cohen, I. B. (1976), *Sci. Am.* **234**(5):98–107.]

Hooke, R. (1665) *Micrographia*, London. (Excerpts reprinted by The Alembic Club, Edinburgh, No. 5, 1902.)

Lavoisier, A. (1789) *Traite elementaire de la chimie*, Paris.

Lechevalier, H. (1976) Louis Joblot and his microscopes, *Bacteriol. Rev.* **40**:241–258.

Priestley, J. (1772) Observations on different kinds of air, *Phil. Trans. Royal Soc.*, London, **62**:147.

Priestley, J. (1775) *Experiments and Observations on Different Kinds of Air*, London. (Excerpts reprinted by The Alembic Club, Edinburgh, No. 7, 1923.)

Rey, J. (1630) *Essays de Jean Rey, docteur en medecine, sur la Recherche de la cause pour laquelle l'Estain et le Plomb augmentent de poids quand on les calcine*, Bazas. (Reprinted by The Alembic Club, Edinburgh, No. 11, 1904).

Scheele, C. W. (1777) *Chemische Abhandlung von der Luft und der Feuer*, Upsala und Leipzig. (Reprinted by The Alembic Club, Edinburgh, No. 8, 1923).

Additional Reading

Comroe, J. H., Jr., ed. (1976) *Bench Mark Papers in Human Physiology/5*, Part I, Dowden, Hutchinson and Ross, Stroudsburg, Pennsylvania.

Loomis, W. E. (1960) Historical introduction, in *Encyclopedia of Plant Physiology, Vol. 5*, Part I (W. Ruhland, ed.), Springer-Verlag, Berlin, pp. 85–114.

Parkes, G. D., ed. (1951) *Mellor's Modern Inorganic Chemistry*, Longmans, Green and Co., London. (This book contains excellent discussions of the early development of chemistry.)

Photosynthesis Bicentennial Symposium (1971) *Proc. Natl. Acad. Sci. USA* **68**:2875–2897. (This symposium emphasizes the major discoveries made during the 20th century.)

Priestley, J. (1962) *Selections from His Writings* (I. V. Brown, ed.), The Pennsylvania State University Press, University Park, Pennsylvania.

Rabinowitch, E. I. (1945) *Photosynthesis and Related Processes*, Vol. I, Interscience Publishers, New York. (This book includes an excellent, detailed chapter on the development of photosynthesis.)

Scott, A. F. (1984) The invention of the balloon and the birth of chemistry, *Sci. Am.* **250**:(1)126–137.

2

Structure of the Chloroplast

I. Importance of the Membrane

The mechanism by which an organelle such as the chloroplast carries out its functions is understood only to the extent that the physical, chemical, and topographical properties of the organelle are known. The actual mechanisms are, of course, described by chemical reactions, which are determined by the physical–chemical properties of the reactants. For reactions in free solution, a knowledge of these properties is usually sufficient. However, a chloroplast is foremost a system of membranes, and a membrane adds another feature to chemical reactions. The end result of a reaction between components of a membrane is determined not only by the inherent properties of the reactants, but also by their locations within the membrane. The physical existence of a membrane can profoundly affect the course and consequences of a reaction. Of importance in this regard is the fact that a membrane acts as a barrier separating two compartments and also provides a means to achieve vectoral reactions between the compartments, in which products are separated from substrates. Therefore, a nonequilibrium distribution of substances, a state containing potential free energy, can occur. Such conditions have great importance in the overall function of the membrane system.

The unique function of chloroplasts is the transformation of light energy first into electrical energy and then into chemical energy. Electrical energy in biological systems appears both as reducing agents and as voltage gradients resulting from a separation of charges and/or concentrations of ions across a membrane. The simplest structural requirement for the production and maintenance of an ionic gradient is a compartment bounded entirely by a semipermeable membrane capable of pumping ionic species vectorially across the membrane. In the chloroplast this requirement is met by the system of internal membranes (Fig. 2.1). These membranes exist as extended, flattened vesicles or sacs, called **thylakoids** by the German botanist W. Menke. (The word *thylakoid* is derived from a Greek word meaning "like an empty pouch.") This shape of the thylakoids allows a large surface-to-volume ratio. The green pigments in plants, the chlorophylls, are located within these membranes. The absorption of light by these pigments drives an intermolecular transfer of electrons, which

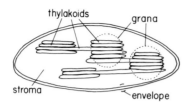

FIG. 2.1. Diagram of a typical chloroplast in a higher plant leaf. The chloroplast is surrounded by an envelope composed of two membranes. The fluid or soluble phase in the chloroplast is the stroma. Thylakoids are individual, flattened vesicles of the photosynthetic membrane that lie within the stroma, not connected to the envelope. Thylakoids are differentiated into the areas where adjacent membranes become tightly appressed into regular aggregates or stacks to form the grana and the connecting segments between grana, or the unstacked stromal thylakoids. Within the stroma are soluble enzymes involved in CO_2 fixation. The chloroplast DNA, ribosomes, and other components of genetic expression also reside in the stroma.

is coupled to the production of a gradient of H^+ ions across the membrane. The flow of H^+ ions down this electrochemical gradient performs work, as evidenced by the production of high-energy chemical compounds (see Chapter 4 for details).

II. Structure of the Thylakoid Membrane

Thylakoid membranes contain lipids (substances soluble in organic solvents) and proteins in a weight ratio of about 1:1. In this respect they are similar to other cellular membranes. Before the structure of thylakoid membranes is discussed in detail, a brief description of general membrane structure is appropriate. Most lipids in membranes are arranged in a manner such that they form a double-layered structure, with polar, hydrophilic groups on the membrane surface in contact with the aqueous environment and the hydrophobic (water-insoluble) portions projecting into the interior of the membrane, away from the water (Fig. 2.2). The hydrophobic interior of the membrane is responsible for the inability of most ions and water-soluble substances, including water, to freely diffuse across the membrane. Embedded in this lipid matrix are specific functional proteins and lipoprotein complexes. These "membrane" proteins are folded into tertiary and quaternary structures that allow portions of their surface to contain hydrophobic amino acid residues, the side chains of which "dissolve" into the nonaqueous interior of the lipid matrix. The hydrophobic surface areas, therefore, determine that these proteins will reside in the mem-

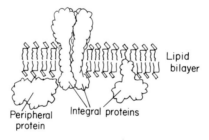

FIG. 2.2. Diagrammatic representation of the basic features of the structure of a membrane. The matrix is a lipid bilayer, in which amphipathic lipids are oriented in each layer with their polar groups (◖) in contact with the aqueous environment and the nonpolar fatty acyl groups inward. Embedded into the membrane, some on one side only and others extending through the bilayer, are integral proteins. Peripheral proteins, although associated with the membrane, can be removed with salt solutions and thus are not physically part of the membrane structure. Interaction of integral proteins with the membrane is determined largely by the extent and distribution of hydrophobic areas on the protein surface, whereas peripheral proteins are held primarily by electrostatic or other types of polar interactions.

brane. Other portions of the membrane proteins extend out into the aqueous environment and are similar to "soluble" proteins in that they contain hydrophilic groups on the surface. A few proteins span the membrane and have portions of the molecule exposed to water on both sides.

Although membrane components are mutually miscible in each other, the forces of attraction between these components are only the weak, noncovalent, gravitational forces. The thermodynamic stability of the membrane is not provided by these forces but rather by the strong dipole attractive forces between the surrounding water molecules. Any hydrophobic or nonpolar group, with which water cannot interact via ionic or dipole interactions, is forced as much as possible out of the aqueous environment. Therefore, the membrane is actually held together by the surrounding lattice of bulk water.

The structure of thylakoid membranes follows this general structural pattern. The matrix of the membrane is a lipid bilayer composed primarily of carbohydrate-containing glycolipids (see Chapter 3). The hydrophilic carbohydrate residues are positioned on the surfaces of the membrane, whereas the hydrophobic fatty acid portions are oriented toward the interior of the structure. Within this matrix, lipoprotein complexes are embedded at a rather high density. These lipoprotein complexes contain most, if not all, of the chlorophyll in the membrane and act as the functional units involved in the initial events of photosynthesis.

A technique that has been of great value in revealing the positions of some of the lipoprotein complexes within membranes is that of **freeze-fracture** (or freeze-etching). Cells frozen in ice at low temperatures can be cracked open by chopping at the ice block with a knife blade. As the ice breaks, the fracture plane travels along paths of low resistance. The regions of lowest resistance to fracture occur along the middle of membranes, within the oily interior formed by the hydrophobic, hydrocarbon portions of the lipids. Thus, with this technique the membrane is split at the interior junctions of the lipids, exposing the inner faces of the halves of the bilayer. Large proteins that span the membrane or lipoprotein complexes that are embedded deeply within the interior of the membrane become exposed (Fig. 2.3). Typically, in green plants and algae (the Chlorophyta) the face provided by the outer half of the thylakoid membrane, called the **P** or protoplasmic half, which is in contact with the chloroplast stroma, exhibits an array of particles 8–12 nm in diameter (Fig. 2.4). The other inner face, provided by the half of the membrane in contact with the thylakoid interior, the **E** or exoplasmic half, exhibits larger particles, some that are 16–18 nm in diameter.

In most other classes of algae the appearance of the P face is similar to that of membranes in the chlorophytes. However, the E face of these other algae, in particular the chromophytes, exhibits considerably fewer particles and has nearly a smooth appearance. In higher plants structural variations occur within the same chloroplast, with granal membranes showing a high density of particles on both faces, as illustrated in Fig. 2.4. whereas stromal membranes have fewer of the large particles on the E face. These structural variations seem to be correlated with differences in the photochemical activities between the two portions of the thylakoid membrane. The functional unit of photosystem 1 apparently is associated with 10- to 12-nm particles on the P face, distributed

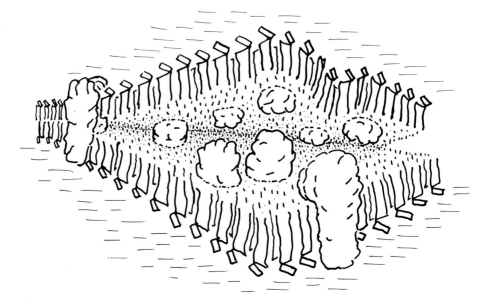

FIG. 2.3. Illustration of splitting a membrane during the technique of freeze-fracture. The sample containing the membrane is frozen and then fractured at low temperature with a knife blade. The polar groups of the membrane remain embedded in the ice. When the fracture plane encounters a membrane, it travels through its interior thus splitting the membrane in half. Proteins deeply embedded into or traversing the membrane remain with one half as determined by the properties of each protein. As the halves of the membrane are separated, these proteins appear as protrusions on the smooth background of the inner surface of the single lipid layer. Vaporized platinum and carbon then are allowed to deposit on the exposed surface to form a thin coating. The underlying tissue is digested away, and the replica is examined with the electron microscope.

⟶

FIG. 2.4. Illustration of freeze-fracture of thylakoid membranes. A: When viewed by transmission microscopy in cross section, spinach chloroplasts exhibit the stacked (grana) and interconnecting unstacked (stroma) membranes. Magnification, ×105,000. B: When the thylakoid membranes are cleaved by freeze-fracture, the internal faces of the membranes are exposed. The diagram in (C) illustrates the path of the fracture plane through the membranes in cross section. In the diagram, the heavy, solid line represents the fracture plane, which travels through the interior of the membranes, but occasionally jumps to adjacent membranes. After the upper, dotted portion is removed, along with the upper half of the ice block, the inner faces of the membrane halves in the lower portion of the ice are exposed on the surface. A thin film of carbon and platinum is deposited on the surface, the underlying tissue is digested away, and the "replica" is viewed face on with the electron microscope. Panel B shows such a micrograph.

The two halves of the membrane are by convention referred to as **P** (protoplasmic, toward the cytoplasm) or **E** (exoplasmic, toward the cell exterior). To determine which is a **P** or **E** half of an intracellular membrane, the membrane can be considered in reference to the cytoplasm of the original cell. Since chloroplasts may have arisen by endocytosis of a prokaryotic cyanobacterium, the stroma then is derived from the endosymbiont's cytoplasm. With thylakoid membranes, the **P** half is in contact with the stroma and the **E** half is adjacent to the thylakoid lumen.

As the interior faces are exposed, features are revealed that differ according to whether the membrane resided within a stacked, granal region (PFs or EFs) or in unstacked, stromal regions (PFu or EFu). These different regions are seen clearly in panel B. Characteristically, the EF contains larger intramembranous particles (lipoprotein complexes) than the PF. The particles on the EF are more abundant within stacked regions, whereas the opposite is true for the PF. Magnification for panel B, ×85,000. (From Staehelin *et al.*, 1980. Reprinted with permission.)

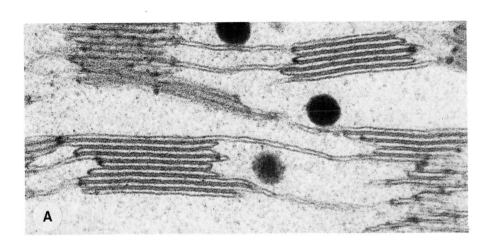

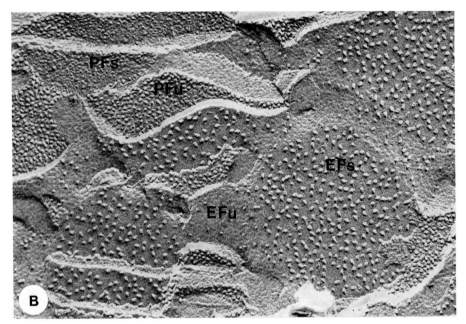

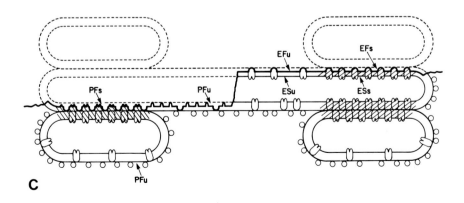

primarily on stroma thylakoids. Smaller, 8- to 9-nm particles on the P face are generated by a portion of the light-harvesting complexes and other protein-lipid complexes. The large 16- to 18-nm particles on the E half of the membrane, occurring at higher density on grana membranes, contain the functional unit of photosystem 2 and its accompanying light-harvesting chlorophyll–protein complexes. A schematic interpretation of the structure of the thylakoid membrane is shown in Fig. 2.5. (For more details of the structure of the membrane see Chapter 4.)

III. Morphological Aspects of Photosynthetic Membranes

Within, or associated with, all photosynthetic membranes are four major functional assemblies: (1) the light-harvesting complexes composed of proteins and pigments, whose role is absorption of light energy; (2) the reaction centers, to which energy from light-harvesting complexes is funneled and where the primary photochemical reactions occur; (3) the carriers for transport of electrons and protons across the membrane; and (4) the coupling factor responsible for synthesis of chemical energy as ATP. As stated at the beginning of this chapter, these assemblies must be part of a membrane system. However, although similarities exist in how the functional parts are integrated within the mem-

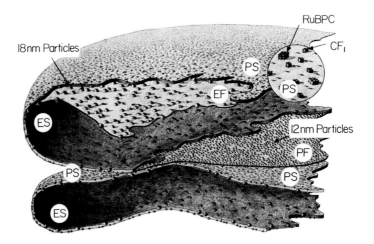

FIG. 2.5. A diagrammatic representation of two thylakoids showing some of the structural features of the membrane. Although not depicted, each thylakoid is a completely closed vesicle. As described in Fig. 2.4, freeze–fracture cleaves the membrane to expose interior faces. The EF contains large 18-nm particles, which are markers of photosystem 2. The PF contains smaller particles (8–12 nm), only some of which contain photosystem 1 centers. The outer surfaces of the membranes contain peripheral proteins. Two well-characterized peripheral proteins on the P surface (PS) are the coupling factor (CF_1) and ribulose 1,5-bisphosphate carboxylase (RuBPC) (see Chapters 4 and 5). The luminal surface of the thylakoid is designated the E surface (ES). (Adapted from Arntzen *et al.*, 1969. Drawing courtesy of C. J. Arntzen.)

brane, the morphological arrangements of the membranes within photosynthetic cells are quite variable.

The simplest morphological features are found in the prokaryotic photosynthetic bacteria. In green photosynthetic bacteria (Chlorobiaceae), the energy-transducing activities are contained within specific, localized regions of the cell membrane. Light-harvesting complexes, however, are contained in nonmembranous particles, called **chlorosomes,** that are bound to the cytoplasmic surface of these specialized regions of the membrane.

In most of the purple photosynthetic bacteria (Rhodospirillaceae and Chromatiaceae), photosynthetic activities are localized in invaginations of the cell membrane. These intracytoplasmic membranes, although continuous with the cell membrane, are highly differentiated chemically and functionally. Because of their high pigment content, these structures are called **chromatophores.** In cross-sectional views of the cell, chromatophores often appear as vesicles (Fig. 2.6A). Freeze-fracture of the membranes shows that one internal face of the chromatophores is highly particulate, as is the **P** face of the cell membrane (Fig. 2.6A). Considerable variations occur among these organisms, however, with respect to the details of their morphology, function, and differentiation patterns.

The most primitive cells that contain free-standing thylakoids are the **cyanobacteria** (prokaryotic blue-green algae). In these cells, thylakoids course through the cytoplasm, usually near the periphery (Fig. 2.6B). Thylakoids are not compartmentalized in these prokaryotic cells. Attached to the surface of the thylakoids are accessory pigment complexes responsible for the blue or red coloration of these organisms (Fig. 2.6C). These complexes, called **phycobilisomes,** are described in Chapter 3.

In all **eukaryotic** algae and higher plants, thylakoids are enclosed within a double-membrane envelope, forming a subcellular organelle, the **chloroplast.** Although the basic structural unit of chloroplasts is the thylakoid, the arrangement of these units among the different types of plant cells is quite variable. These variations are dependent not only on the *type* of cell but also on the *physiological state* of the cell.

Table 2.1 lists several distinguishing features of algal chloroplasts, as documented and compiled by Sarah Gibbs (1981b). The simplest arrangement of thylakoids within chloroplasts is found in the red algae (Rhodophyta), in which the thylakoids exist individually and lie generally in parallel with each other (Fig. 2.7). In the red algae, adjacent thylakoids appear to have few, if any, points of contact. Attached to the outer surface of the membranes are large particles, the phycobilisomes, whose pigment components provide the red color of these cells. The phycobilisomes are 40–50 nm on their long axis and 30–32 nm on their short axis. Some cryptophytes also contain accessory pigments associated with thylakoid membranes, but in these organisms the phycobiliproteins appear to be located within thylakoids rather than on the surface and do not seem to be organized into large particles.

Most classes of algae belong to the division Chromophyta, of which the Bacillariophyceae (diatoms), the Chrysophyceae (yellow-green algae), and the Phaeophyceae (brown algae) are examples. In the chloroplasts of these organisms the thylakoids are usually grouped in bands of three, with each thylakoid

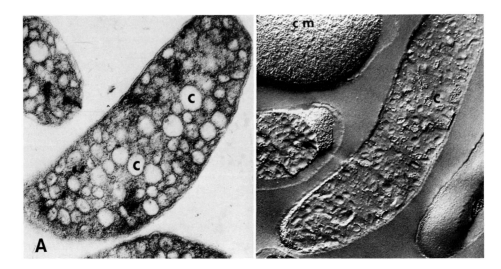

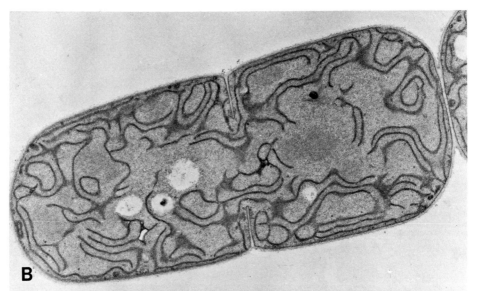

FIG. 2.6. Ultrastructure of prokaryotic photosynthetic cells. A: The purple photosynthetic bacterium *Rhodospirillum rubrum*. As seen in thin section with transmission microscopy (left), the cells are filled with chromatophores (c), which appear as vesicles in the cytoplasm but are continuous with the cell membrane. When viewed by freeze–fracture (right), both the cell membrane (cm) and the intracytoplasmic chromatophores exhibit highly particulate interior faces. (×35,000. Micrographs courtesy of P. Siekevitz.)

B: The cyanobacterium *Anabena cylindrica*. The photosynthetic membranes, as single thylakoids, extend throughout most of the cytoplasm and apparently are not continuous with the cell membrane. Moreover, the thylakoids are not enclosed within an envelope in these prokaryotic cells. Phycobilisomes are associated with thylakoid membranes in cyanobacteria (see panel C), but for this micrograph, cells were fixed with $KMnO_4$ to emphasize the membranes. This treatment did not reveal phycobilisomes. (×23,000. Micrograph courtesy of T. H. Giddings and L. A. Staehelin.)

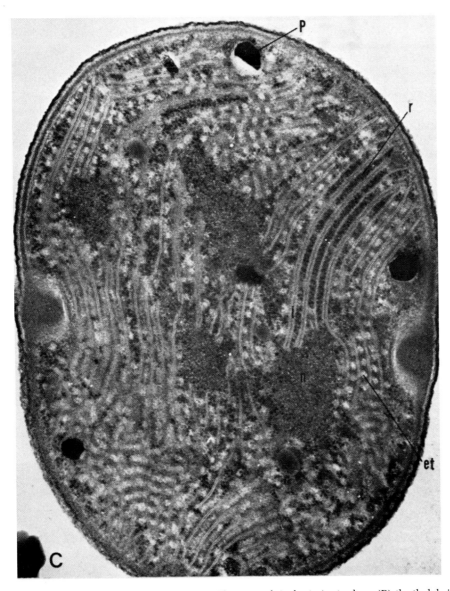

C: The single-celled, coccoid cyanobacterium *Gleocapsa alpicola*. As in *Anabena* (B), the thylakoids extend throughout most of this prokaryotic cell. The cell was fixed with glutaraldehyde and osmium tetroxide, which preserved the phycobilisomes on the surface of thylakoids. The thylakoid membranes and attached phycobilisomes appear as electron transparent structures (et). Also indicated are ribosomes (r), the nucleoplasm (n), and polyphosphate granules (p). (×72,000. From Allen, 1968. Reprinted with permission.)

near, but not in contact with, its neighbor (Figs. 2.8 and 2.9). The thylakoids extend for a considerable distance through the chloroplast, most from pole to pole. In some organisms, there are also several concentric thylakoids just inside the chloroplast envelope that encircle the bulk of the stroma. These structures are called **girdle bands.**

An additional feature of these chloroplasts in the chromophytes is their relationship to an external membrane system. As shown in Figs. 2.8 and 2.10, these chloroplasts are surrounded by a double-membrane structure in addition to the chloroplast envelope. These additional membranes are called the **chloroplast endoplasmic reticulum** (see Section IV). A sketch of a typical member of this division, *Ochromonas*, which illustrates this structure, is shown in Fig. 2.11.

In the Chlorophyta (green algae), chloroplasts are not surrounded by a chloroplast endoplasmic reticulum. Within the organelle, adjacent thylakoids are fused together into definite stacks called **grana** (Fig. 2.12). The number of thylakoids per granum varies depending on the species and the conditions of growth. In these respects, the chloroplasts of the green algae are similar to those of higher plants.

In the green algae, fusion of membranes generally occurs over most of the thylakoid surface (Fig. 2.12). However, in higher plants the membranes are dif-

TABLE 2.1. Some Characteristics of Chloroplasts in Different Algal Classes

Class	Usual number of thylakoids per band	Girdle bands present	Grana present	Degree of apposition of thylakoids	Number of membranes enclosing chloroplast
Cyanophyta					
Cyanophyceae (blue-green algae)	1	No	No	None	None[a]
Rhodophyta					
Rhodophyceae (red algae)	1	No	No	None	2
Cryptophyta					
Cryptophyceae (cryptomonads)	2	No	No	Loose	4
Dinophyta					
Dinophyceae (dinoflagellates)	3	No	No	Variable	3
Prymnesiophyta					
Prymnesiophyceae	3	No	No	Variable	4
Chrysophyta					
Raphidophyceae (chloromonads)	3	Yes	No	—	4
Chrysophyceae	3	Yes	No	Variable	4
Bacillariophyceae (diatoms)	3	Yes	No	Variable	4
Xanthophyceae	3	Variable	No	Variable	4
Phaeophyta					
Phaeophyceae (brown algae)	3	Yes	No	Loose	4
Euglenophyta (euglenoids)	3	No	No	Tight	3
Chlorophyta (green algae)					
Prasinophyceae	2–4	No	Yes	Tight	2
Chlorophyceae	2–6	No	Yes	Tight	2

[a]The prokaryotic Cyanophyta do not contain subcellular chloroplasts. Indeed, many similarities exist between these cells and the chloroplasts in eukaryotic cells. (Adapted from Gibbs, 1970, and 1981b.)

ferentiated into localized regions, those where fusion into small, discrete grana occurs, and unfused membranes that interconnect the grana. The later are called **stromal** thylakoids (Fig. 2.13). Within some plant leaves, cells of different layers of the tissue exhibit structural variations. The **mesophyll** cells of the so-called C_4 plants (See Chapter 4) contains thylakoids that are segregated into granal and stromal regions, whereas the **bundle sheath** cells of some, but not all, such plants contain chloroplasts with undifferentiated, normally unfused membranes. These structural modifications are accompanied by biochemical differences as well (see Chapter 5).

When seeds of many higher plants are germinated in complete darkness, the plastid develops into an **etioplast,** which lacks the green pigment chlorophyll and has no photosynthetic activity. Within the etioplast, a complex lipid-rich structure develops, which forms the **prolamellar body** (Fig. 2.14). Extending from the prolamellar body are rudimentary membranes called **prothylakoids.** When these dark-grown seedlings are transferred to light, the prolamellar body disperses and the etioplast develops into the mature chloroplast (see Chapter 7).

IV. The Chloroplast Endoplasmic Reticulum and the Periplastidal Compartment

A particularly interesting structure in the majority of the classes of algae (primary exceptions are the red and green algae) is the chloroplast endoplasmic reticulum (see Figs. 2.8–2.11). In these organisms the cell's chloroplasts are completely surrounded by a double-membrane structure that also is continuous with the nuclear envelope. Consequently, the chloroplasts and nucleus are enclosed within a compartment separate from the remainder of the cell. The existence of *four* membranes surrounding these chloroplasts undoubtedly has significant ramifications concerning fluxes of material into and out of the organelle.

The chloroplast endoplasmic reticulum has connections with the nuclear envelope similar to features found with the endoplasmic reticulum itself. Ribosomes are attached to the cytosolic (outer) surface, but not to the surface adjacent to the plastid (Fig. 2.10). The question has been raised by Gibbs (1981a) as to whether these ribosomes are involved in synthesizing nuclear-coded proteins that are destined for the chloroplast. An additional structure that also might be involved in transfer of material into and/or out of the chloroplast is a layer of tubules and vesicles that lies between the chloroplast envelope and the chloroplast endoplasmic reticulum (Fig. 2.10). This layer is called the **periplastidal reticulum** and resides within the **periplastidal compartment.** Unfortunately, biochemical studies on these structures are in their infancy and essentially no information is available on their physiological functions.

In algae of the class Cryptophyceae, which are flagellated algae containing phycobilin pigments, the periplastidal compartment has some additional interesting features. Although around the outside of the cup-shaped chloroplast (toward the cell membrane) the periplastidal compartment is quite thin, on the inner side of the chloroplast this compartment is wide and contains unique structures not found in other algae. Ribosomes are sparsely distributed and are similar in size to cytoplasmic ribosomes. Most unusual, however, is the presence of a small body, surrounded by a double membrane, called a **nucleomorph** because of its resemblance to a nucleus (Figs. 2.15 and 2.16). Gibbs (1981a) found that this structure also contains DNA. Furthermore, within the

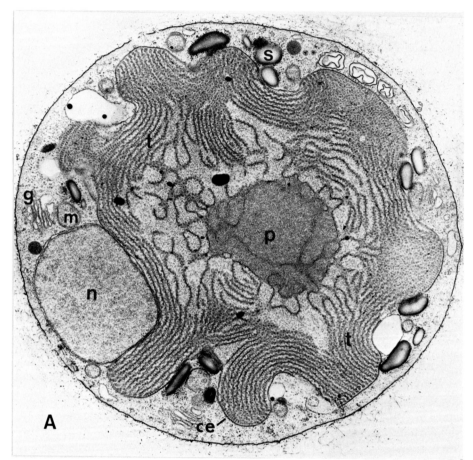

FIG. 2.7. Ultrastructure of *Porphyridium cruentum*, a single-celled, eukaryotic red alga. A: In this micrograph of the whole cell, the single, large chloroplast is delimited by a double-membrane envelope (ce). Within the chloroplast are generally parallel-arranged thylakoids (t) and a pyrenoid body (p). The thylakoids are not fused together to form grana or connected with the chloroplast envelope. Identifiable structures in the remainder of the cell are the nucleus (n), mitochondria (m), and Golgi bodies (g). Scattered throughout the cytoplasm are starch grains (s). (×14,000. From Gantt and Conti, 1965. Reprinted with permission of The Rockefeller University Press.) B: An enlargement of a portion of the chloroplast of *Porphyridium cruentum* that shows the alternating arrangement of the phycobilisomes on the stromal surface of thylakoid membranes. (×115,000. From Gantt and Conti, 1965. Reprinted with permission of the Rockefeller University Press.)

nucleomorph is a dense, granular region resembling a nucleolus. There is only one nucleomorph per chloroplast (and per cell) and it divides just prior to cell division. Several investigators have suggested that the chloroplasts in this class of algae, and perhaps other classes except the red and green algae, originated through capture of a red, eukaryotic alga by an unpigmented flagellated eukaryotic cell. Thus, the periplastidal compartment would represent the cytoplasm of the endosymbiont and contain the vestigial nucleus still capable of making ribosomes. This concept is described in more detail in Section V.

V. Evolutionary Ontogeny of Chloroplast Structure

The diverse and, to some extent, bizarre structural features of algal chloroplasts may hold clues to the evolutionary development of the organelle. Several investigators have drawn attention to the possible origins of the multiple membranes surrounding plastids. The scheme illustrated in Fig. 2.17, developed largely by Jean Whatley (1981), attempts to systematically correlate evolutionary events with chloroplast structure.

The progenitor of the chloroplast is generally thought to have been an organism similar to modern prokaryotic cyanobacteria. The lineage of this alga

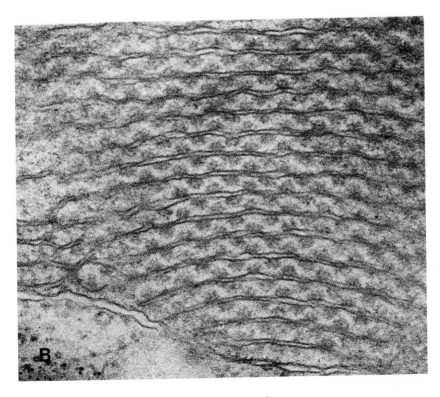

FIG. 2.7 (Continued)

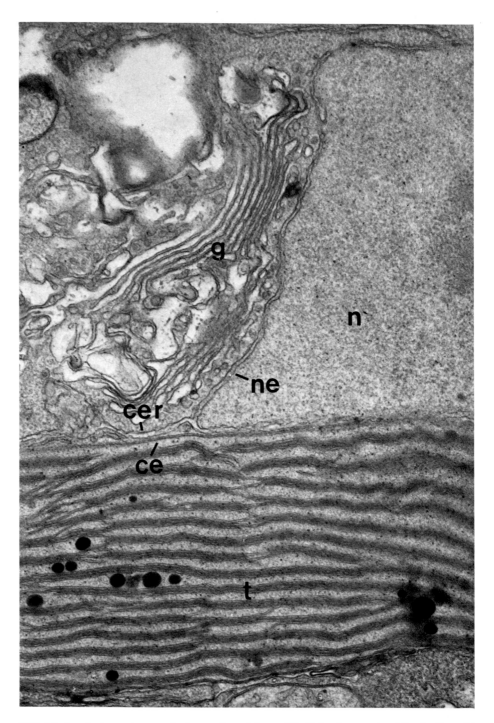

FIG. 2.8. An electron micrograph of a portion of the multicellular brown alga *Chorda filum*. The close association between the chloroplast and nucleus is apparent. The chloroplast is enclosed by a double-membrane envelope (ce) and also by the two membranes of the chloroplast endoplasmic

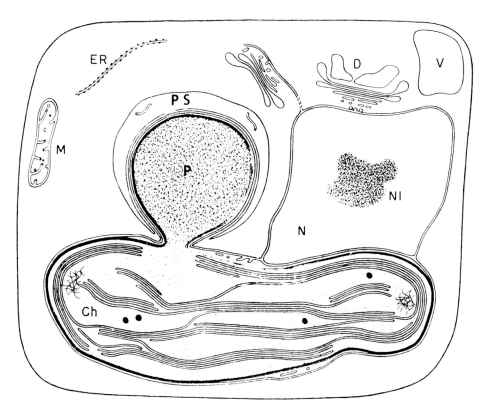

FIG. 2.9. Diagram of a hypothetical brown algal cell that illustrates some of the organelle associations shown in Fig. 2.8. The chloroplast envelope encloses the chloroplast (ch) and extends around the protruding pyrenoid body (P). The chloroplast endoplasmic reticulum also encloses the chloroplast and is continuous with the membranes of the nuclear envelope. Golgi bodies (also called dictyosomes, D) are usually closely associated with the nucleus (N) or the sparse endoplasmic reticulum (ER). A pyrenoid sac (PS) surrounds the pyrenoid body, as a cap outside the chloroplast endoplasmic reticulum, and may contain reserve polysaccharide material. M, mitochondrion; V, vacuole; Nl, nucleolus. (Adapted from Bouck, 1965. Diagram courtesy of G. B. Bouck.)

to the red algae and the chromophytes is traced through the presence of the phycobilisomes. The red algae possibly arose by endosymbiotic capture of a cyanobacterium by a eukaryotic (or protoeukaryotic) host. In the red algae, which have plastids surrounded by two membranes, the inner membrane would then have been derived from the prokaryotic cell membrane, whereas the outer plastid membrane would have formed by invagination of the host's plasma membrane. Fission of the membrane to form an algal-containing vacuole resulted in the first "chloroplast."

reticulum (cer). The chloroplast endoplasmic reticulum is continuous with the outer membrane of the nuclear envelope (ne). Within the chloroplast thylakoids are organized usually into bands of three but are not fused together into grana. A large Golgi apparatus (g) is close to the nucleus (n). (×30,000. From Bouck, 1965. Reprinted with permission of The Rockefeller University Press.)

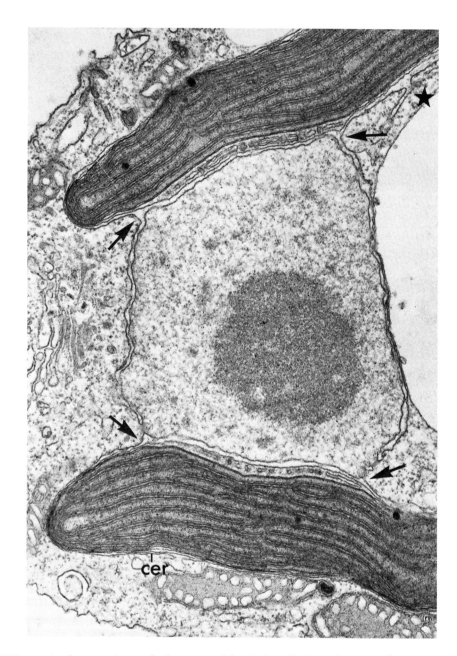

FIG. 2.10. An electron micrograph of a portion of the single-cell, chrysophycean (yellow-green) alga *Ochromonas danica*. The section shows the nuclear region of a light-grown cell. Thylakoid membranes in the chloroplast are grouped in bands of three but are not entirely fused together into grana. Girdle bands of thylakoids surround the stroma just inside the chloroplast envelope. Most of the remaining thylakoid bands also course concentrically around the chloroplast. Continuity of the chloroplast endoplasmic reticulum (cer) with the nuclear envelope is clearly seen at the arrows. Ribosomes are present on the outer surface of the chloroplast endoplasmic reticulum, which in places also is continuous with the endoplasmic reticulum (at star). (×36,500. From Gibbs, 1970. Reprinted with permission.)

However, the development of the progenitor of the chloroplasts in green algae and higher plants involved both the loss of the ability to make phycobilisomes and the acquisition of reactions for chlorophyll *b* synthesis. (It is also possible that the reverse was true.) The recent discovery of the prokaryotic alga *Prochloron* has been hailed as providing the link between the blue-green and green algae. *Prochloron* lives in symbiotic association with ascidians (sea squirts) in marine environments. In this photosynthetic alga, thylakoids are not enclosed within an intracellular membrane, but the membranes are quite similar in structure and composition to those in the green algae. Although most of the thylakoids in *Prochloron* are unstacked, in localized areas they are differentiated into stacked structures closely resembling grana. The freeze–fracture images of these stacked and unstacked regions are essentially indistinguishable from those of chloroplasts of higher plants. Stanier and Cohen-Bazire (1977) suggested that *Prochloron*, which contains both chlorophylls *a* and *b*, may be the modern counterpart of the organism that developed into the chloroplast of the green plants. Capture of this organism by a eukaryotic host would have given rise to two surrounding membranes, as described above for genesis of the red algae, with the prokaryotic inner membrane and the eukaryotic outer membrane forming the "chloroplast" envelope.

The chloroplasts in the algal species *Euglena* are surrounded by three membranes. The proposal presented for the origin of the third, outer membrane suggests that a eukaryotic host cell, through the process of endocytosis, acquired a complete chloroplast from a green algal donor (see Fig. 2.17). The outer chloroplast membrane would then be derived from the plasma membrane of the host cell.

The presence of additional membranes around the chloroplasts of cryptomonads and chromophytic algae has generated the hypothesis illustrated on the left side of Fig. 2.17. Since these algae contain chlorophyll *c* in addition to chlorophyll *a* (see Chapter 3), it is thought that either an ancestral prokaryotic alga or a red alga developed this ability and thus became a now-extinct prechromophyte alga. Dinoflagellates, which have three membranes surrounding the chloroplasts, possibly arose by transfer of a chloroplast from this prechromophyte into a eukaryotic host, an origin similar to that of the chloroplasts in euglenoids. However, a scenario that could explain development of the four membranes surrounding the chloroplasts in modern cryptomonads and other chromophytes includes entrance of the complete, eukaryotic prechromophyte alga into another eukaryotic host. Reduction of the original symbiont over time would have led to vestigial extraplastid structures such as the nucleomorph (see Fig. 2.15). Thus, the chloroplasts in the resulting cryptomonad algae are surrounded, from the inner to the outer, by membranes derived from the original prokaryotic alga, the endocytic vacuolar membrane of the first eukaryotic host cell, the plasma membrane of this first host cell, and finally the endocytic vacuolar membrane of the second eukaryotic host cell. The cytoplasm of the first host has been reduced to the periplastidal compartment. Further reduction of the first host cell, with the loss of the nucleomorph and other periplastidal components, apparently led to the variety of chromophytic algae.

The scheme of sequential endosymbiosis shown in Fig. 2.17 is supported by freeze–fracture analysis of the polarity of the membrane surrounding the

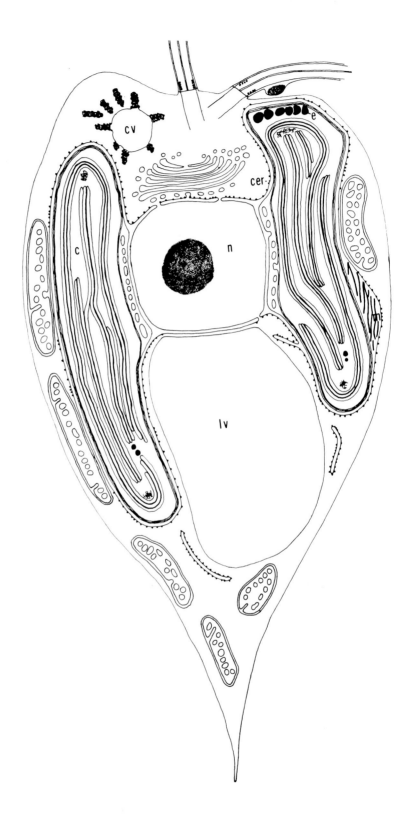

chloroplasts. This analysis was based on the general observation that the PF or protoplasmic face of a fractured membrane, presented by the half of the membrane in contact with the cytoplasm or protoplasm, contains a higher density of particles, often of smaller size, than the EF or exoplasmic ("extracellular") face.

A puzzling feature of this evolutionary explanation is the site of storage of the products of photosynthesis. Green algae accumulate the glucose polymer starch within the chloroplast, whereas in red algae starch is stored in the cytosol. Euglenoids and dinoflagellates, which contain chloroplasts surrounded by three membranes, also store polymeric forms of glucose in the cytosol. The putative prechromophyte probably stored starch in its cytosol as well, since cryptomonads accumulate starch in the periplastidal compartment. However, further reduction to the chromophyte algae resulted in the cytosol of the second host as the site of storage of polysaccharides. Although its significance is not known, it is interesting to note that in the euglenoids and chromophyte algae the storage products are glucose units joined by β-1,3 linkages, whereas in all other plant cells the glucose monomers are joined by α-1,4 linkages (Fig. 2.18).

VI. The Pyrenoid Body

A localized, granular region called the **pyrenoid body** is present within the chloroplasts of many but not all varieties of algae. However, pyrenoid-containing species are distributed among all major groups of eukaryotic algae. The pyrenoid body is centrally located in some types, whereas in others, as for example the brown algae, the pyrenoid body occurs as an appendage that protrudes from the main body of the plastid (Figs. 2.8 and 2.9). Starch, the usual storage form of carbohydrate in the cell, surrounds the pyrenoid body within the chloroplast in most green algae (Fig. 2.12), but in others the storage polysaccharides are outside the chloroplast. In some algae starch grains form a cytoplasmic cap surrounding the protruding pyrenoid body (Fig. 2.9), whereas in some species the starch is scattered throughout the cytoplasm (Fig. 2.7). Although higher plants generally do not have a pyrenoid body, they nevertheless store starch within the chloroplast.

The available evidence indicates that the dense, granular matrix of the pyrenoid body results from a high concentration of the enzyme ribulose 1,5-bisphosphate carboxylase, the major CO_2-fixing enzyme in these plant cells.

←

FIG. 2.11. A diagrammatic representation of a cell of *Ochromonas danica*, which illustrates some of the features shown in Fig. 2.10. The chloroplast endoplasmic reticulum (cer) is continuous with the nuclear envelope. The double-membrane chloroplast envelope surrounds the plastid inside the chloroplast endoplasmic reticulum. The chloroplast (c) is thus separated from the cytoplasm by four membranes. Ribosomes are attached to the cytoplasmic side of the chloroplast endoplasmic reticulum except where it overlays the eyespot (e) and where mitochondria are adjacent to the chloroplast. The contractile vacuole (cv), the leucosin vacuole (lv), the nucleus (n), mastigonemes (m) within the plastid, and several mitochondria also are shown. (From Gibbs, 1981a. Reprinted with permission.)

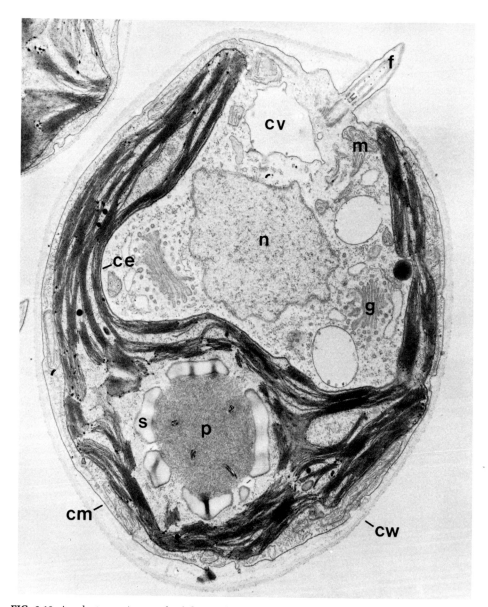

FIG. 2.12. An electron micrograph of the single-cell, green alga *Chlamydomonas reinhardtii*. The cell contains a single, cup-shaped chloroplast that occupies nearly half of the cell volume. Two membranes surround the chloroplast and comprise the chloroplast envelope (ce). In this light-grown cell, the chloroplast is filled with long thylakoids that are fused along most of their surfaces into grana, without clear differentiation into granal and stromal thylakoids. In the posterior region of the chloroplast, the pyrenoid body (p) is surrounded by starch particles (s). Ribosomes are abundant within the chloroplast and are smaller than those in the cytoplasm. Also shown are the nucleus (n), Golgi bodies (g), the contractile vacuole (cv), and mitochondria (m). A portion of one of the two flagella (f) is seen at the anterior end of the cell. cm, Cell membrane; cw, cell wall. (× 16,000. From Ohad *et al.*, 1967. Reproduced with permission of The Rockefeller University Press.)

The function of this enzyme is discussed in Chapter 5. This enzyme is present at very high concentrations in all algal and higher plant chloroplasts, with the exception of those in the mesophyll cells of C_4 plants, which seem to lack the protein. From all indications, this enzyme is the most abundant protein in nature. What causes an aggregation of the carboxylase to form the pyrenoid body is not known. Also, the functional advantage, if any, of this local accumulation of the enzyme is not known.

VII. Chloroplast Size and Number

Among other highly variable features of chloroplasts are their shapes and the quantity per cell. Mature chloroplasts in higher plants generally are regular in shape, occurring as biconcave or lens-shaped structures 1–3 μm across by 5–

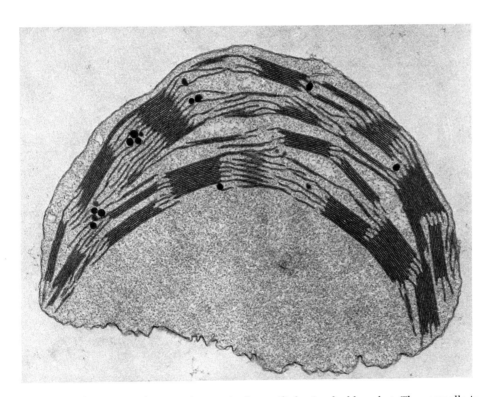

FIG. 2.13. A thin-section electron micrograph of a purified spinach chloroplast. The organelle is limited by a double-membrane envelope. Thylakoids are distinctly differentiated into stacked regions **(grana)** and the unstacked **stromal thylakoids.** Such clearly defined grana stacks are characteristic of most higher plant thylakoid membranes. Several densely stained lipid globules are present. Within the matrix or **stroma** are soluble proteins involved in catalyzing biosynthetic reactions. Ribosomes in the stroma are visible as small granules. (×37,500. Micrograph courtesy of K. R. Miller.)

7 μm on their long axis. Higher plant cells typically contain between 50 and 200 plastids per cell. In the younger areas of a leaf, near the base, cells contain fewer plastids than those in the outer, mature areas.

In the algae, although the internal volume of the plastid is roughly the same as in higher plants, the shape varies from lens-shaped to lobed, cup-shaped to even long, spiral, or irregular shapes. The number of chloroplasts per algal cell ranges from one in some species such as *Chlorella* and *Chlamydomonas* to over 100 in some large-celled species.

VIII. Summary

The morphology of photosynthetic organisms is highly variable. However, one feature common to all photosynthetic organisms is the presence of a specialized membrane that surrounds a compartment. In photosynthetic bacteria, this compartment is the cell itself. In cyanobacteria, algae, and higher plants, this compartment is provided by the thylakoid, which is the basic morphologi-

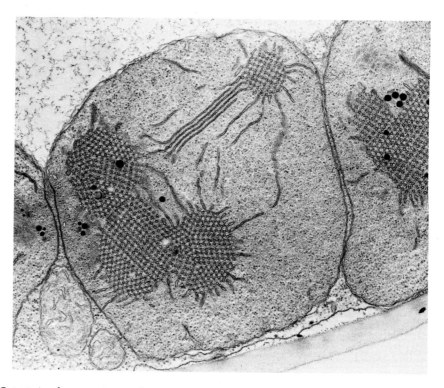

FIG. 2.14. An electron micrograph of a developed etioplast in a primary leaf of an oat *(Avena sativa)* seedling grown in the dark for 11 days. The plastids in dark-grown plant cells accumulate a lipid-rich, organized three-dimensional arrangement of interconnecting tubules, called the **prolamellar body.** The time course of etioplast development is variable among different species, but usually ranges from 1 to 2 wk for maximal production of prolamellar bodies. Extending from these structures are rudimentary **prothylakoid** membranes. (From Lütz, 1981. Reprinted with permission.)

cal unit of photosynthesis in these organisms. Within the membrane are the pigments and proteins that serve as the functional components of photosynthesis. These components are embedded within the lipid bilayer of the membrane in specific orientations determined by their physical–chemical properties. The characteristics of these components will be described in subsequent chapters.

Chloroplast structure is highly plastic, which is an indication of the durability and strength of the basic design developed through the evolutionary pro-

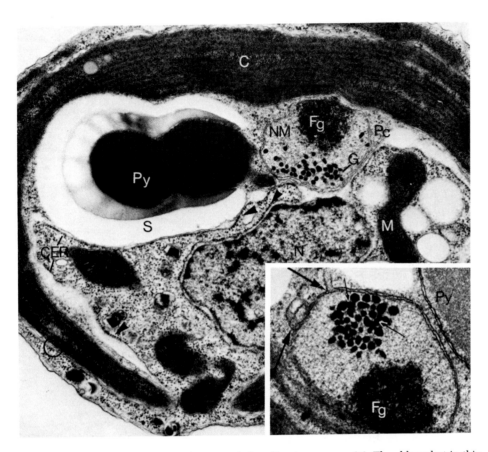

FIG. 2.15. The ultrastructure of a cryptomonad alga, *Cryptomonas* sp. (φ). The chloroplast in this alga is surrounded by a chloroplast endoplasmic reticulum (CER). On the outer surface of the chloroplast, the space between the chloroplast endoplasmic reticulum and the chloroplast envelope is small (circle). However, on the inner surface, the periplastidal compartment (Pc) is wide and contains a starch cap (S) over the pyrenoid body (Py), ribosomes, and a nucleomorph (NM). The arrowheads point to ribosomes attached to the cytoplasmic surface of the chloroplast endoplasmic reticulum. (×25,000.) Inset: An enlargement of the nucleomorph. Within the nucleomorph are fibrillogranular material (Fg) and densely staining globules (G). The double-membrane nucleomorph envelope contains porelike interruptions and is continuous with periplastidal tubules (at the large arrows). This arrangement is reminiscent of the association of the nuclear envelope and endoplasmic reticulum in many eukaryotic cells. Some of the nucleomorph globules appear to be connected (small arrows). (From Gillott and Gibbs, 1980. Reprinted with permission.)

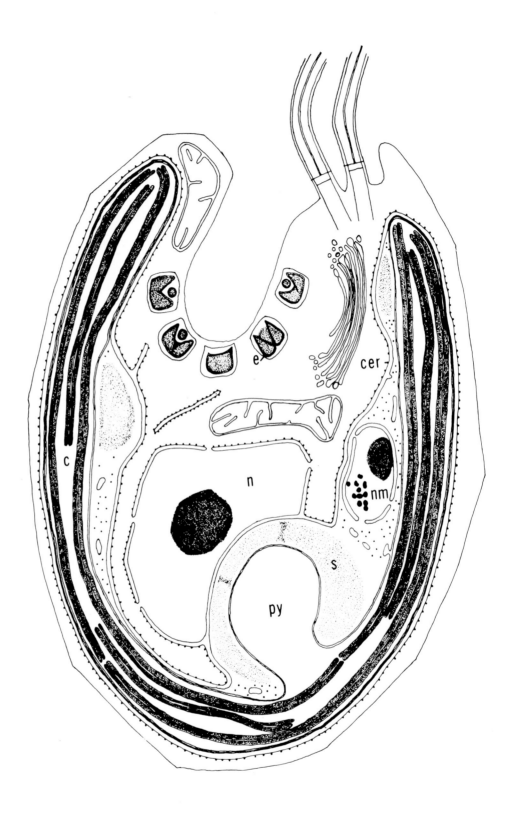

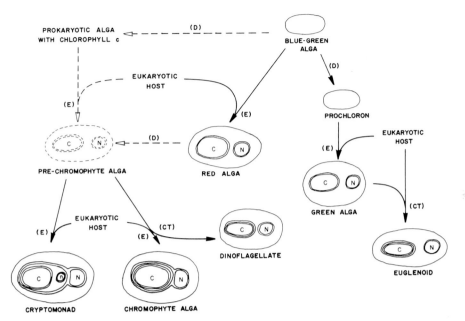

FIG. 2.17. A diagrammatic scheme for the evolutionary relationships of chloroplast structure in algae. C, chloroplast; N, nucleus; n, nucleomorph. Membranes surrounding the chloroplast are indicated, but thylakoid membranes within chloroplasts or prokaryotic cyanobacteria (blue-green algae) are not. Dashed lines refer to formation of organisms for which there is no known modern counterpart. Developmental processes thought to be involved are (D), differentiation; (E), endosymbiosis; and (CT), chloroplast transfer. (Adapted from Whatley, 1981a).

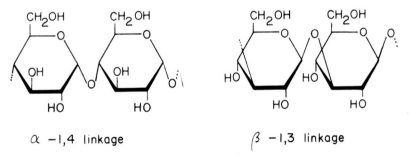

α –1,4 linkage β –1,3 linkage

FIG. 2.18. Structures of glycosidic linkages between glucose units in the storage forms of carbohydrate produced by photosynthesis, Starch, made in most types of algae and higher plants, contains α-1,4 linkages, whereas the glucan made in *Euglena* and chromophyte algae contains β-1,3 linkages.

FIG. 2.16. A diagram of a cell of *Cryptomonas*, illustrating features shown in Fig. 2.15. The nucleomorph (nm) is shown within the periplastidal compartment, which is enclosed by the chloroplast endoplasmic reticulum (cer). The chloroplast envelope separates the pyrenoid body (py) inside the chloroplast (c) from the starch grains (s) in the periplastidal compartment. n, Nucleus; e, ejectosomes. (From Gibbs, 1981a. Reprinted with permission.)

cess. The basic features of this design were in place early, since many similarities exist with the prokaryotic cyanobacteria (blue-green algae), the apparent progenitors of the chloroplasts in eukaryotic cells. The functional importance of several interesting chloroplast-associated structures, such as the chloroplast endoplasmic reticulum and the pyrenoid body, remains to be elucidated.

Literature Cited

Allen, M. M. (1968) Ultrastructure of the cell wall and cell division of unicellular blue-green algae, *J. Bacteriol.* **96**:842–852.

Arntzen, C. J., Dilley, R. A., and Crane, F. L. (1969) A comparison of chloroplast membrane surfaces visualized by freeze–etch and negative staining techniques, and ultrastructural characterization of membrane fractions obtained from digitonin-treated spinach chloroplasts, *J. Cell Biol.* **43**:16–31.

Bouck, G. B. (1965) Fine structure and organelle associations in brown algae, *J. Cell Biol.* **26**:523–537.

Gantt, E., and Conti, S. F. (1965) The ultrastructure of *Porphyridium cruentum, J. Cell Biol.* **26**:365–381.

Gantt, E., and Conti, S. F. (1966) Granules associated with the chloroplast lamellae of *Porphyridium cruentum, J. Cell Biol.* **29**:423–434.

Gibbs, S. P. (1970) The comparative ultrastructure of the algal chloroplast, *Ann. N. Y. Acad. Sci.* **175**:454–473.

Gibbs, S. P. (1981a) The chloroplast endoplasmic reticulum: Structure, function, and evolutionary significance, *Int. Rev. Cytol.* **72**:49–99.

Gibbs, S. P. (1981b) The chloroplasts of some algal groups may have evolved from endosymbiotic eukaryotic algae, *Ann. N. Y. Acad. Sci.* **361**:193–208.

Gillott, M. A., and Gibbs, S. P. (1980) The cryptomonad nucleomorph: its ultrastructure and evolutionary significance, *J. Phycol.* **16**:558–568.

Kaplan, S., and Arntzen, C. J. (1982) Photosynthetic membrane structure and function, in *Photosynthesis*, Vol. 1: *Energy Conversion by Plants and Bacteria* (Govindjee, ed.), Academic Press, New York, pp. 65–151.

Lütz, C. (1981) Development and ageing of etioplast structures in dark grown leaves of *Avena sativa* (L.), *Protoplasma* **108**:83–98.

Lütz, C. (1981) On the significance of prolamellar bodies in membrane development of etioplasts, *Protoplasma* **108**:99–115.

Ohad, I., Siekevitz, P., and Palade, G. E. (1967) Biogenesis of chloroplast membranes, *J. Cell Biol.* **35**:521–584.

Staehelin, L. A., Carter, D. P., and McDonnel, A. (1980) Adhesion between chloroplast membranes: experimental manipulation and incorporation of the adhesion factor into artificial membranes, in *Membrane-Membrane Interactions* (N. B. Gilula, ed.), Raven Press, New York, pp. 179–193.

Whatley, J. M. (1981) Chloroplast evolution—Ancient and modern, *Ann. N. Y. Acad. Sci.* **361**:154–164.

Additional Reading

Anderson, J. M. (1975) The molecular organization of chloroplast thylakoids, *Biochim. Biophys. Acta* **416**:191–235.

Arntzen, C. J., and Briantais, J.-M. (1975) Chloroplast structure and function, in *Bioenergetics of Photosynthesis* (Govindjee, ed.), Academic Press, New York, pp. 51–113.

Gantt, E. (1980) Structure and function of phycobilisomes: Light harvesting pigment complexes in red and blue-green algae, *Int. Rev. Cytol.* **66**:45–80.

Giddings, T. H., Jr., Withers, N. W., and Staehelin, L. A. (1980) Supramolecular structure of stacked and unstacked regions of the photosynthetic membrane of *Prochloron* sp., a prokaryote, *Proc. Natl. Acad. Sci. USA* **77**:352–356.

Kirk, J. T. O., and Tilney-Bassett, R. A. E. (1978) *The Plastids: Their Chemistry, Structure, Growth and Inheritance*, 2nd ed., Elsevier/North Holland, Amsterdam.

Lang, N. J. (1968) The fine structure of blue-green algae, *Annu. Rev. Microbiol.* **22**:15–46.

Menke, M. (1962) Structure and chemistry of plastids, *Annu. Rev. Plant Physiol.* **13**:27–44.

Miller, K. R. (1979) The photosynthetic membrane, *Sci. Am.* **241**(4):102–113.

Miller, K. R., and Cushman, R. A. (1979) A chloroplast mutant lacking photosystem II, *Biochim. Biophys. Acta* **546**:481–497.

Miller, K. R., Miller, G. J., and McIntyre, K. R. (1977) Organization of the photosynthetic membrane in maize mesophyll and bundle sheath chloroplasts, *Biochim. Biophys. Acta* **459**:145–156.

Ohad, I., and Drews, G. (1982) Biogenesis of the photosynthetic aparatus in prokaryotes and eukaryotes, in *Photosynthesis, Vol. 2: Development, Carbon Metabolism and Plant Productivity* (Govindjee, ed.), Academic Press, New York, pp. 89–140.

Schiff, J. A., ed. (1982) *On the Origins of Chloroplasts*, Elsevier, New York.

Stanier, R. Y., and Cohen-Bazire, G. (1977) Phototrophic prokaryotes: The cyanobacteria, *Annu. Rev. Microbiol.* **31**:225–274.

Tanford, C. (1980) *The Hydrophobic Effect: Formation of Micelles and Biological Membranes*, 2nd ed., John Wiley & Sons, New York.

Characteristic Components of Chloroplast Membranes

I. Purification of Chloroplasts

To determine its chemical composition as well as the mechanism of function, the chloroplast must be prepared in a relatively pure state. An important consideration during purification is the osmolarity of the surrounding medium. In solutions with osmolarities less than that inside the organelle, bulk water is drawn in until the chloroplast swells to the limits of elasticity of its envelope, beyond which it bursts. This condition permits escape of nearly all water-soluble components contained in the stroma of the chloroplast. Such drastic consequences can be avoided, however, by adding to the medium a solute to which the chloroplast envelope is relatively impermeable, at a concentration sufficient to balance the internal osmotic pressure. The two most widely used substances for this purpose are sucrose and sorbitol (Fig. 3.1).

By 1900, from the work of Englemann, it was recognized that isoosmotic sugar solutions would preserve chloroplast activities outside the living cell. By 1950 two media were in common usage, 0.35 M NaCl and 0.5 M sucrose or glucose. With the development of the electron microscope, it became possible to critically evaluate the ability of the isolation procedures to preserve intactness of the organelle. The impression of intactness obtained by viewing preparations with the light microscope was not always confirmed when they were examined at higher magnifications. As work continued, solutions containing sucrose became the medium of choice. Nevertheless, modifications are continually being made. Sorbitol, a sugar alcohol that is not metabolized by these systems, was introduced as an effective substitute for sucrose. Also some investigators add to the isolation medium, in addition to sucrose or sorbitol, polymers such as Ficoll (a polymer of sucrose), dextran (a polymer of glucose), or bovine serum albumin (a protein). These polymeric substances tend to stabilize the chloroplast envelope. Other important components of the isolation medium are a buffer to control the pH, a divalent cation such as Mg^{2+} for membrane stability, and a reducing agent such as isoascorbic acid. The various experimental modifications are designed to increase both the recovery of intact chloroplasts

roplasts and the particular activity or function of the chloroplast the investigator is interested in. Thus, specific details of the procedures often depend on the intent of the experiment.

A persistent problem in the isolation of chloroplasts has been due to the fact that, to release them from the cell, rather strenuous procedures must be employed to break, the plant cell wall. A number of mechanical means, such as grinding or blending, have been used with variable success. But in every case some of the chloroplasts are broken along with the cells. Yet the investigator must proceed, first filtering the chloroplasts and other small subcellular structures through a cloth mesh to remove large fragments of leaf tissue. The chloroplast fraction then is obtained by centrifugation of the broken cell preparation.

A recently developed approach to purification of chloroplasts and other subcellular organelles involves enzymatic digestion of plant tissue with cellulase and pectinase. These enzymes selectively digest the cell wall of higher plants and allow the release of protoplasts. After collection by a centrifugation step, the protoplasts are gently sheared open in a medium that preserves the integrity of the intracellular structures, as described in the last paragraph. Chloroplasts or other organelles then are isolated by centrifugation procedures.

The initial step in purifying chloroplasts is usually **differential centrifugation.** In this procedure, the preparation of broken cells is centrifuged first at 500 times the force of gravity for several minutes to sediment large debris and unbroken cells. The supernatant fluid from this first centrifugation then is subjected to 1000–2000 times the force of gravity for 5–10 min. After the second

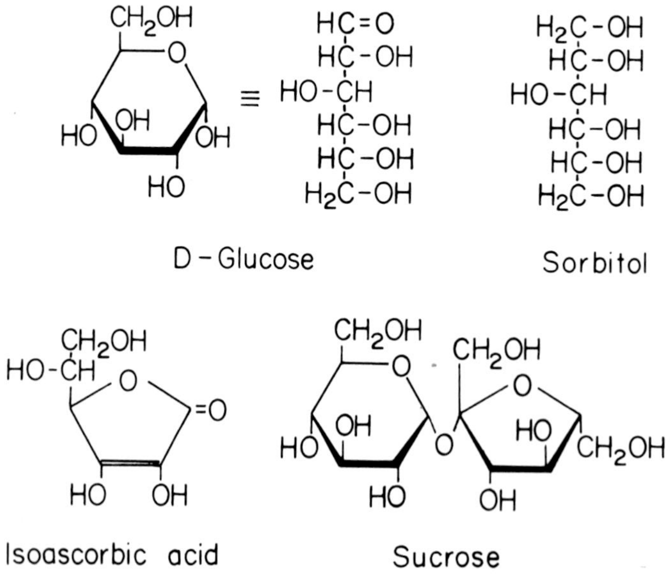

FIG. 3.1. Structures of α-D-glucose, sorbitol (the alcohol derivative of glucose), isoascorbic acid, and sucrose (D-glucose[α,1 → β,2]-D-fructose).

centrifugation the chloroplast fraction is obtained as a pellet at the bottom of the centrifuge tube (Fig. 3.2). Cellular components smaller than chloroplasts remain in the second supernatant fluid. This technique, therefore, separates cellular components on the basis of their rate of sedimentation in a centrifugal field, which is related to the size of the particle. But intact and broken chloroplasts cannot be separated by this technique. They can, however, be resolved by an additional centrifugation step that takes advantage of differences in the density of particles that are to be separated. For example, the plastid fraction can be further fractionated by sedimentation through a concentration gradient of sucrose or of silica sol, a technique called **density gradient centrifugation.** The rate of sedimentation becomes zero when the particle reaches a position in the gradient at which the density of the medium is equal to its own density. Since intact chloroplasts, containing the stroma as well as the membranes, have a density different from that of broken chloroplasts, with only membranes, this procedure yields preparations of highly pure, intact, and functional chloroplasts (Fig. 3.3).

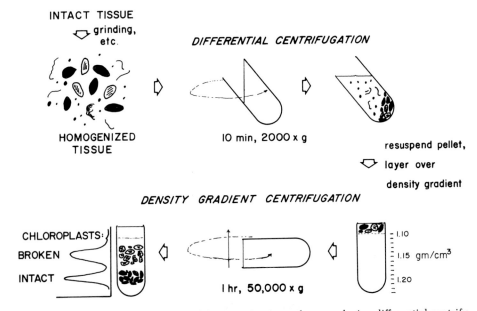

FIG. 3.2. Schematic illustration of a cell fractionation procedure employing differential centrifugation and density gradient centrifugation. The cellular tissue first is homogenized by a procedure that disrupts the cells while causing minimal damage to subcellular organelles. The differential centrifugation step, which separates structures on the basis of size, can be done with a rotor that holds tubes at an angle (as shown) or with a swinging-bucket rotor. Gradient centrifugation generally is done with a swinging-bucket rotor, which allows the tubes to swing into a horizontal position during centrifugation. The contents of the tubes are collected by puncturing the bottom and either allowing the gradient to drip out the bottom or displacing it upward into a collection device with a liquid of greater density. Individual fractions then are analyzed to determine the distribution of cellular components in the gradient. (The density scale corresponds to a 1–2 M sucrose gradient.)

A critical factor in the study of many of the chloroplast components and functions is the length of time required for the isolation procedure. Isolated chloroplasts have a limited survival time *in vitro*, with degradation of components and loss of activities increasing with time. For some experiments, short-cuts can be taken in the isolation procedure, balancing some loss of purity against an increase in activity. But for the critical analysis of chemical composition, satisfactory results can be achieved only with highly purified preparations.

The soluble fraction or stroma of the chloroplasts can be obtained by suspending the intact, purified organelles in a dilute buffer to deliberately achieve lysis. After centrifugation of the sample of lysed chloroplasts, the supernatant fluid is assayed for the presence of various soluble enzymes and other components. The membranes, collected in the pellet after centrifugation, contain both the thylakoid membranes, which are the predominant type, and the membranes of the envelope. Further fractionation of these membranes on the basis of their difference in density allows the preparation of pure membrane types.

II. Components of the Thylakoid Membrane

Before proceeding to the mechanisms of photosynthesis, it is well to consider the properties of the reactants engaged in the process. Foremost are the

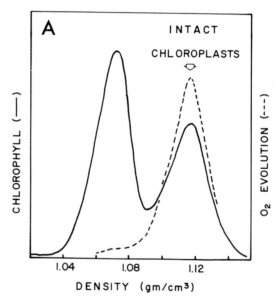

FIG. 3.3. A: A profile of chlorophyll and oxygen evolution obtained after sedimentation of a crude chloroplast fraction into a density gradient of silica sol. Intact chloroplasts (density about 1.12 g/cm^3) were separated from broken chloroplasts (density about 1.06 g/cm^3). Intact chloroplasts retain components that are required for transferring electrons photosynthetically from water to carbohydrate intermediates. Thus, only intact chloroplasts that can fix CO_2 also can evolve O_2. B: An electron micrograph of the crude preparation, containing both intact and broken chloroplasts. C: The purified, intact chloroplasts obtained after separation on gradients of silica sol. The chloroplasts were obtained from spinach leaves. (From Morgenthaler *et al.*, 1975.)

components of the thylakoid membranes. Although current knowledge is far from complete in terms of the participants in every reaction of photosynthesis and of the properties of these participants, a significant amount of information is available, particularly on the major constituents. Table 3.1 lists the major types of lipids in thylakoid membranes and their amounts in relation to chlorophyll. Each of these will be discussed in turn in the following sections. The properties of several major protein components will be described after the section on the lipids.

A. Chlorophyll

Most conspicuous of the chloroplast's chemical components, although not the major ones, are the green chlorophylls. The most important of these lipid

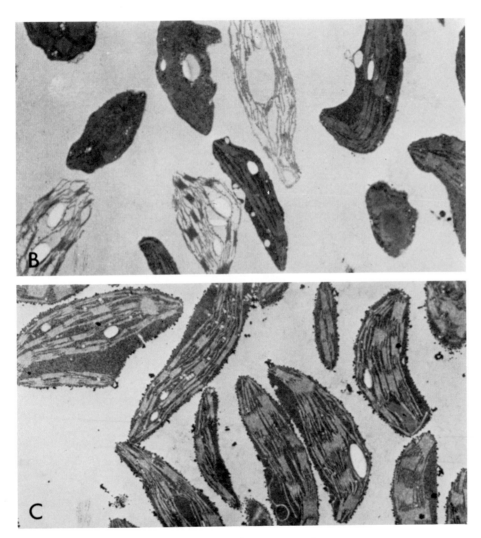

FIG. 3.3 (Continued)

TABLE 3.1. Major Lipid Components of Thylakoid Membranes[a]

Lipid	Percent by weight	Moles per 100 moles of chlorophyll
Chlorophylls	20–22	100
Chlorophyll a		70
Chlorophyll b		30
Carotenoids	3	22
Lutein		10
β-Carotene		6
Violaxanthin		3
Neoxanthin		3
Galactolipids	40–50	200–300
Monogalactosyl diglyceride		140–200
Digalactosyl diglyceride		60–100
Sulfoquinovosyl diglyceride	ca. 5	20–25
Phosphatidyl glycerol	ca. 5	20–30
Quinones	ca. 3	ca. 20
Plastoquinone A		7
Plastoquinone B		4
Others		ca. 9
Unidentified[b]	ca. 12	

[a]Data from Lichtenthaler and Park, 1963, and Vernon and Shaw, 1971.
[b]An ether lipid has recently been discovered in *Chlamydomonas*, in which diacyl glycerol is linked through an ether oxygen to trimethyl homoserine (see Fig. 3.11). This lipid accounts for 10–13% of the membrane lipid.

pigments is chlorophyll *a*, found in all photosynthetic plant cells. Chlorophyll *a* is the only green pigment in some algae, but in many plant cells it is accompanied by either chlorophyll *b* or *c* (Table 3.2). Chlorophyll *b* is the minor form only in the higher plants and the green algae and is usually present in a ratio of one chlorophyll *b* molecule to two to four molecules of chlorophyll *a*. Chlo-

TABLE 3.2. Distribution of Chlorophylls among Photosynthetic Organisms

Organism	Chlorophyll		
	a	b	c
Cyanophyta	+	−	−
Rhodophyta	+	−	−
Cryptophyta	+	−	+
Dinophyta	+	−	+
Chrysophyta			
Xanthophyceae	+	−	−
Chrysophyceae	+	−	+,−
Bacillariophyceae	+	−	+
Phaeophyta	+	−	+
Euglenophyta	+	+	−
Chlorophyta	+	+	−
Higher Plants	+	+	−

(+) Indicates the presence and (−) indicates the absence of the specific form of chlorophyll.

rophyll c occurs in some species of the Chromophyta (Table 3.2). Very little is known about the synthesis of these minor forms of chlorophyll, but a curious correlation arises from a comparison of Table 3.2 with Table 2.1 in Chapter 2. The algae that contain chlorophyll c are among those whose chloroplasts are surrounded by the added membranes of the chloroplast endoplasmic reticulum. These algae do not contain chlorophyll b. Another minor form, chlorophyll d, has been obtained from extracts of some Rhodophyceae.

The three principal chlorophyll analogues were first separated by Stokes in 1864, who also separated the two major carotenoid components of chloroplasts. The first extensive studies of the structure of chlorophyll were carried out by Richard Willstätter and his colleagues (1928) in Germany during the first decade of this century. A significant event that also occurred during this period was the development of column chromatography for the separation of plant pigments, described in 1905 by Michael Tswett, a botanist of Russian origin who worked in Geneva. However, Willstätter chose not to use this powerful new tool, preferring instead to use solvent extractions to achieve separation of the pigments.

Another great German chemist, Hans Fischer, who also extensively studied the structure of heme, nearly completed the determination of the structure of chlorophyll a by 1940. The final positional and stereochemical assignments of several hydrogen atoms were made by Linstead's group in the early 1950s (Ficke et al., 1956). As the ultimate proof of structure, the complete chemical synthesis of chlorophyll a was accomplished by Robert Woodward and his colleagues in 1961.

The structures of the chlorophylls are shown in Fig. 3.4. Chlorophyll a and b are amphipathic molecules, whose properties are determined by their two distinct parts. The presence of a long, isoprenoid alcohol, **phytol,** esterified to the carboxyl group of the propionic acid residue on ring D, affords the molecule its lipid solubility. This property apparently determines the membrane as the chlorophylls' site of residence and function. Since chlorophyll c lacks the phytol moiety, it is a naturally occurring **chlorophyllide.** (The ending -ide indicates the porphyrin without the alcohol side chain.) Chlorophyll c is not very soluble in lipids or in organic solvents at neutral pH, and its role in cells that contain it is not known. All chlorophylls contain the porphyrin ring structure, with its conjugated arrangement of double bonds, and it is this portion of the molecule that gives them the ability to absorb light and to function in the process of photosynthesis.

The chlorophylls, as porphyrins, are cyclic tetrapyrrole compounds, which differ from each other by the substituents on the pyrrole rings and by the number of double bonds within the cyclic, conjugated system. Chlorophylls a and b differ only in that the methyl group on pyrrole ring B of chlorophyll a is oxidized to a formyl group in chlorophyll b. This alteration changes slightly the polarity of the molecule and also shifts the absorption spectrum. Fig. 3.5 shows the absorption spectra of these two chlorophylls. Chlorophyll a has absorption maxima (in ether) at 428.5 and 660 nm, whereas those for chlorophyll b are 452.5 and 642 nm. Chlorophyll c_1 shows maximal absorption at 444 and 628 nm. However, the in vivo spectra of these pigments have different maxima than those found in organic solvents and are more complex, as shown in Fig. 3.6. By care-

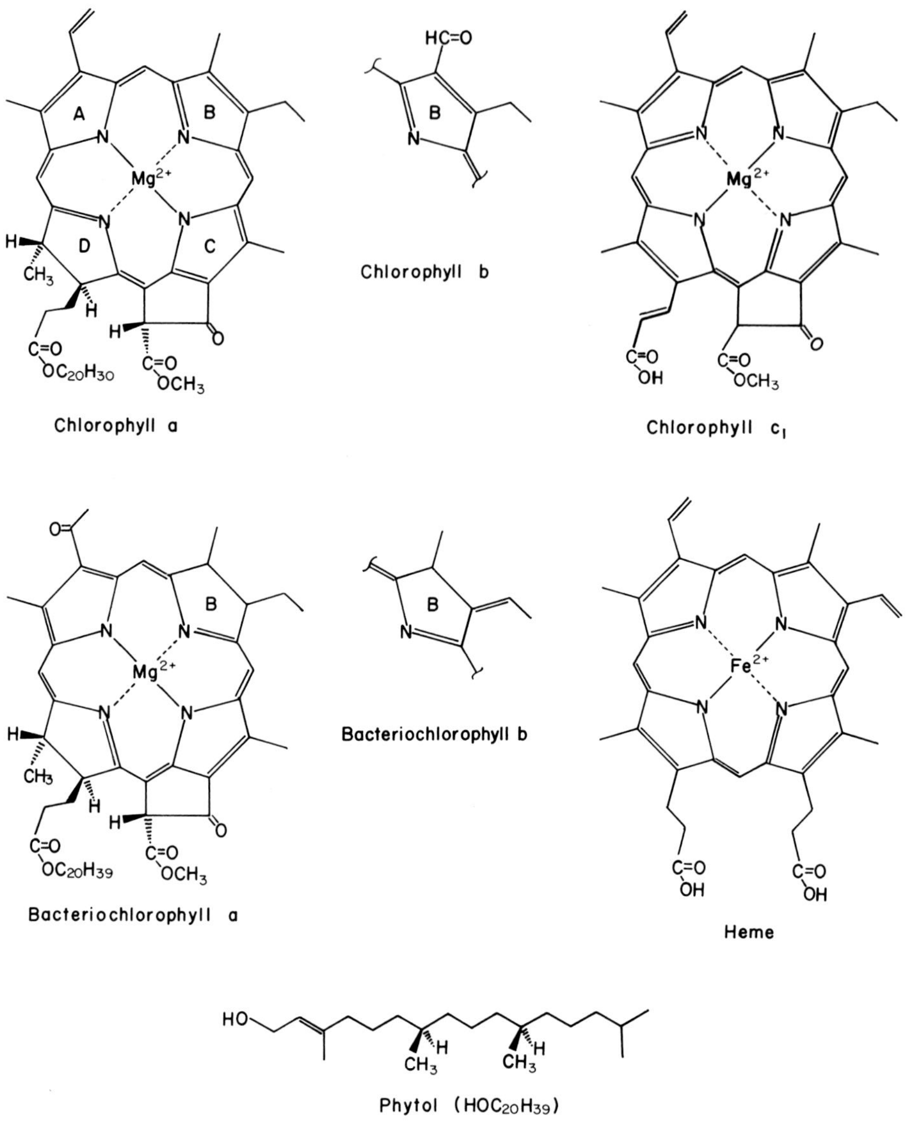

Chlorophyll b

Chlorophyll a

Chlorophyll c_1

Bacteriochlorophyll b

Bacteriochlorophyll a

Heme

Phytol $(HOC_{20}H_{39})$

FIG. 3.4. Structures of the major porphyrins found in nature. Chlorophyll *a* exists predominantly in the monovinyl form, as shown, but the divinyl form, with a vinyl group in place of the ethyl group on ring B, also occurs. Chlorophyll *b* is identical to chlorophyll *a* except for the presence of the formyl group on ring B. Chlorophyll c_2 differs from chlorophyll c_1 in having a vinyl side chain on ring B rather than the ethyl group. Bacteriochlorophylls have both ring B and D reduced. Bacteriochlorophyll *b* contains an ethylidene side chain on ring B in place of the ethyl group. In bacteriochlorophylls *c*, *d*, and *e*, the carbonyl group in the side chain on ring A is reduced to the α-hydroxyethyl substituent, and the $-COOCH_3$ group is not present on the isocyclic ring. Bacteriochlorophyll *e* also contains a formyl group on ring B, similar to chlorophyll *b*. Chlorophylls generally are esterified to the C_{20} alcohol phytol, but the bacteriochlorophylls also may contain farnesol or geranylgeranol, more highly unsaturated derivatives. All chlorophylls contain a Mg^{2+} ligand coordinated with the nitrogen atoms in the pyrrole rings. Heme is protoporphyrin IX with a Fe^{2+} ligand.

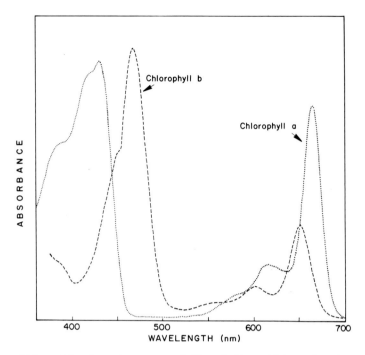

FIG. 3.5. The absorption spectra of chlorophyll *a* and *b* in methanol.

FIG. 3.6. Composite absorption spectrum of purified chloroplast thylakoid membranes. The heavy line is the observed spectrum, measured at −196°C, of membranes from spinach that contain both photosystem 1 and 2. The dotted curves indicate spectra of Gaussian subcomponents of the overall spectrum. Each of these subcomponents is considered to be a different form of chlorophyll. The analysis revealed several forms of chlorophyll *a* (Chl *a*661, Chl *a*669, Chl *a*677, Chl *a*684, Chl *a*691, and Chl *a*700–720). The environment of the chlorophyll molecules and the proteins to which they are bound influence the spectral properties. The forms absorbing at the longest wavelengths (690–720 nm) are components of photosystem 1 and are usually present in very small amounts. Chlorophyll *b* exists in two forms (Chl *b*640 and Chl *b*650). (Adapted from French, 1971.)

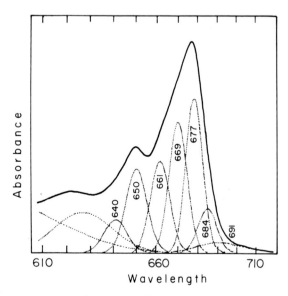

ful analysis of the spectra of thylakoid membranes or whole cells, numerous spectral forms of the pigments can be detected. The variable spectral properties of chlorophyll *a in vivo* apparently result both from aggregation of the pigment within the membrane and from combinations with other membrane components, particularly specific proteins. As many as six forms of chlorophyll *a* exist within the chloroplast, whereas chlorophyll *b* appears to exist in perhaps two forms. Recent studies have indicated that most, and perhaps all, of the chlorophyll in the membrane is noncovalently associated with proteins.

The chlorophylls all contain a central Mg^{2+} ion as a chelate complex, in which the metal is held by coordination bonds with the nitrogens of the pyrrole rings. The Mg^{2+} is lost readily if chlorophyll is exposed to an acidic environment. The porphyrins lacking the metal ion are called **pheophytins.** The pheophytins apparently are not present at any significant level within cells, but care must be taken during the extraction and analysis of chlorophyll that the conditions do not promote generation of these derivatives. Because the spectral properties of the pheophytins vary from those of the chlorophylls, spectrophotometric assays are subject to error if loss of the Mg^{2+} occurs.

By way of contrast, the other major type of porphyrin structure within cells, the **heme,** contains an iron atom in place of Mg^{2+} (Fig. 3.4). This change leads to profound differences in the manner with which these two types of metalloporphyrins function. Both types of porphyrins are involved in one-electron oxidation–reduction reactions. In the case of chlorophyll, the absorption of light leads to excitation of an electron in the conjugated ring system and the subsequent transfer of a ring π electron to an acceptor. On the other hand, the hemes carry out oxidation–reduction reactions as part of the cytochromes by changing the valence state of the central iron ion. In fact, the hemes are not light-sensitive reagents, since the paramagnetic iron atom is an effective quencher of the type of excited states involved in the photochemical reactions of chlorophyll.

Photosynthetic bacteria contain still other chlorophyll derivatives, the **bacteriochlorophylls.** Bacteriochlorophyll *a* differs from chlorophyll *a* by having pyrrole ring B in addition to ring D in the reduced form (Fig. 3.4). Bacteriochlorophyll *b* is similar to bacteriochlorophyll *a* but has an ethylidene group on ring B instead of an ethyl group (Fig. 3.4). The absorption spectrum of bacteriochlorophyll *a* is shown in Fig. 3.7. The bacteriochlorophylls absorb light principally in the near infrared and near ultraviolet regions.

The primary function of the chlorophylls in photosynthesis is the absorption of light energy. The energized chlorophyll molecules then drive the endergonic reactions involved in the synthesis of chemical energy. This ability of chlorophyll makes possible the process of photosynthesis.

Chlorophyll is not the predominant chemical component of thylakoid membranes but it is the most easily measured. Since it is found only in these membranes, it is often used as a marker for chloroplast development. Mature chloroplasts contain about 1–2 pg, or 5 to 10×10^8 molecules, of chlorophyll per plastid.

B. Carotenoid Pigments

About 3% of the weight of the thylakoid lipids is contributed by the intensely yellow or orange pigments, the **carotenoids.** The four principal mem-

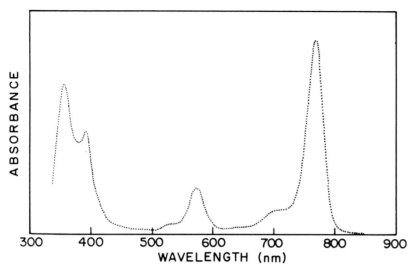

FIG. 3.7. Absorption spectrum of bacteriochlorophyll *a* in diethyl ether.

bers of this group in green plants are listed in Table 3.1. The major carotenoid is β-carotene, along with its products of oxidation. The carotenoids contain an extensive system of conjugated double bonds, as shown by their structures in Fig. 3.8. As a result, they strongly absorb blue light. The absorption spectrum of β-carotene is shown in Fig. 3.9. However, the efficiency of the transfer of absorbed energy from these components to the reaction centers of photosynthesis is quite low. It is for this reason that the efficiency of photosynthesis in blue light is often lower than that in red light, even though the energy contained in a photon of blue light is greater. Therefore, the main purpose of the carotenoids does not seem to be that of light harvesting. Rather, they play a very important *protective* function within the chloroplast membranes.

This protective function can best be exemplified by describing the process of photooxidation. Oxidative reactions are facilitated by a large number of colored compounds that absorb light energy. Examples of such "photosensitizers" are chemically synthesized dyes (methylene blue and rose bengal) and naturally occurring pigments (chlorophyll). In the light, these compounds can catalyze the oxidation of most cellular materials. Thus, the combination of an endogenous natural pigment or an exogenously supplied dye, an aerobic environment, and light can have disastrous effects on cellular structures.

The mechanism of photooxidation is as follows. The energy of the excited, triplet state of the dyes, attained by the absorption of light, is effectively transferred to molecular oxygen to yield excited, singlet-state oxygen, indicated by $^1O_2$ in the following reactions (excited states are indicated by *):

$$D + hv \rightarrow D_s^* \qquad \text{(initial generation of singlet-state dye)} \qquad [3.1]$$

$$D_s^* \rightarrow D_T^* \qquad \text{(intersystem crossing of singlet- to triplet-state dye)} \quad [3.2]$$

$$D_T^* + {}^3O_2 \rightarrow D + {}^1O_2^* \quad \text{(quenching by oxygen)} \qquad [3.3]$$

By these reactions, the dye (D) acts as a "photosensitizer" of oxidation by activating molecular oxygen. Transfer of energy from the triplet state of the dye to oxygen transforms ground-state, triplet oxygen to the excited-state, singlet oxygen. Singlet oxygen is highly reactive and will attack an easily oxidized, electron-rich structure, such as the double bonds in membrane lipids and in the side chains of several amino acids in proteins. If the oxidative damage is extensive enough, the cell cannot survive. Plant cells, existing as they do in a self-made aerobic environment and exposed to intense light from the sun, have the ingredients necessary to destroy themselves, since chlorophyll is an excellent photosensitizing dye. An additional aspect of the potentially damaging

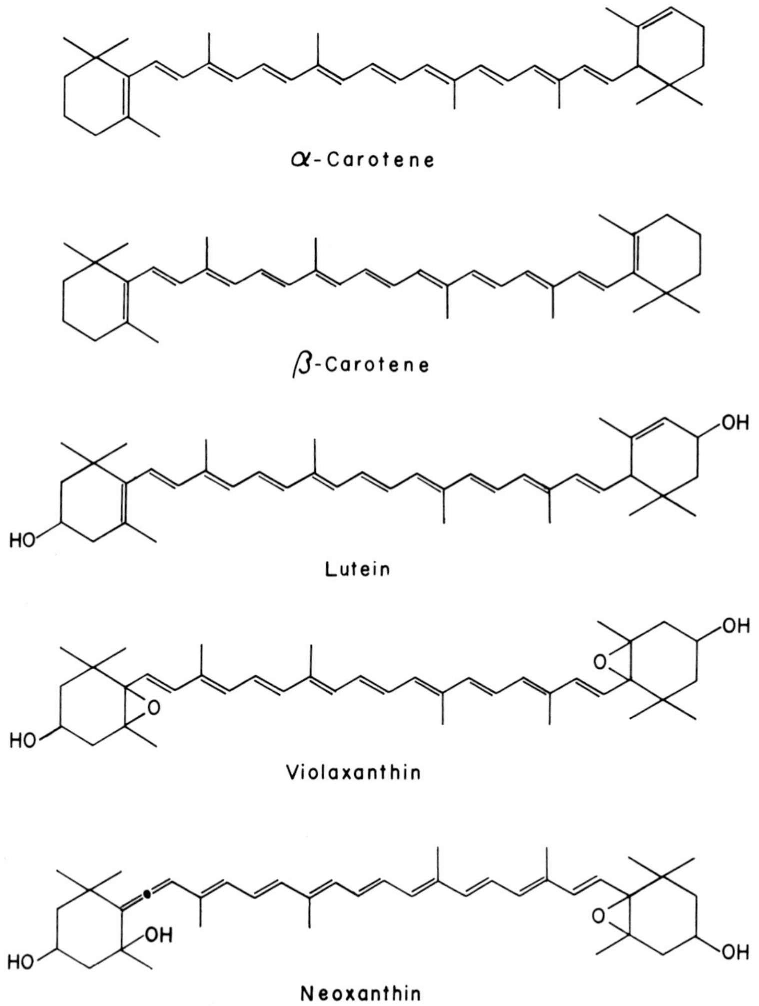

FIG. 3.8. Structures of the major carotenoids in chloroplasts. α- and β-carotene differ only in the position of the double bond in one ring. The other carotenoids are products of oxidation of the carotenes.

effect of chlorophyll-sensitized photooxidation is the solubility property of oxygen. Since it is five to ten times more soluble in hydrocarbon solvents than in water, the relatively high concentration of oxygen within the thylakoid membranes, where it is produced, also would promote destruction of these structures.

The substances that protect plant cells from chlorophyll-sensitized destruction are the carotenoids, lipid substances located in the membranes along with chlorophyll. Chlorophyll generally reacts in functional photoreactions in the singlet state (see Chapter 4), but if the initial excited state is not immediately trapped, it will achieve the triplet state by intersystem crossing. Triplet chlorophyll can be quenched directly by carotenoids. This reaction is very rapid and is the major protective function of carotenoids. However, the longer lifetime of the triplet state enables molecular oxygen to also quench the excited chlorophyll. The transfer of energy to oxygen will produce singlet oxygen. The carotenoids also are exceptionally effective in quenching singlet oxygen, without being oxidized themselves.

$$^1O_2^* + Car \rightarrow O_2 + Car^*$$

$$Car^* \rightarrow Car + heat$$

[3.4]

A conjugated system is necessary for this quenching effect, with the most effective system containing 10 or more double bonds. β-Carotene and the other

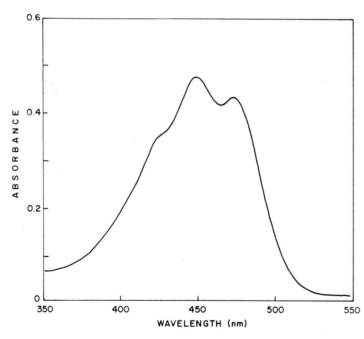

FIG. 3.9. Absorption spectrum of the carotenoid fraction from etiolated cells of *Chlamydomonas reinhardtii* y-1. The major pigment in this fraction is β-carotene.

carotenoids are well suited for this role in protection of the chloroplast and of the cell.

A poorly understood function of the carotenoids is their role as major constituents of the **eyespot** or **stigma** within the chloroplasts of many flagellated algae. As shown in Fig. 3.10, the eyespot is an organized arrangement of primarily carotenoid globules. It has been suggested that the eyespot is involved in phototaxis in these algae, which respond by moving either toward or away from a light source. However, the role of the eyespot in this process is neither understood nor established.

C. Galactosyl Diglycerides

The most abundant chloroplast lipids, which constitute the bulk of the bilayer matrix of the thylakoid membrane, are the **mono-** and **digalactosyl diglycerides.** These glycolipids (carbohydrate-containing lipids) make up about 50% of the total lipid of the membrane. The monogalactosyl lipid is the predominant form, and the ratio of the mono- to the digalactosyl forms is usually 2:1 in mature chloroplasts. The structures of these lipids are shown in Fig. 3.11.

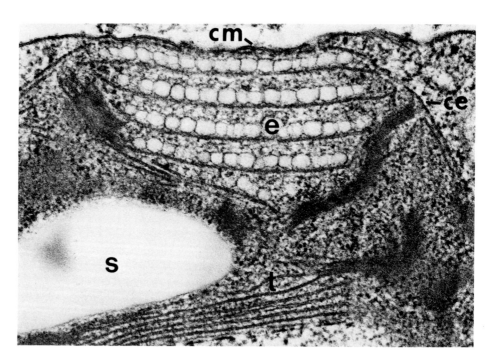

FIG. 3.10. Electron micrograph of the eyespot within the chloroplast of *Chlamydomonas reinhardtii.* The eyespot (e) lies just inside the chloroplast envelope (ce) and is composed of four layers of pigment globules. These globules are rich in carotenes. Each layer of globules is covered on its inner surface by a single thylakoid. Over the position of the eyespot, the chloroplast is very near the cell membrane (cm). Thylakoids (t) and a starch grain (s) also are seen. (×53,000. Micrograph courtesy of L. A. Staehelin.)

The galactosyl diglycerides are very rich in unsaturated fatty acids, particularly in the triunsaturated linolenic acid (Table 3.3). These lipids are, like other membrane lipids, *amphipathic* structures, which means that they contain both polar, water-soluble and nonpolar, water-insoluble portions. Since they contain no phosphate, they are uncharged. Chloroplast membranes function in close apposition to each other, particularly within the grana, and the absence of ionic groups in these lipids may be important in preventing electrostatic repulsion of adjacent membrane surfaces.

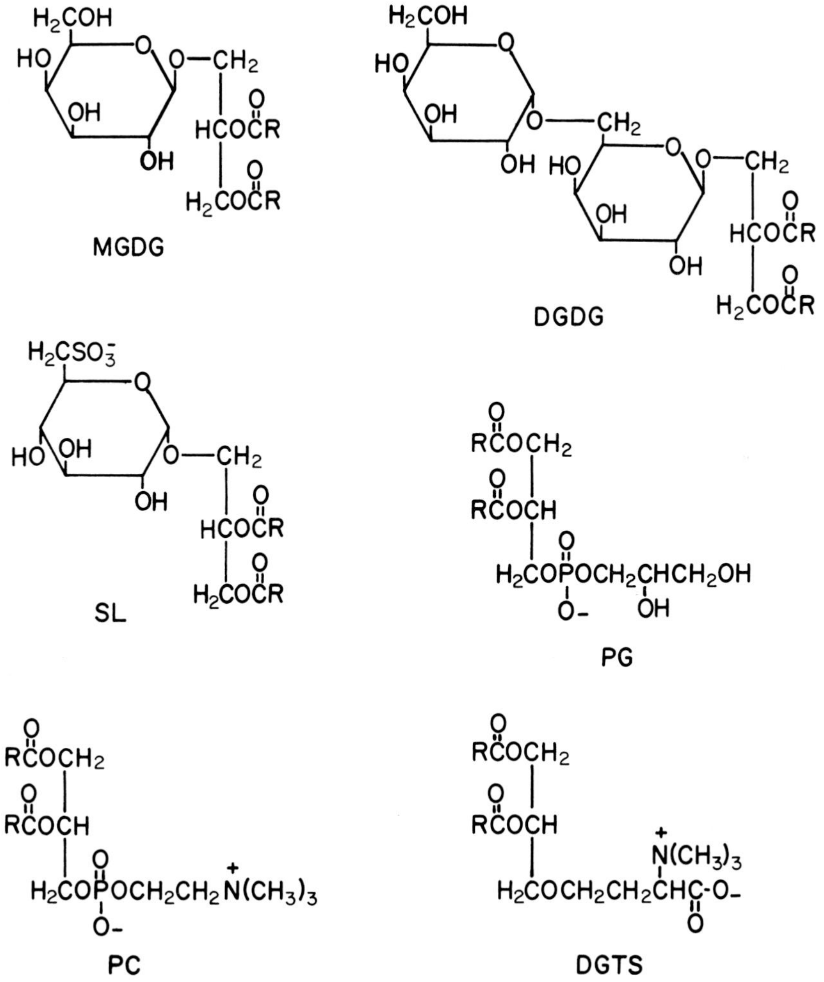

FIG. 3.11. Structures of the polar lipids of thylakoid membranes. MGDG, monogalactosyl diglyceride [1,2-diacyl-3-O-β-D-galactopyranosyl-sn-glycerol]; DGDG, digalactosyl diglyceride [1,2-diacyl-3-O-(α-D-galactopyranosyl-($1''\rightarrow6'$)-O-β-D-galactopyranosyl)-sn-glycerol]; SL, sulfoquinovosyl diglyceride [1,2-diacyl-3-O-(6-sulfo-α-D-quinovopyranosyl)-sn-glycerol]; PG, phosphatidyl glycerol; DGTS, diacylglceryl trimethyl homoserine [1,2-diacylglyceryl-3-O-4'(N,N,N-trimethyl) homoserine]; PC, phosphatidyl choline.

TABLE 3.3. Fatty Acid Composition (%) of Individual Lipid Components in Spinach Chloroplast Thylakoid Membranes[a]

Component	16:0	Δ³-trans-16:1	16:3	18:0	18:1	18:2	18:3
Monogalactosyl diglyceride	Tr	—	25	—	Tr	2	72
Digalactosyl diglyceride	3	—	5	—	2	2	87
Sulfolipid	39	Tr	Tr	Tr	1	7	53
Phosphatidyl glycerol	11	32	2	Tr	2	4	47

[a]Small amounts of unsaturated C-14, C-20, and C-22 fatty acids also are present. Tr, trace. (Adapted from Kates, 1970.)

D. Minor Lipids

As Table 3.1 shows, chloroplast membranes contain a small amount of a unique **sulfoglycolipid.** This lipid, comprising about 5% of the total lipid fraction, contains the sulfated derivative of 6-deoxyglucose. This sugar, also called **quinovose,** is attached to the diacyl glycerol portion of the molecule by a glycosidic linkage. The structure of the sulfolipid is shown in Fig. 3.11.

Phospholipids, the major lipid type in other cellular membranes, comprise less than 10% of the lipid of thylakoid membranes. The predominant phospholipid in the chloroplast is **phosphatidyl glycerol** (Fig. 3.11). This lipid is unique in that it is the only one that contains a significant amount of the naturally occurring *trans*-3-hexadecenoic acid (see Table 3.3), which is located specifically at the 2-position of glycerol. All other unsaturated fatty acids in the chloroplast contain *cis* double bonds. Furthermore, in plant cells this *trans* unsaturated fatty acid has been found only in thylakoid membranes.

Recently, another thylakoid lipid has been found in the algae *Ochromonas* and *Chlamydomonas.* This lipid contains an N-trimethyl homoserine residue in ether linkage to diacyl glycerol (Fig. 3.11). Whether this lipid is common to all plants is not known, but it has not been found in nonplant cells. In *Chlamydomonas* it accounts for 10–13% of the membrane lipid. The zwitterionic nature of this unusual lipid may have an important, but as yet not understood, function within the membrane.

E. Quinone Compounds

A group of quinone derivatives, called **plastoquinones** because of their location, are found in thylakoid membranes. The structures of the two predominant derivatives, plastoquinones A and B, are shown in Fig. 3.12. Quinones undergo oxidation–reduction reactions as components within the electron transport chain of the membrane. Since the oxidation and reduction of the quinones also involve the release and uptake of H^+ ions, these compounds are thought to be part of the system that transports protons across the membrane as electrons are transferred through the electron transport chain (see Chapter 4).

F. Accessory Pigments

Some plant cells have predominant colors other than green. A pale green or yellow color often is caused simply by a deficiency of chlorophyll. But some

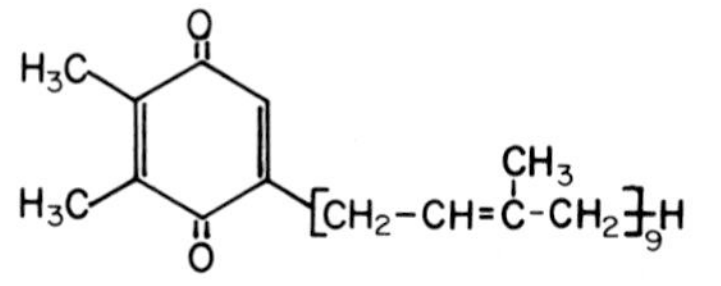

FIG. 3.12. Structures of the major quinone compounds in thylakoid membranes.

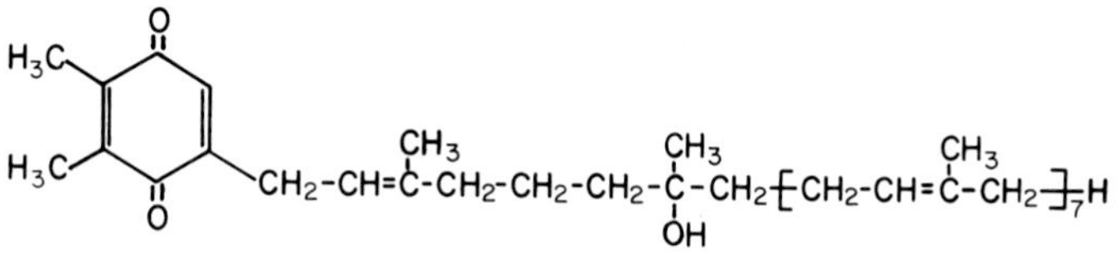

cells, particularly the cyanobacteria (Cyanophyta), the red algae (Rhodophyta), and the cryptomonad algae (Cryptophyta), appear blue or red because of the presence of additional pigments within the chloroplast. Although not part of the chloroplast membrane *per se*, these pigments are found as complexes with proteins on the surface of the thylakoid membranes. In cyanobacteria and the red algae these chromoproteins, called **phycobiliproteins,** are aggregated into large particles, the **phycobilisomes.** In Chapter 2, Fig. 2.7 illustrates the chloroplast of a red alga and shows the relationship of the phycobilisomes with the membranes.

Although there are several types of phycobiliproteins (Table 3.4), only two major types of chromophores, called **phycobilins,** are found in these pigment–protein complexes. These are **phycocyobilin,** which absorbs light of about 630 nm and imparts a blue color to cells, and **phycoerythrobilin,** which absorbs

TABLE 3.4. Composition of the Phycobiliproteins[a]

Biliprotein	Subunit	Subunit mol. wt.	Phycobilins per subunit	Color of subunit
C-phycoerythrin $(\alpha,\beta)_6$	α	19,700	2 PEB	Red
	β	22,000	4 PEB	Red
C-phycocyanin $(\alpha,\beta)_6$	α	18,500	1 PCB	Blue
	β	20,500	2 PCB	Blue
R-phycoerythrin $(\alpha,\beta)_6\gamma$	α	19,500	? PEB	Red
	β	19,500	? PEB, PUB	Orange
R-phycocyanin $(\alpha,\beta)_3$	α	18,500	1 PCB	Blue
	β	20,500	1 PCB, 1 PEB	Red
Allophycocyanin $(\alpha,\beta)_3$	α	15,500	1 PCB	Blue
	β	15,500	1 PCB	Blue

[a]C-phycobiliproteins are found in Cyanophyta and R-phycobiliproteins are found in Rhodophyta. Allophycocyanins of similar spectral properties are found in both. The colors of intact organisms are provided by the biliproteins in combination with the green and yellow colors of chlorophyll and the carotenoids. PEB, phycoerythrobilin; PCB, phycocyanobilin; PUB phycourobilin.

principally light of about 550 nm and is the predominant form in the red algae. **Phycourobilin** is a minor chromophore in some biliproteins (Table 3.4) and absorbs maximally light of 498 nm. The phycobilins have an open tetrapyrrole structure, as shown in Fig. 3.13. An inspection of the side chains of these compounds suggests that they are not degradation products of chlorophyll but are formed directly from another porphyrin precursor. By analogy to heme degradation in animal cells, it is thought that heme or protoporphyrin IX may be the precursor of the phycobilins. In this conversion, the methylidene bridge between rings A and B of the porphyrin is oxidized, causing the macrocyclic structure to open (refer to Fig. 3.4). The bridge carbon is lost as CO. Additional modifications also occur, but little is known about these reactions.

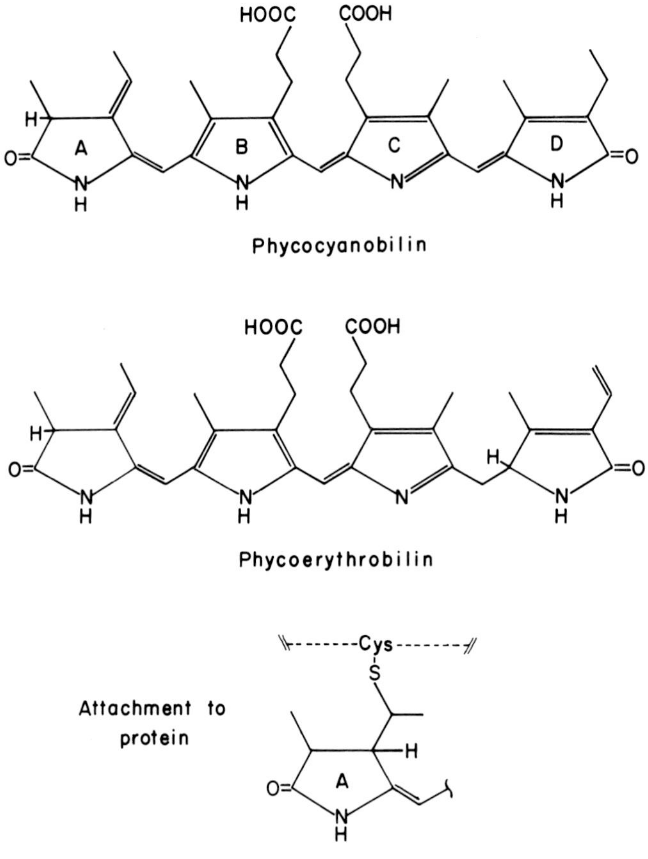

FIG. 3.13. Structures of phycocyanobilin and phycoerythrobilin. These two open tetrapyrrole chromophores differ in the bridge between rings C and D (a methylidene bridge in phycocyanobilin and a methylene bridge in phycoerythrobilin) and in the side chain on ring D (an ethyl group in phycocyanobilin and a vinyl group in phycoerythrobilin). The chromophores are covalently attached to the protein through addition of a sulfhydryl group of a cysteine residue in the protein across the double bond of the ethylidine side chain on pyrrole ring A. The propionate side chain on pyrrole ring C also may be esterified with a hydroxyl group of a serine residue in the protein, but evidence for this second linkage is not conclusive.

The chromophores are linked covalently to the proteins by addition of the sulfhydryl group of a cysteine residue across the double bond of the ethylidene group on pyrrole ring A. A second attachment also may occur through an ester linkage between the propionic acid side chain on ring C and a serine residue in the protein. Differences in the proteins apparently cause the differing spectral properties within each group of phycobiliproteins (Fig. 3.14). Each type of protein contains two different subunits, an α and a β subunit, and each one bears from one to four chromophores. Allophycocyanin contains one chromophore per subunit polypeptide, whereas phycoerythrin may contain as many as four chromophores per subunit polypeptide. The polypeptide subunits have molecular weights in the range of 15,000–22,000. The smallest stable unit of the

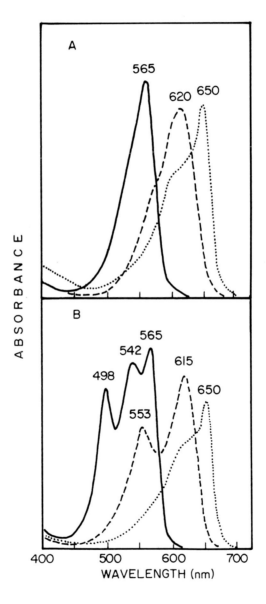

FIG. 3.14. Absorption spectra of phycobiliproteins. A: The phycobiliproteins of cyanobacteria. _____, C-phycoerythrin; ----------, C-phycocyanin; . . . . , allophycocyanin. B: The phycobiliproteins of red algae. _____, R-phycoerythrin; ----------, R-phycocyanin; . . . . , allophycocyanin. The differences in the spectra of the phycocyanins and allophycocyanins are not caused by the chromophore, which is the same, but by the polypeptide portions of the complexes.

phycobiliproteins usually is a trimer $(\alpha, \beta)_3$, but the basic building block of the phycobilisome is the hexamer $(\alpha, \beta)_6$.

Phycobilisomes recently have been shown to be specific heteroaggregates composed of three phycobiliproteins (Fig. 3.15). The core of the phycobilisome consists of allophycocyanin. Extending radially from this core is first a layer of phycocyanin, to which is attached stacked rods composed of phycoerythrin. In organisms lacking phycoerythrin, the rods contain phycocyanin. The complete phycobilisome in red algae is roughly hemispherical in shape, with pheripheral rods extending in all directions.

The role these accessory pigment–protein complexes play in the chloroplast is that of additional light-harvesting complexes. The phycobiliproteins absorb light of wavelengths that are poorly absorbed by chlorophyll and thereby considerably enhance the photosynthetic capability of the cells that contain them (Fig. 3.16). The light energy captured by these complexes is transferred to photosystem 2 with an efficiency greater than 90%.

Excitation energy can be transferred from one molecule to another only when their absorption spectra overlap, and then only in the direction of decreasing energy levels (i.e., longer wavelengths). Thus, the order of energy transfer within the phycobilisome is from the shorter-wavelength–absorbing phycoerythrin to the phycocyanins in the sequence: phycoerythrin → phycocyanin → allophycocyanin → photosystem 2. This sequence also correlates with the arrangement of the phycobiliproteins within the phycobilisome structure (Fig. 3.15). (See Chapter 4 for a description of the photosystems.)

It is interesting that some organisms, such as *Fremyella* and *Tolypothrix* (cyanobacteria), respond to different wavelengths of incident light by changing the composition of their phycobilisomes. Thus, growth in red light produces cells containing phycocyanin but little phycoerythrin, whereas phycoerythrin predominates in cells grown in blue or green light. This process, called **complementary chromatic adaptation,** is brought about by changes in the rates of synthesis of the phycobilins and of their respective polypeptide subunits. The switch in synthesis of one phycobiliprotein to the other is mediated by the wavelength of light, but the mechanism of this control process is still to be determined.

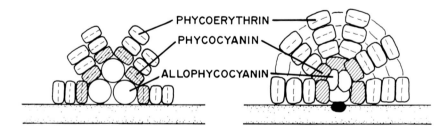

PHYCOERYTHRIN

PHYCOCYANIN

ALLOPHYCOCYANIN

FIG. 3.15. Arrangement of the phycobiliproteins within the phycobilisome. The model on the left represents thin, hemidiscoidal-shaped phycobilisomes in the cyanobacteria *Rhodella* and *Pseudoanabaena*. The model on the right, for the phycobilisome in the red alga *Porphyridium cruentum*, shows the internal arrangement of this hemispherical structure. A protein (large black dot) that is shared by both phycobilisome and thylakoid membrane is thought to anchor the phycobilisome to the membrane. (From Gantt, 1981. Reprinted with permission of Annual Reviews, Inc.)

A variety of colors are produced in some plants by another group of pigments, the **anthocyanins.** In a few species that have been examined, the anthocyanins were recovered with intact vacuoles of the plant cell and not with the chloroplasts.

G. Integral Membrane Proteins

Our knowledge of the protein fraction of thylakoid membranes has lagged behind that of the lipid fraction and is far from complete. The protein fraction is complex, comprising all the proteins needed for the membrane to perform its functions. To consider membrane proteins, however, requires a definition. There are some proteins that function in association with a membrane but can be removed relatively easily from the membrane by washing with dilute solutions of salts, buffers, or chelating agents (such as ethylenediaminetetraacetic acid). Examples of such loosely bound proteins, called **peripheral proteins,** are the phycobiliproteins; the chloroplast coupling factor 1, a protein required for photophosphorylation; and ribulose 1,5-bisphosphate carboxylase, the CO_2-fixing enzyme (see Chapter 4). But after the removal of most loosely bound pro-

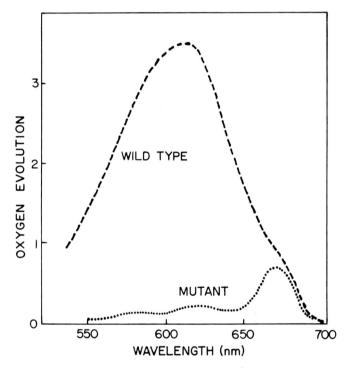

FIG. 3.16. Demonstration of the utility of phycobiliproteins in photosynthetic light harvesting. The graph shows the action spectra for photosystem 2 in wild-type cells of the red alga *Cyanidium caldarium,* which contains principally C-phycocyanin, and in a mutant strain lacking phycocyanin. The difference in the spectra illustrates the effectiveness with which light between 550 and 650 nm, absorbed by the phycobilisomes, is transferred to photosystem 2 reaction centers. In the mutant strain, photosystem 2 activity is driven only by the light absorbed by chlorophyll in thylakoid membranes. Photosystem 2 activity was measured by evolution of oxygen. (Redrawn from Diner, 1979.)

teins, additional proteins can be removed by more drastic procedures. Concentrated solutions of urea or guanidine chloride, both agents commonly used to denature proteins, remove all but the most tightly bound, **integral proteins.** Integral proteins generally can be solubilized only with detergents, agents that also destroy the structure of the membrane as well. By these criteria, about 70–80% of the protein associated with isolated thylakoid membranes is recovered in the integral protein fraction.

The most useful procedure for estimating the number of different polypeptides and the relative amounts of each in the membrane has been electrophoresis through a porous gel in the presence of a detergent. The gel is formed by polymerization of a mixture of acrylamide and N,N'-methylene-*bis*-acrylamide (the cross-linking agent) to form a semirigid, insoluble gel. Most membrane proteins have a low solubility in water, which necessitates the use of detergents as solubilizing agents. The detergent most commonly used is sodium dodecyl sulfate. In addition to surrounding each polypeptide chain with polar groups, thus making the complex soluble in water, the detergent also provides a large num-

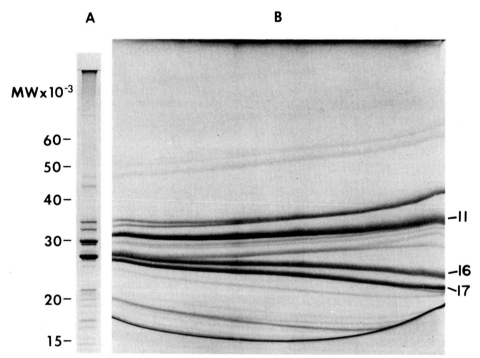

FIG. 3.17. Electrophoretic analysis of the integral polypeptides in thylakoid membranes from *Chlamydomonas reinhardtii*. Purified membranes were washed with 8 M urea containing 1 M KCl to remove polypeptides loosely associated with the membrane. For panel A, a polyacrylamide gradient slab gel (10–20%) was used to separate polypeptides at pH 8.7 in the presence of sodium dodecyl sulfate. The most prominent bands correspond to polypeptides of 29,500 and 26,000 in molecular weight. For panel B, the membrane polypeptides were layered over a slab gel containing a pH gradient of 9.0–8.0 (left to right). Across the gel, bands merge or split depending on the effect of conditions on individual polypeptides. The major polypeptides are numbered as 11 (molecular weight 29,500), 16, and 17 (both apparently 26,000 in molecular weight, from panel A) according to the designation of Chua *et al.* (1975). (Adapted from Hoober *et al.*, 1980.)

ber of negative charges. As a result, all protein–detergent complexes move through the gel toward the anode when an electric field is applied. The amount of detergent that binds to each polypeptide is approximately proportional to its length and is not influenced (except in extreme cases) by the amino acid composition. This property is extremely useful, since the rate of migration through the gel then is a function only of the size of the complex. Smaller polypeptides move more rapidly than larger ones. The molecular weight of any polypeptide, consequently, can be estimated by relating its rate of migration to those of proteins of known size.

Fig. 3.17 shows an example of this type of analysis of the integral polypeptides in thylakoid membranes from the green alga *Chlamydomonas reinhardtii*. In a one-dimensional separation, under a defined set of conditions, polypeptides were separated on the basis of their size (panel A). Two major polypeptide fractions were resolved, with molecular weights of 29,500 and 26,000. However, as shown in panel B, some fractions can be further resolved into more than one polypeptide. A polyacrylamide gel was prepared to contain a gradient in pH from 9.0 to 8.0 (left to right) and thereby subject the polypeptides to different conditions within the same gel. Three major polypeptides were resolved, with the 26,000-molecular-weight fraction splitting into two polypeptides, 16 and 17.

Thylakoid membranes in higher plants are similar in many respects to those in green algae. Figure 3.18 illustrates the electrophoretic analysis of thylakoid polypeptides from bean *(Vicia faba)* and barley *(Hordeum vulgare)*. The major polypeptide again has a molecular weight near 26,000.

This type of analysis is valuable primarily as a tool for determining the complexity of the protein fraction and the sizes of the polypeptide components. A further refinement has been the development of a two-dimensional analysis in which polypeptides first are separated by their isoelectric points within a porous gel containing a gradient of pH (a technique called **isoelectric focusing**). This gel then becomes the sample layer for a second polyacrylamide gel, into which the polypeptides migrate, according to their size, during electrophoresis in the presence of sodium dodecyl sulfate. The number of different polypeptides detected by this method is at least double that in one-dimensional procedures.

It is not possible to identify the function of any polypeptide simply by examining electrophoretic patterns. Several polypeptides have been purified, however, and their positions in the electrophoretic pattern established. In Fig. 3.18 the bands that correspond to the largest subunits, α and β, of the coupling factor CF1 are marked. The position of cytochrome f, an electron transport protein (see Chapter 4) also is marked.

H. Chlorophyll–Protein Complexes

Of great interest have been polypeptides that migrate during electrophoresis in association with chlorophyll. Included in this group are two major polypeptides, one 64,000–68,000 and the other 26,000–30,000 in molecular weight. After treatment of thylakoid membranes under mild conditions, with sufficient detergent to just solubilize the membrane, both major polypeptide groups can be purified as complexes with chlorophyll. Studies on these complexes indicate that they are present also *in vivo* and are not artifacts of the procedure.

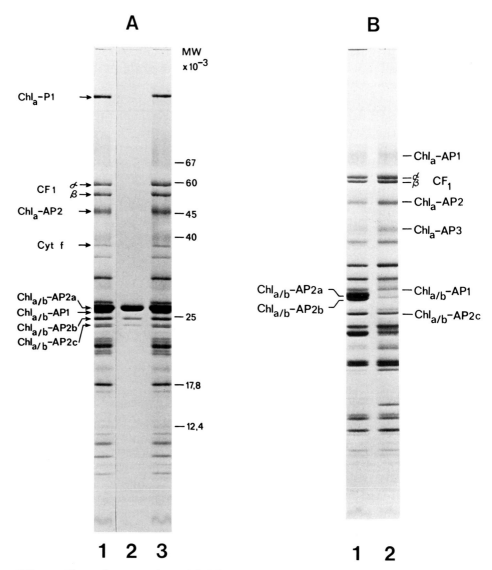

FIG. 3.18. Electrophoretic analysis of thylakoid polypeptides from higher plants. A: In lanes 1 and 3 is shown the pattern of total polypeptides in membranes from bean (*Vicia faba*). The major polypeptide has a molecular weight of about 26,000 and is designated $Chl_{a/b}$-AP2a [apoprotein (AP) a of the light-harvesting chlorophyll a/b-protein complex 2]. This polypeptide is the major component in highly purified preparations of this complex (lane 2). Other minor polypeptides in this complex are $Chl_{a/b}$-AP1, -AP2b, and -AP2c. Chl_a-P1 is the chlorophyll-protein complex that contains the reaction center of photosystem 1. This band is not present after membrane samples are heated in boiling water before electrophoresis, since all chlorophyll-protein complexes are destroyed by these conditions.

B: In lanes 1 and 2 are shown patterns of total polypeptides in membranes for barley (*Hordeum vulgare*). Two major polypeptides of the light-harvesting complex are present at approximately equal amounts ($Chl_{a/b}$-AP2a and -AP2b) in the wild-type strain (lane 1). In the *chlorina-f2* mutant strain (lane 2), which phenotypically lacks chlorophyll b and thus the light-harvesting complex, these two polypeptides also are absent. Although the light-harvesting complexes are associated primarily with photosystem 2, the mutant that lacks these complexes still has activity of photosystem 2 reaction centers (see also Fig. 3.19). The polypeptide marked Chl_a-AP3 has been identified as a component of these reaction centers. (Adapted from Machold, 1981. Figure courtesy of O. Machold.)

The larger polypeptide (Chl$_a$–AP1 in Fig. 3.18, about 67,000 molecular weight) is complexed with chlorophyll *a*. This complex is found in thylakoid membranes from all photosynthetic, chlorophyll-bearing plant cells and is referred to as the **P700 chlorophyll *a*–protein complex** or Chl$_a$–P1 (chlorophyll *a*–protein complex 1 or simply CP1) (see Fig. 3.19). The specialized chlorophyll molecules at the reaction center of photosystem 1 have an absorption maximum at 700 nm and are recovered with the purified Chl$_a$–P1 complex.

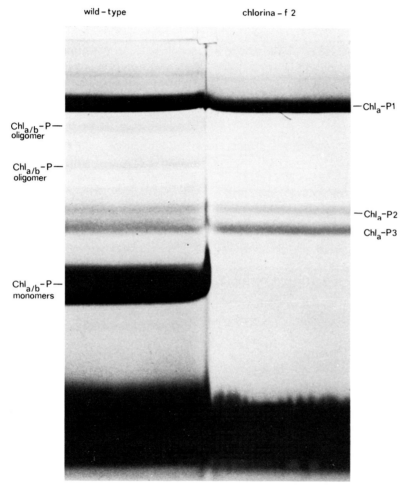

FIG. 3.19. Electrophoretic pattern of the chlorophyll–protein complexes in thylakoid membranes from barley *(Hordeum vulgare).* Membranes were solubilized in sodium dodecyl sulfate at room temperature and then subjected to electrophoresis in a polyacrylamide gradient (10–20%) gel at 5°C. The low temperature stabilized the complexes, which are marked according to whether they contain only chlorophyll *a* or chlorophyll *a* and *b*. The major complexes are Chl$_a$–P1, the complex that contains the reaction center for photosystem 1, and Chl$_{a/b}$–P, the major light-harvesting complex. In lane on the left are shown the complexes separated from wild-type membranes. The lane on the right shows complexes in the *chlorina-f2* mutant strain, which lacks chlorophyll *b* (see Fig. 3.18B). The mutant strain contains chlorophyll *a*–protein complexes, including Chl$_a$–P2 and Chl$_a$–P3, the most probable candidates for the reaction center for photosystem 2, but none of the chlorophyll *a/b*–protein complexes. Oligomers (probably dimers and trimers) of the Chl$_{a/b}$–P complex are indicated. The large band at the bottom of the gel is free chlorophyll. (Adapted from Machold, et al., 1979. Figure courtesy of O. Machold.)

The predominant polypeptides in the membrane, 26,000–30,000 in molecular weight, are complexed with equal amounts of both chlorophyll *a* and *b*. Although the amount has not been definitely established, at least three molecules of each type of chlorophyll are bound to each polypeptide in the complex. About half of the total chlorophyll, and most of the chlorophyll *b* in the membrane, is contained in this complex, which is called the **light-harvesting chlorophyll *a*/*b*–protein complex** ($Chl_{a/b}$–P in Fig. 3.19 or simply CP2). This complex has been found only in those organisms that contain chlorophyll *b*.

Highly purified preparations of the light-harvesting chlorophyll–protein complex (LHC) from *Vicia faba* contain three polypeptides, as shown in Fig. 3.18, with a 26,000-molecular-weight polypeptide as the major one. This polypeptide comigrates with the major polypeptide in the membrane. Further evidence that this polypeptide is indeed the apoprotein of the light-harvesting complex is provided in the patterns for *Hordeum vulgare* in Fig. 3.19. A mutant strain, *chlorina f2*, lacks chlorophyll *b* and thus the light-harvesting complex. Thylakoid membranes of this plant also lack the 26,000-molecular-weight polypeptide.

As the electrophoretic procedures become more refined, more chlorophyll–protein complexes are resolved. Figure 3.19 shows the large number of complexes resolved from thylakoids of barley *(Hordeum vulgare)*. Most of the minor complexes contain both chlorophyll *a* and *b*. However, reaction center complexes apparently contain only chlorophyll *a*, and Chl_a–P3 has properties of the reaction center for photosystem 2. An apoprotein (Chl_a–AP3) of this complex has a molecular weight of about 41,000 (see Fig. 3.18).

I. Photosystem 1 and 2 Complexes

Two major functional complexes occur in photosynthetic membranes in plant cells. These complexes carry out the activities associated with photosystems 1 and 2 (see Chapter 4). Each of the photosystems has been purified under conditions that retain the "native" complement of proteins that function in association with the reaction centers. An analysis of the polypeptides in such preparations from pea *(Pisum sativum)* chloroplasts is shown in Fig. 3.20. Photosystems 1 and 2 contain at least 13 and 10 polypeptides, respectively. The polypeptides that bind chlorophyll to form light-harvesting complexes (LHC I or LHC II) or the reaction center complexes (RC I or RC II) are identified. The function of most of the other polypeptides has not been established.

Within the membrane, the light-harvesting complexes surround the reaction centers of photosynthesis and act as "antennae." Light energy absorbed by these complexes is funneled into the reaction centers, and as a result, more energy is collected to drive photosynthetic reactions. The efficiency of photosynthesis consequently is greater than what could be provided by the relatively few chlorophyll molecules in the reaction centers alone.

J. Minor Membrane Polypeptides

The remaining integral polypeptides of the membranes are each present in relatively small amounts. Included in this population of minor polypeptides are components of the electron transport chain. The functions of these com-

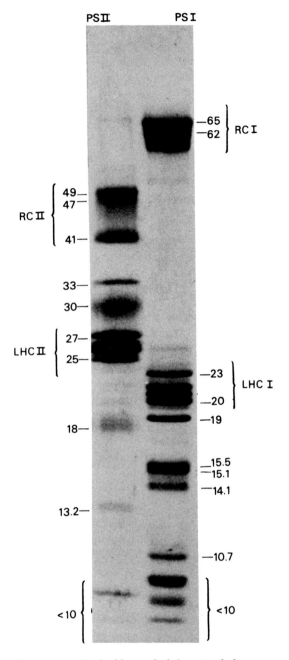

FIG. 3.20. The polypeptides contained in highly purified, functional photosystem 1 (PS I) and photosystem 2 (PS II) complexes from pea *(Pisum sativum)* thylakoid membranes. The electrophoresis was performed with a polyacrylamide slab gel containing 0.1% sodium dodecyl sulfate and 8 M urea. The numbers indicate the molecular weight ($\times 10^{-3}$) of each polypeptide. Bands labeled LHC I and LHC II are polypeptides associated with light-harvesting complexes for PS I and PS II, respectively. The major polypeptides in the PS I and PS II reaction centers are labeled RC I and RC II, respectively. (From Kaplan and Arntzen, 1982. Reprinted with permission).

ponents are discussed in the context of their roles in photosynthesis in Chapter 4. These minor polypeptides are present in the membrane at levels of 1 molecule per 200–1000 molecules of chlorophyll. However, most of the minor polypeptide fractions, observed after electrophoresis as illustrated in Fig. 3.18, have not been identified. Because of the difficulties of working with these polypeptides, which generally are insoluble in water in the absence of detergents, few of them have been purified and characterized.

III. Components of the Chloroplast Stroma

The **stroma** or **matrix** contains those components of the chloroplast that are soluble in water and are recovered in the supernatant fluid after centrifugation of lysed chloroplasts. The principal constituent of this soluble fraction is the CO_2-fixing enzyme **ribulose 1,5-bisphosphate carboxylase.** This protein was discovered in 1947 by S. Wildman, who analyzed soluble proteins of plants by electrophoresis and centrifugation. One fraction had a sedimentation coeffecient of 18 S (Svedburg units) (Fig. 3.21). Because of the homogeneity of this fraction, it initially was called Fraction I protein, and the remainder of the soluble proteins was referred to as Fraction II. Later studies revealed the function of the Fraction I protein, which is described in detail in Chapter 4. The protein

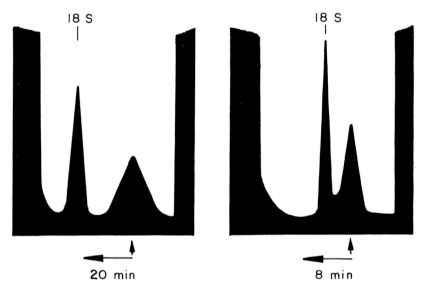

FIG. 3.21. Sedimentation pattern during centrifugation of soluble proteins from tobacco leaves. Arrows indicate the direction of sedimentation and the position of the starting boundary in the centrifuge tube. Time is indicated after the rotor had attained a speed of 51,100 rpm. The schlieren pattern represents a second derivative analysis of the refractive index within the centrifuge tube. The large amount of ribulose 1,5-bisphosphate carboxylase in the soluble fraction of the leaf cell is indicated by the boundary with a sedimentation coefficient of 18 S. This protein was initially referred to "Fraction I." The other soluble proteins ("Fraction II") did not sediment appreciably under the conditions of centrifugation, since their sizes were much smaller. (Adapted from Cohen et al., 1956.)

is composed of two types of subunits, a polypeptide with a molecular weight of 55,000 and one with a molecular weight of 12,000–16,000, depending on the source. The whole protein is composed of eight subunits of each type and has a molecular weight of about 560,000.

Up to 50% of the soluble protein in the chloroplast is accounted for by the Fraction I protein, ribulose 1,5-bisphosphate carboxylase. Some estimates suggest that this is the most abundant protein in nature.

The enzymes involved in the synthesis of glucose and starch are found in the stroma. The functions of some of these enzymes are discussed in Chapter 5. Synthesis of fatty acids for the membrane lipids also occurs with enzymes located in the stroma. In addition to the enzymes involved in CO_2 fixation, those that carry out the assimilation of sulfur (as sulfate) and of nitrogen (as nitrate) are located here. The principal requirement for these elements is in the synthesis of amino acids and the nucleic acids of the cell.

Within the stroma of the chloroplast is the entire machinery for the synthesis of proteins. DNA is found here along with ribosomes and the enzymes and protein factors required for protein synthesis. These components are described in Chapter 6.

IV. Summary

Thylakoid membranes are typical in that they contain lipids and proteins. However, they are unusual in that their major lipids are glycolipids (the galactosyl diglycerides) and the pigment chlorophyll. Most and perhaps all proteins are unique to the plastid membrane. Chlorophyll exists as complexes with proteins, and the two major complexes are the P700 chlorophyll *a*–protein complex associated with photosystem 1 and the light-harvesting chlorophyll *a/b*–protein complex associated with photosystem 2. The major integral proteins of the membrane are those in the light-harvesting complexes. The major peripheral proteins are the phycobiliproteins, in cyanobacteria and red and cryptomonad algae, and the coupling factor 1 in green plants.

The major protein found in the stroma of the chloroplasts, and in fact the major protein in most chloroplasts, is ribulose 1,5-bisphosphate carboxylase. In some species of algae this protein is condensed in a localized region to form a pyrenoid body. The reaction that fixes CO_2 in most plant cells is catalyzed by this protein.

The chloroplast stroma also contains the enzymes involved in the synthetic pathway of carbohydrates and of the fatty acids of membrane lipids. The stroma is the location of an important protein-synthesizing system of the plant cell.

Literature Cited

Chua, N.-H., Martin, K., and Bennoun, P. (1975) A chlorophyll-protein complex lacking in photosystem 1 mutants of *Chlamydomonas reinhardtii*, *J. Cell Biol.* **67**:361–377.
Cohen, M., Ginoza, W., Dorner, R. W., Hudson, W. R., and Wildman, S. G. (1956) Solubility and color characteristics of leaf proteins prepared in air and nitrogen, *Science* **124**:1081–1082.

Diner, B. A. (1979) Energy transfer from the phycobilisomes to photosystem II reaction centers in wild type *Cyanidium caldarium, Plant Physiol.* **63**:30–34.

Ficken, G. E., Johns, R. B., and Linstead, R. P. (1956) Chlorophyll and related compounds, IV. The position of the extra hydrogens in chlorophyll. The oxidation of pyropheophorbide-a, *J. Chem. Soc.* **1956**:2272–2280.

Fischer, H. (1940) Progress of chlorophyll chemistry, *Naturwissenschaften* **28**:401–405.

French, C. S. (1971) The distribution and action in photosynthesis of several forms of chlorophyll, *Proc. Natl. Acad. Sci. USA* **68**:2893–2897.

Gantt, E. (1981) Phycobilisomes, *Annu. Rev. Plant Physiol.* **32**:327–347.

Hoober, J. K., Millington, R. H., and D'Angelo, L. P. (1980) Structural similarities between the major polypeptides of thylakoid membranes from *Chlamydomonas reinhardtii, Arch. Biochem. Biophys.* **202**:221–234.

Janero, D. R., and Barnett, R. (1981) Cellular and thylakoid-membrane glycolipids and phospholipids of Chlamydomonas reinhardtii 137+, *J. Lipid Res.* **22**:1119–1130.

Janero, D. R., and Barnett, R. (1982) Isolation and characterization of an ether-linked homoserine lipid from the thylakoid membrane of *Chlamydomonas reinhardtii* 137+, *J. Lipid Res.* **23**:307–316.

Lichtenthaler, H. K., and Park, R. B. (1963) Chemical composition of chloroplast lamellae from spinach, *Nature (London)* **198**:1070–1072.

Kaplan, S., and Arntzen, C. J. (1982) Photosynthetic membrane structure and function, in *Photosynthesis, Energy Conversion by Plants and Bacteria* (Govindjee, ed.), Vol. 1, Academic Press, New York, pp. 65–151.

Kates, M. (1970) Plant phospholipids and glycolipids, *Adv. Lipid Res.* **8**:225–265.

Machold, O. (1981) Chlorophyll$_{a/b}$-proteins and light-harvesting complex of *Vicia faba* and *Hordeum vulgare, Biochem. Physiol. Pflanzen* **176**:805–827.

Machold, O., Simpson, D. J., and Moller, B. L. (1979) Chlorophyll-proteins of thylakoids from wild-type and mutants of barley (*Hordeum vulgare* L.), *Carlsberg Res. Commun.* **44**:235–254.

Morgenthaler, J. J., Marsden, M. P. F., and Price, C. A. (1975) Factors affecting the separation of photosynthetically competent chloroplasts in gradients of silica sols, *Arch. Biochem. Biophys.* **168**:289–301.

Vernon, L. P., and Shaw, E. R. (1971) Subchloroplast fragments: Triton X-100 method, *Meth. Enzymol.* **23**:277–289.

Willstätter, R., and Stoll, A. (1928) *Untersuchungen über Chlorophyll*, J. Springer, Berlin. (English edition, *Investigations on Chlorophyll*, Science Press, Lancaster, Ohio.)

Woodward, R. B. (1961) The total synthesis of chlorophyll, *Pure Appl. Chem.* **2**:383–404.

Woodward, R. B., Ayer, W. A., Beaton, J. M., Bickelhaupt, F., Bonnett, R., Bushschacher, P., Closs, G. L., Dutler, H., Hannah, J., Hauck, F. P., Ito, S., Langemann, A., Le Goff, E., Leimgruber, W., Lwowski, W., Sauer, J., Valenta, Z., and Volz, H. (1961) The total synthesis of chlorophyll, *J. Am. Chem. Soc.* **82**:3800–3802.

Additional Reading

Barr, R., and Crane, F. L. (1971) Quinones in algae and higher plants, *Meth. Enzymol.* **23**:372–408.

Benson, A. A. (1963) The plant sulfolipid, *Adv. Lipid Res.* **1**:387–394.

Black, C. C., Jr., and Rouhani, I. (1980) Isolation of leaf mesophyll and bundle sheath cells, *Meth. Enzymol.* **69**:55–68.

Glazer, A. N. (1982) Phycobilisomes: Structure and dynamics, *Annu. Rev. Microbiol.* **36**:173–198.

Glazer, A. N. (1983) Comparative biochemistry of photosynthetic light-harvesting systems, *Annu. Rev. Biochem.* **52**:125–157.

Goodwin, T. W. (1976) Distribution of the carotenoids, in *Chemistry and Biochemistry of Plant Pigments* (T. W. Goodwin, ed.), Vol. 1, Academic Press, New York, pp. 225–261.

Jackson, A. H. (1976) Structure, properties and distribution of chlorophylls, in *Chemistry and Biochemistry of Plant Pigments* (T. W. Goodwin, ed.), Vol. 1, Academic Press, New York, pp. 1–63.

Ó Carra, P., and Ó hEocha, C. (1976) Algal biliproteins and phycobilins, in *Chemistry and Biochemistry of Plant Pigments* (T. W. Goodwin, ed.), Vol. 1, Academic Press, New York, pp. 328–376.

Poincelot, R. P. (1980) Isolation of chloroplast envelope membranes, *Meth. Enzymol.* **69:**121–128.

Reeves, S. G., and Hall, D. O. (1980) Higher plant chloroplasts and grana: General preparative procedures (excluding high carbon dioxide fixation ability chloroplasts), *Meth. Enzymol.* **69:**85–94.

Scheer, H., and Katz, J. J. (1975) Nuclear magnetic resonance spectroscopy of porphyrins and metalloporphyrins, in *Porphyrins and Metalloporphyrins* (K. M. Smith, ed.), Elsevier, Amsterdam, pp. 399–524.

Strain, H. H., Cope, B. T., and Svec, W. A. (1971) Analytical procedures for the isolation, identification, estimation and investigation of the chlorophylls, *Meth. Enzymol.* **23:**452–476.

Thornber, J. P., and Markwell, J. P. (1981) Photosynthetic pigment-protein complexes in plant and bacterial membranes, *Trends Biochem. Sci.* **6:**122–125.

Walker, D. A. (1980) Preparation of higher plant chloroplasts, *Meth. Enzymol.* **69:**94–104.

4

The Process of Photosynthesis

The Light Reactions

I. Introduction

The oxidation of foodstuffs to carbon dioxide and water provides the energy required by all aerobic cells to maintain their state of life and to grow. Energy stored in these foodstuffs is lost irreversibly as heat as the work of life is performed. But since the foodstuffs themselves were once parts of living cells, eventually all living matter reverts to carbon dioxide, water, and ammonia or free nitrogen. To prevent life from consuming itself, the energy-bearing compounds must be regenerated, which requires an input of external energy at least equal to that lost as heat. The only source of readily available external energy for living systems on earth is the sun. Given the appropriate mechanisms for performing the process, light energy can be converted to electrical or chemical energy. However, cells in the animal kingdom do not have this capacity. Plant cells, on the other hand, because they contain photosynthetic structures, can accomplish this feat. Chlorophyll-bearing cells are the primary* sector of the earth's population that are capable of production, in this sense, and they alone enable all cells in the animal kingdom to exist. Green plant cells can synthesize their fabric *de novo* from carbon dioxide, water, and other inorganic compounds. Thus, they are **autotrophic** (self-fed, requiring no organic matter for growth), whereas all other cells are **heterotrophic** (fed by others, requiring preformed organic matter for growth). Plant cells accumulate energy, whereas others dissipate it. The synthesis of carbohydrates from carbon dioxide and water, with the aid of sunlight, is without doubt the most fundamental of all biological activities.

Not only do chloroplasts provide the energy-bearing compounds on which living organisms depend, they also provide the oxygen required for maximal recovery by oxidative metabolism of the energy stored in these compounds. The relationship between the fixation of carbon dioxide into carbohydrate with the

---

*The *Halobacteria*, which live in waters with very high salinity, contain bacteriorhodopsin, a membrane protein containing retinal (vitamin A aldehyde) as a chromophore, which functions in a light-driven proton pump. The resulting proton gradient is used by these bacterial cells to produce ATP. These cells, therefore, are also capable of converting light energy into chemical energy.

concomitant evolution of oxygen, on the one hand, and the reversal of this process, on the other, is shown by the overall reaction written as

$$CO_2 + H_2O + 4.7 \times 10^5 \text{ J (per mole of } CO_2) \rightleftarrows \frac{1}{6}(C_6H_{12}O_6) + O_2 \quad [4.1]$$

Although the overall process can be written as a reversible equation, in reality it is better represented by a cycle that indicates that the forward and reverse processes occur in separate cellular compartments, as shown in Fig. 4.1. The forward reaction, photosynthesis in chloroplasts, is driven by the absorption of light energy. The reverse reaction, oxidative respiration in bacteria and in the mitochondria of eukaryotic cells, is pulled by the release of energy as heat. This continuous cycling of carbon atoms is an expression of the interdependence of plant and animal life.

The total amount of carbon introduced into carbohydrates by all photosynthetic organisms on earth is estimated to be currently about 200×10^9 metric tons/year, at a cost of 8×10^{21} J/year of light energy. However, since about 3×10^{24} J of total radiant energy strike the earth's surface per year, only between 0.2 and 0.3% of this energy is actually trapped by plant cells. A major factor that contributes to this low yield is that much of the land is arid or otherwise unhospitable to plant life. Yet even in heavily vegetated areas, several factors must be taken into account in connection with energy conversion, as listed in Table 4.1. Plant pigments absorb light with wavelengths only between 380 and 700 nm (Fig. 4.2). Fortunately, within this range lie the wavelengths of greatest

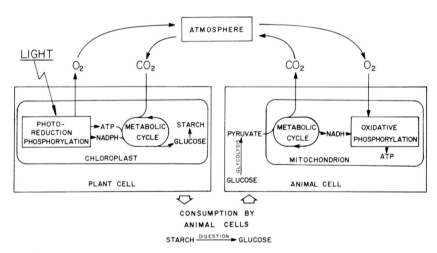

FIG. 4.1. The carbon cycle. Carbon as CO_2 is "fixed" into organic compounds (primarily carbohydrates) by photosynthesis in chloroplasts. The bond energy stored in the carbohydrate molecules is released as they are metabolized. Oxidative metabolism in aerobic cells results in the most complete metabolism and thus the maximal release of energy. The important metabolic processes, through which the stored energy in carbohydrates is made available, are glycolysis, the tricarboxylic acid cycle, and oxidative phosphorylation. In eukaryotic cells glycolysis is localized in the cytosolic compartment but the tricarboxylic acid cycle and oxidative phosphorylation are compartmentalized within the mitochondria. The end products of oxidative metabolism are CO_2 and H_2O, the primary chemical substrates for photosynthesis in chloroplasts.

TABLE 4.1. Photosynthetic Efficiency and Energy Levels[a]

	Available light energy (%)
Total radiant energy at sea level	100
Wavelengths between 400 and 700 nm are photosynthetically usable (50% loss)	50
Reflection, nonproductive absorption, and transmission by leaves (20% loss)	40
Quantum efficiency for CO_2 fixation in 680-nm light (assuming 10 quanta/CO_2)[b] (77% loss)	9.2
Respiration (40% loss)	5.5

The energy made available through photosynthesis for anabolic processes (constructive metabolism) in plant and animal cells is thus about 5.5% of the total radiant energy.

[a]Adapted from Hall (1976).
[b]If the quantum efficiency is 8 quanta per CO_2, this loss factor becomes 72%, rather than 77%, and thus the overall efficiency would be 6.7% instead of 5.5%.

solar intensity (Fig. 4.2), so that nearly one-half of the total spectral energy reaching the earth's surface is available for photosynthesis. Various other losses also occur, as listed in Table 4.1; so the average practical maximal efficiency of photosynthetic energy conversion is on the order of 5–7%. The actual efficiency varies among different plants and between geographical locations. Furthermore, photosynthetic production is not uniform throughout the year and in most nontropical climates occurs only 3–4 mo of each year. When averaged over a

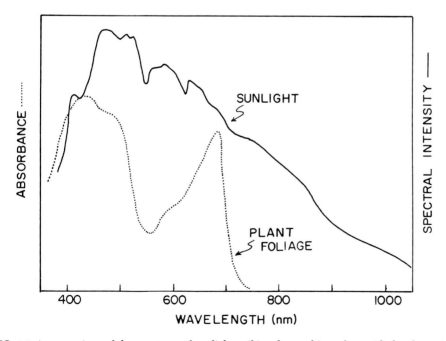

FIG. 4.2. A comparison of the spectrum of sunlight striking the earth's surface with the absorption spectrum of green plant foliage. It is quite apparent that only a portion of the spectrum of light emitted from the sun can be used for photosynthesis by these plants.

whole year, photosynthetic conversion efficiencies are between 0.5 and 1.3% of total available light energy in temperate zones and 0.5 and 2.5% in subtropical regions.

The least energetic light that performs photosynthesis efficiently in plant cells has a wavelength of about 680 nm, which corresponds to a peak in the absorption spectrum for chlorophyll *a in vivo* (see Fig. 3.6). The energy of a mole quantum (an einstein) of red light at 680 nm is 1.76×10^5 J. From a consideration of the energy factor in equation 4.1, the minimal number of einsteins required for the fixation of 1 mole of carbon dioxide is $4.7 \div 1.76$, or 2.7. As described below, experimental measurements have shown that in reality more than 2.7, and at least 8–10, einsteins of red light must be absorbed by plant cells for the fixation of 1 mole of carbon dioxide. Therefore, with respect to a plant's ability to use the energy it absorbs, the *efficiency* for carbon dioxide fixation in red light is about 30%. Since the same number of quanta is required regardless of the wavelength, the maximal efficiency in blue light, which contains nearly twice the energy per photon as red light, is less.

II. Primary Mechanisms of Photosynthesis

For a photochemical reaction to occur, light must be absorbed by a pigment to produce an excited or high-energy state of the pigment. The electrons in these molecules are normally in orbitals having the lowest possible energy, a condition referred to as the ground state. An input of energy is required to cause the transition of an electron from a lower to a higher orbital. Within an organic molecule that is capable of absorbing light, only discrete, discontinuous energy levels are allowed by the electronic orbital structure of the molecule, and therefore, energy transitions between orbitals occur in definite steps. Light is absorbed only if the photon contains the right amount of energy to raise an electron to one of the higher energy levels. Quanta of electromagnetic vibrations with wavelengths between 100 and 1300 nm, that is, light in the ultraviolet, visible, and near infrared range, have the appropriate amounts of energy to span the range between the ground state and the lower excited states of most organic compounds. Thus, the energy of a **photon,** a unit of energy of a given wavelength of light, can be used to promote an electron in the pigment molecule from its ground state to an excited state.

The active pigment in photosynthesis is **chlorophyll.** The absorption of a photon by a chlorophyll molecule results in redistribution of the energy levels of the electrons within the conjugated porphyrin ring system. The molecule is then in an excited state.

$$h\nu + \text{Chl} \rightarrow \text{Chl}^* \qquad [4.2]$$

The energy needed to raise the electron from an energy level corresponding to the ground state into this excited state is provided by that in the absorbed photon, an amount that is described by the relationship

$$\Delta\epsilon = \epsilon_1 - \epsilon_0 = h\nu = hc/\lambda \qquad [4.3]$$

where ν is the frequency of the absorbed radiation, λ is its wavelength, h is Planck's constant (6.62×10^{-27} erg sec), and c is the velocity of light (2.997×10^{10} cm sec^{-1}).

The absorption process, described by reaction 4.2, has certain restrictions. Within the chlorophyll molecule is a set of allowable energy levels, each with a characteristic energy and electron distribution (see Fig. 4.3). A photon, as a unit, is absorbed only if it has the correct energy to produce one of the allowable levels above the ground state. Photons of intermediate energy are transmitted by the compound, leaving it unaffected. The *probability* that a photon of a certain energy content will be absorbed and cause an energy transition is described by the **absorption spectrum** of the molecule.

The fact that chlorophyll is the primary photoreactive agent in photosynthesis was deduced from a comparison of an action spectrum for photosynthesis with the absorption spectrum of chlorophyll (Fig. 4.4). An action spectrum is defined as the ability of absorbed light of various wavelengths to achieve the photochemical reaction under investigation. In the determination of an action spectrum, plant cells or chloroplast preparations are illuminated with a series of monochromatic or narrow-band wavelengths of light. Careful measurements

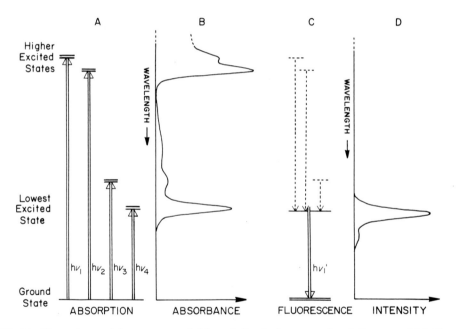

FIG. 4.3. Absorption and fluorescence of chlorophylls. A: An energy-level diagram, which illustrates the electronic transitions (vertical arrows) that result from the absorbance of photons of light. B: The absorption spectrum describes the probability for the occurrence of the energy transitions shown in A. The spectrum is turned 90° from the usual orientation to show the relationship to the energy levels. C: The diagram shows the radiationless decay (dashed arrows) from the higher excited states to the lowest excited state and the reemission of energy as fluorescence accompanying the further decay to the ground state. D: A fluorescence emission spectrum that corresponds with the energy transitions shown in C. A small red shift of the fluorescence spectrum as compared with the absorption spectrum is due to vibrational relaxation in the excited state prior to fluorescence and in the ground state after emission. (Adapted from Sauer, 1975.)

are made of both incident and transmitted light. The relation between the difference—the absorbed light—and the magnitude of the resulting photochemical reaction is then determined. The most favored expression of the action spectrum is a plot of the amount of light of a particular wavelength that must be absorbed to cause a half-maximal response.

The time span for the absorption of a photon of light and a consequent energy transition is on the order of 10^{-15} sec. The excited state then has several possible fates. The initial excited condition, described as the **singlet state,** has an electron in a higher energy level but still maintaining the direction of spin it had when paired in the ground state with an electron of opposite spin (Fig. 4.5). The excited electron can return to the ground state, reemitting the energy as light (fluorescence) and/or as heat. These decay events occur within the range of 10^{-12} (pico-)–10^{-9} (nano-)sec. Alternatively, a portion of the energy may be used to achieve the **triplet state,** reached by a mechanism called **intersystem crossing.** In the triplet state the spin of the excited electron is reversed, resulting in a pair of electrons with the same directional spin. The magnetic fields of these electrons oppose each other and thus the triplet state is more stable than the singlet state. Triplet states have lifetimes in the range of 10^{-3} (milli-)sec to several seconds. Again, the energy of the triplet state may be dissipated by the emission of light (phosphoresence) or by a radiationless return of the electron to the ground state (Fig. 4.5).

For the absorption of light to accomplish a photochemical reaction, the energy in the excited state must be trapped. An excited electron is held less tightly by the parent compound and is readily transferred to a second com-

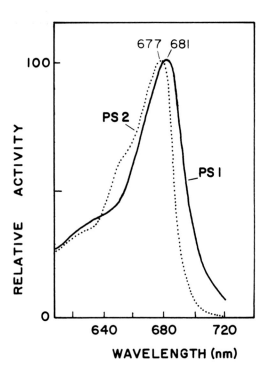

FIG. 4.4. Action spectra of photosystems 1 and 2 for spinach chloroplasts. The activity of photosystem 1 was measured with methylviologen as the electron acceptor by superimposing monochromatic light of various wavelengths on a continuous background of 650-nm light to maintain an excess of photosystem 2 activity. Photosystem 2 was measured in a similar manner, with NADP$^+$ as the electron acceptor and background light of 720-nm light to maintain an excess of photosystem 1 activity. The spectra correspond primarily to the absorbance characteristics of the light-harvesting complexes. (Adapted from Joliet, 1972.)

pound, resulting in a chemical reduction of the electron acceptor (Fig. 4.5). The initial step in photosynthesis can therefore be described as a one-electron oxidation–reduction reaction driven by the absorption of light.

$$D \cdot Chl \cdot A \leftrightarrow D \cdot Chl^* \cdot A \leftrightarrow D \cdot Chl^+ \cdot A^- \leftrightarrow D^+ \cdot Chl \cdot A^- \qquad [4.4]$$

In this reaction D is an electron donor and A is the acceptor. The initial charge separation occurs with the singlet state of chlorophyll within a time span of 5 psec following the absorption of a photon. The triplet state of chlorophyll is achieved only if the energy in the initially excited state is not immediately trapped.

To derive useful work from this initial reaction, the charges must be separated to prevent a back reaction and the consequent loss of energy. The reac-

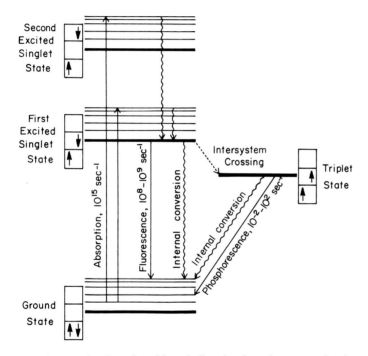

FIG. 4.5. Diagram of energy levels within chlorophyll molecules. Absorption of a photon raises the energy level of an electron either to the first or second excited singlet state, depending on the wavelength (energy content) of the photon (see Fig. 4.3). The second excited singlet state decays to the first by radiationless vibrational relaxation. From the first excited singlet state, the energy can be reemitted as fluorescence or as heat (vibrational) energy. Alternately, some of the energy can be used to reverse the direction of spin of the electron, thereby achieving the triplet state by intersystem crossing. The triplet state can decay by a relatively slow reemission of light, phosphorescence, or as heat. The small arrows in the boxes indicate directions of spins of a pair of electrons in the different energy levels. In a reaction center, absorption of light energy raises an electron in a chlorophyll molecule to a higher energy level, from which the electron can be transferred to an acceptor. The acceptor molecule consequently becomes reduced. The oxidized chlorophyll molecule then abstracts an electron from a donor to satisfy its electron deficiency, resulting in oxidation of the donor. The chlorophyll molecule acts catalytically in this reaction. (Adapted from Govindjee and Govindjee, 1975.)

tion described by reaction 4.4 occurs within the thylakoid membrane of the chloroplast, and the organization of components in the membrane allows the primary acceptor of the electrons from chlorophyll to in turn reduce other components in an electron transport chain. The resultant Chl^+, a strong oxidizing agent, then abstracts electrons from a donor, indicated by D, on the opposite side of the membrane from where the initial reduction occurred. Evidence for the efficiency of this reaction is the fact that the primary photochemical reactions occur with a quantum yield near 1. The ability of excited chlorophyll molecules to give up or accept electrons is the fundamental property that makes photosynthesis possible.

III. Chemical Currency

The primary role of photosynthesis is the conversion of light energy to electrical and chemical energy. As with any photoreaction, the process of photosynthesis can be divided into three stages. First, the absorption of a photon produces an electronically excited molecule. Second, the excited molecule participates in the primary photochemical reaction. This reaction, as described above, converts light energy into electrical energy. The third stage is characterized by secondary or "dark" reactions, which occur as the result of the primary photochemical reactions. During this stage the electrical energy is converted into chemical energy. These forms of energy then can be used to synthesize organic compounds.

Electrical energy is transferred by oxidation–reduction reactions. The principal metabolic pool of reducing equivalents in cells is the reduced **nicotinamide adenine dinucleotides,** NADH and NADPH (Fig. 4.6). The redox (oxidation–reduction) potential, $E_{m,7}$ (E_0 at pH7) of NADPH (and NADH) is -0.32

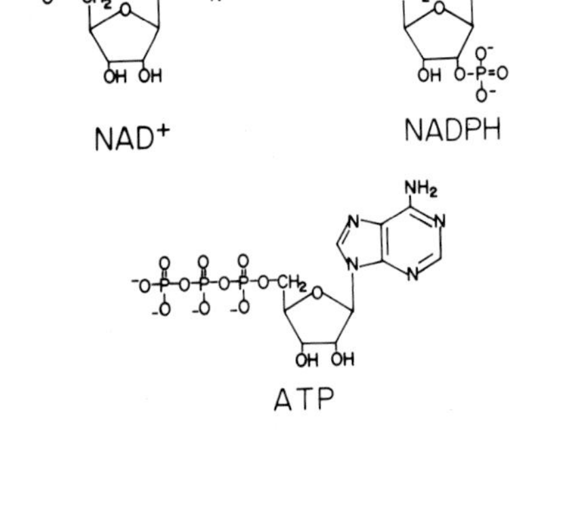

NAD$^+$ NADPH   ATP

FIG. 4.6. Structures of NAD^+, NADPH, and ATP. Reduction of NAD^+ and $NADP^+$ adds two electrons and a proton (as an electron and a hydride radical) to the pyridine ring of the nicotinamide portion of the molecule. NAD^+, nicotinamide adenine dinucleotide (oxidized); NADPH, nicotinamide adenine dinucleotide phosphate (reduced); ATP, adenosine triphosphate.

the substance exerts on an inert electrode. The value for E can be measured with appropriate apparatus. The more negative the value of E, the more "electron pressure" the substance exerts and the better a reducing agent it is. Therefore, NADPH will reduce, i.e., transfer electrons to, another compound with a more positive value of $E_{m,7}$ and likewise will be reduced by other compounds with a more negative value of $E_{m,7}$.

The principal currency of *chemical energy* in cells is **adenosine triphosphate** (ATP) (Fig. 4.6). Electrical energy is expended to synthesize ATP. The minimal amount of energy required for the production of ATP is determined by the energy released during its hydrolysis. In reality, considerably more energy is expended to drive the synthesis of ATP against an unfavorable equilibrium in the reaction

$$\text{ATP} + \text{H}_2\text{O} \rightleftarrows \text{ADP} + \text{inorganic phosphate} + 3.1 \times 10^4 \text{ J/mole} \quad [4.5]$$

ATP is the immediate source of chemical energy for most energy-requiring biological reactions. The energy released by cleavage of the "high-energy" pyrophosphate bonds in ATP is captured during enzymatic reactions and used for the production of new chemical bonds. Nearly all biosynthetic reactions require an introduction of energy into the system in the form of ATP.

The primary role of the "light" reactions in photosynthesis is the production of NADPH and ATP. Both these energy-rich compounds then are used in subsequent "dark" reactions in the biosynthesis of carbohydrates and other cellular materials.

IV. The Flow of Electrons during Photosynthesis

The ultimate *donor* of electrons in photosynthesis in green plants is water, which is oxidized to molecular oxygen. The production of each oxygen molecule requires the abstraction of four electrons from two water molecules, which also liberates four H^+ ions. The final *acceptor* of electrons is $NADP^+$, which is reduced to NADPH in a two-electron transfer step. Thus, for each molecule of oxygen produced, two molecules of NADPH are formed. This overall process can be described by the reaction

$$2 \text{ H}_2\text{O} + 2 \text{ NADP}^+ \rightleftarrows \text{O}_2 + 2 \text{ NADPH} + 2 \text{ H}^+ \quad [4.6]$$

Although this reaction involves the transfer of four electrons, careful measurements have shown that at least 8 quanta of light must be absorbed per molecule of oxygen produced. Results from extensive experimental work have been interpreted as evidence that the actual photosynthetic electron transport occurs by a process involving two separate photoreactions connected in series. As shown in Fig. 4.7, at one reaction center, called **photosystem 2,** water is oxidized to oxygen at a redox potential $E_{m,7} = +0.82$ V, whereas the primary electron acceptor Q ($E_{m,7} = -0.035$ V) is reduced. In a second reaction center, referred to as **photosystem 1,** the electron donor plastocyanin ($E_{m,7} = +0.37$ V) is oxidized as the second photochemical electron acceptor X ($E_{m,7} \approx -0.7$ V) is reduced. In photosystem 2, the acceptor Q feeds electrons into a "downhill" electron transport chain consiting of the plastoquinones, an iron–sulfur protein, a cytochrome (cyt f), the copper-containing protein plastocyanin, and then on to the reaction center of photosystem 1. The acceptor X, whose structure is still in doubt, feeds electrons into another "downhill" chain containing the non-heme iron protein ferredoxin and finally NADP$^+$

V. Membrane-Bound Electron Carriers

A. Photosystem 2 (P680)

At the reaction center of photosystem 2 is a small number (probably two) of chlorophyll *a* molecules designated, according to their absorption properties, as **P680.** However, the actual absorbance changes that accompany the photoox-idation of P680 in whole chloroplasts occur at about 687 nm. These chlorophyll molecules exist in a special environment, the properties of which are not fully known, and function to promote the oxidation–reduction reactions described in section IV. Associated with the reaction center chlorophylls are at least two proteins of molecular weight 40,000–50,000.

Adjacent to P680 reaction centers are chlorophyll *a/b*–protein complexes that serve as light-harvesting "antennae." The light energy absorbed by these "antenna" chlorophylls is transferred at the singlet excitation level, through many chlorophyll molecules, to the reaction centers. This transfer, which occurs within a few picoseconds, produces an excited, singlet state of the reaction center chlorophylls, which then become the primary donor of electrons

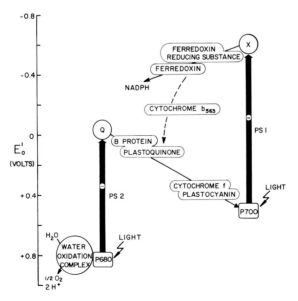

FIG. 4.7. The Z scheme of photosynthetic electron transport. The components of the system are placed in a sequence according to the midpoint of their oxidation–reduction potentials. Positions in regard to the photosystems are determined by whether a component is oxidized or reduced by a photosystem. Thus, if a component experimentally is reduced when photosystem 2 is active, but oxidized when photosystem 1 is active, the conclusion is that it functions in the electron transport chain between the two reaction centers. Components of a cyclic flow of electrons around photosystem 1 also are shown.

into the electron transport chain leading to photosystem 1. The oxidized form of P680 then draws an electron from components on the oxidizing side of photosystem 2.

Few of the components that feed electrons into P680, the side involved in the production of oxygen, have been characterized. However, the effects of several ions have been observed. Chloride ions markedly stimulate the oxidation of water. And it has been recognized for some time that Mn^{2+} ions are essential for photosynthesis with water as the electron donor. Recently, a Mn^{2+}-protein complex has been identified that is located on the inner surface of the thylakoid membrane. Each polypeptide of molecular weight 65,000 binds two Mn^{2+} ions. This complex may be the initial acceptor of electrons from water, with the Mn^{2+} ions as the functional carriers of electrons between water and P680. Studies with mutant strains of algae defective in photosystem 2 activity have shown that there are several different polypeptides involved in the production of oxygen. The oxidation of water occurs on the *inside* surface of the thylakoid membrane.

On the electron accepting side of photosystem 2 is a component that exhibits absorbance changes at 325 and at 550 nm. This component has been designated **Q** because of its ability, in the oxidized form, to quench the fluorescence of chlorophyll in photosystem 2. Q seems to be a plastoquinone in a special environment that serves as a one-electron carrier. Q is considered to be the primary electron acceptor for photosystem 2.

B. B Protein

It has been known since the work of Otto Warburg in the 1920s that CO_2 is required for the evolution of O_2 in chloroplasts. Recently, a protein about 32,000 in molecular weight has been found on the outer, stromal surface of the thylakoid membrane that is involved in electron flow from the initial acceptor of photosystem 2 to the general plastoquinone pool. This protein, simply referred to as the **B protein,** possibly contains a quinone compound as a prosthetic group. However, the activity of this protein is entirely dependent on the presence of a bound CO_2 molecule. Consequently, CO_2-depleted chloroplasts exhibit no electron transport through photosystem 2. This protein also is the site of action of a number of herbicides, notably the triazines and 3-(3,4-dichlorophenyl)-1,1-dimethyl urea (DCMU) (Fig. 4.8), that apparently inhibit electron flow by displacing the bound CO_2.

A striking feature of photosystem 2, and also of photosystem 1 as described in section F, is its transmembrane arrangement. The oxidation of water occurs on the inside surface of the thylakoid membrane, whereas the reduction of electron carriers takes place on the outer, stromal surface. The separation of

FIG. 4.8. Structures of 3-(3,4-dichlorophenyl)-1,1-dimethylurea (Diuron, DCMU) and 2-chloro-4-ethylamino-6-isopropylamino-s-triazine (atrazine), two herbicides that inhibit photosynthetic electron flow by binding to "B" protein on thylakoid membranes.

DCMU

ATRAZINE

these reactions is crucial for the efficiency of photosynthesis and exemplifies one important feature of the membrane structure.

C. Plastoquinone

The **plastoquinones,** whose structures are given in Chapter 3, are reduced by photosystem 2 and are oxidized by photosystem 1. Thus, they function as components in the intersystem electron transport chain. The plastoquinones are present at an appreciable molar excess over the amount of P680. In spinach chloroplasts there are 30–35 moles of plastoquinone per mole of Q. These components provide a large pool of electron acceptors near photosystem 2. Each quinone molecule transfers two electrons during a reduction–oxidation reaction, which must be accompanied by the uptake of two H^+ ions according to the reaction

$$[4.7]$$

The flux of electrons through the plastoquinone pool seems to be the rate-limiting step in photosynthetic electron flow. Reaction 4.7 is sensitive to pH. By mass action, low pH values shift the equilibrium of reaction 4.7 toward reduction. The quinol form has pK values for dissociation of protons of 10.8 and 12.9, so that at physiological pH values (near pH7) the predominant reduced form is QH_2. However, the semiquinone form Q^- which is the initial intermediate in the reaction and may be the predominant form during steady-state photosynthesis, has a pK value of 5.9. Since the internal pH in energized thylakoids is below this value (see p. 103), the semiquinone would also exist primarily in the protonated form, $QH^.$. In chloroplasts, the equilibrium $E_{m,7}$ of the plastoquinone pool is $+0.118$ V

D. Cytochromes

Among the integral polypeptides present in small amounts in thylakoid membranes are electron-carrying hemoproteins, the **cytochromes.** Chloroplasts contain three major cytochromes, a c type and two b types. The cytochromes are characterized by the mode of attachment of the heme to the protein, by the spectral properties of the complex, and by the oxidation–reduction potential of the iron atom in the heme. The c-type cytochrome, designated cytochrome f, was the first hemoprotein observed in chloroplasts. As in other c-type cytochromes, the heme moeity is covalently bound to the protein as the result of addition of two sulfhydryl groups on the protein across two vinyl side chains of the porphyrin. Figure 4.9 shows the absorption spectra of the oxidized and reduced cytochrome f, with the characteristic α-peak for the reduced form at 554 nm. Each molecule transfers a single electron via the heme iron. The redox potential for cytochrome f is $E_{m,7} = +0.36$ V. Studies of its reduction by photosystem 2 and its oxidation by photosystem 1 have permitted an assignment of the functional position of cytochrome f between the two systems but close to photosystem 1 (Fig. 4.7).

Cytochrome f from higher plants and green algae is a relatively hydrophobic protein and is only solubilized with organic solvents or detergents. The molecular weight of the monomeric polypeptide is 33,000–37,000. In higher plants there is one molecule of cytochrome f for about 500 molecules of chlorophyll.

In some algae, generally those that lack plastocyanin, a second c-type cytochrome is found that functions close to cytochrome f in the electron transport chain. This additional cytochrome has an absorption maximum in the reduced form at 552 nm and is small (10,000–13,500 in molecular weight) and water soluble.

Two other cytochromes in the chloroplast, of the b type, are hemoproteins in which the porphyrin moeities are not covalently attached to the protein. These b-type cytochromes are distinguished by their absorption maximum for the α-peak of the reduced form. Cytochrome b-559, with an α-peak at 559 nm, exists in two forms that differ in their oxidation–reduction potential. The high-potential form ($E_{m,7} = +0.37$ V) is the naturally most abundant form, but it can be converted to a low-potential form ($E_{m,7}$ about $+0.02$ V). The precise function of these two species in electron transport is not clear. However, mutant strains of algae that lack cytochrome b-559 are unable to carry out electron transport. Its site of function is not established, but it is tightly associated with the reaction center of photosystem 2 and possibly is part of the H_2O–oxidation complex. This cytochrome contains a small hydrophobic subunit polypeptide with a molecular weight of about 6000.

The second b-type cytochrome is designated b-563 or cytochrome b_6 and has a molecular weight of 18,000–20,000. This cytochrome has the lowest oxidation–reduction potential of the chloroplast cytochromes, with an $E_{m,7} = -0.12$ V. It does not seem to be involved in the electron transport chain between the two photosystems but instead is involved with the cyclic electron flow that is driven by photosystem 1 (Fig. 4.7).

A specific, stoichiometric complex has recently been purified from thylakoid membranes that contains two molecules of cytochrome b_6, one of cytochrome f, and one nonheme, iron–sulfur center. The iron–sulfur center, with a $E_{m,7} = $ ca. $+0.29$ V, has a composition of Fe_2S_2. The sulfur atoms are released as inorganic S^{2-} upon acidification. Additional coordination bonds with the iron atom apparently are provided by sulfhydryl groups of cysteine residues within a polypeptide of molecular weight 20,000. The iron–sulfur center is the functional acceptor of electrons from plastoquinone and subsequently transfers electrons to cytochrome f within the same complex.

E. Plastocyanin

Whereas plastoquinone provides a mobile electron carrier between photosystem 2 and the cytochrome b_6/f complex, plastocyanin serves this purpose between this complex and photosystem 1. Plastocyanin is a copper-containing

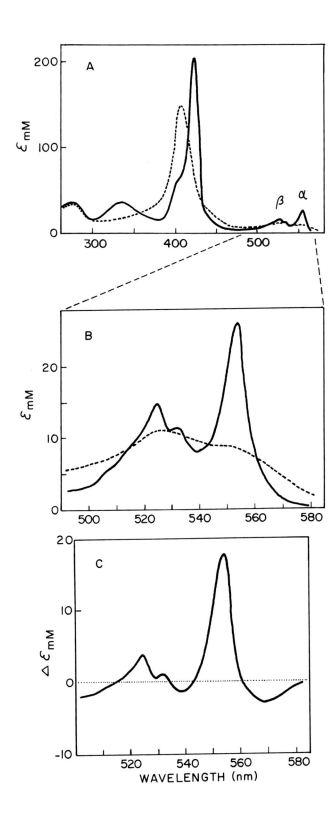

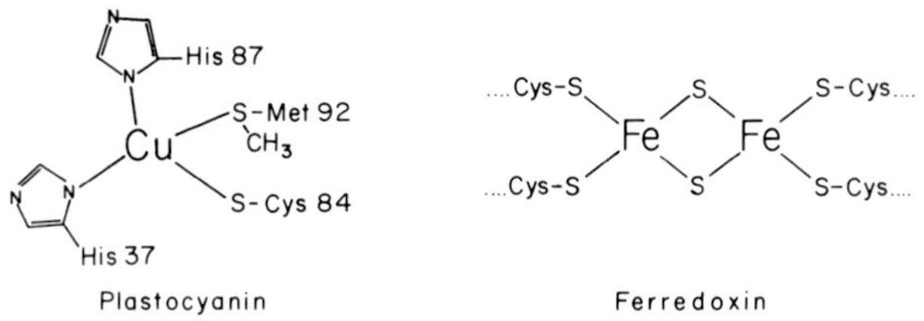

Plastocyanin

Ferredoxin

FIG. 4.10. Left: Structure of the coordinated copper ion in plastocyanin. The amino acid residues involved in the coordination complex are numbered according to their position in the sequence of the 99-amino-acid (molecular weight 10,500) polypeptide. Right: Structure of the iron–sulfur (Fe_2S_2) center in higher plant ferredoxin. Each iron ion is coordinated with two cysteine residues in the polypeptide and two inorganic sulfur atoms that are released as H_2S upon acidification.

protein and occurs in a wide variety of photosynthetic organisms including higher plants and algae but not in photosynthetic bacteria. The oxidation–reduction potential of the copper complex is $E_{m,7} = +0.37$ V. The protein from spinach leaves has a molecular weight of 21,000 and contains two atoms of copper per molecule. However, a similar protein was isolated from bean leaves that has a molecular weight of 10,500 and only one atom of copper per molecule. The copper atom is coordinated to a cysteine, a methionine, and two histidine side chains (Fig. 4.10). Plastocyanin, as a component of the electron transport chain, donates electrons directly to photosystem 1. Inside-out thylakoid membranes can be washed free of plastocyanin rather easily, and the resulting deficient membranes exhibit low electron transport activity. Electron flow through photosystem 1 is restored upon addition of plastocyanin. Thus, studies with this protein indicated that it is located, along with cytochrome f, on the inner surface of the thylakoid membrane.

F. Photosystem 1 (P700)

Photosystem 1, in an analogous fashion to P680, contains a chlorophyll a molecule associated with a specific polypeptide of molecular weight between 65,000 and 70,000 (see p. 71). The environment within this chlorophyll–protein complex is responsible for the shift in maximal absorbance of the chlorophyll to about 700 nm. This form of chlorophyll a, designated **P700**, accounts for 1 of 200–1000 total chlorophyll molecules in thylakoid membranes. P700 has an oxidation–reduction potential of $E_{m,7} = +0.45$ V and is considered to be the reaction center of photosystem 1. It was discovered by a light-induced bleaching of

FIG. 4.9. The absorption spectra of chloroplast cytochrome f. A: The broken line is the spectrum for the oxidized cytochrome and the solid line is the spectrum for the reduced form. B: The spectra between 500 and 580 nm are expanded. C: The difference spectrum for the reduced versus the oxidized forms of cytochrome f. The major long-wavelength peak (the α-peak) in the difference spectrum usually is used for analysis of the cytochromes, with regard to both identification and amount. (Adapted from Bendall et al., 1971.)

the absorbance around 700 nm of a chloroplast preparation. The bleaching is caused by an oxidation of chlorophyll, coupled to the reduction of the primary electron acceptor of photosystem 1, referred to as X. X has a very low oxidation-reduction potential of about $E_{m,7} = -0.73$ V, but its structure has not been established. Electrons then are transferred from X to secondary electron acceptors, which are iron–sulfur proteins (Fe_4S_4 type) with $E_{m,7}$ of -0.6 to -0.7 V.

The reducing side of photosystem 1, i.e., the site of reduction of X, is on the surface of the thylakoid membrane facing the *stroma*. As noted in section E, plastocyanin, the immediate electron donor to photosystem 1, is located on the *luminal* surface of the membrane. Therefore, as with photosystem 2, the oxidizing and reducing reactions of photosystem 1 are located on opposite sides of the membrane.

G. Ferredoxin

The best characterized electron carrier between photosystem 1 and $NADP^+$ is the nonheme, iron–sulfur protein **ferredoxin.** The ferredoxins are small, water-soluble protein molecules. The protein isolated from higher plants and algae has a molecular weight of 12,000. Each molecule contains two iron atoms complexed with two sulfur atoms, which are released as inorganic sulfide when a solution of the protein is acidified. Additional iron–sulfur coordination complexes occur with sulfhydryl groups of protein-bound cysteine residues (Fig. 4.10).

Ferredoxin has a low oxidation-reduction potential ($E_{m,7} = -0.43$ V) and transfers one electron per molecule during a reaction. Electrons are transferred from the primary acceptor of photosystem 1 to ferredoxin by a substance with an even lower potential ($E_{m,7} = -0.55$ V) called the **ferredoxin-reducing substance** (FRS).

H. Ferredoxin–NADP⁺ Reductase

Ferredoxin functions *in situ* in a 1:1 complex with the flavoprotein **ferredoxin–NADP⁺ reductase.** The enzyme catalyzes the transfer of electrons from ferredoxin, a one-electron carrier, to $NADP^+$, a two-electron carrier. Thus, the coenzyme flavin adenine dinucleotide (FAD) must exist in an intermediate form with a single unpaired electron. This form has been identified by electron spin resonance. The reductase contains one FAD moeity per protein molecule of molecular weight 40,000. The enzyme-ferredoxin complex is tightly bound to the stromal surface of the thylakoid membrane. Consequently, the enzyme is accessible to the stromal pool of $NADP^+$

VI. Topographical Arrangement of Electron Carriers

Hill and Bendall proposed in 1960 that the carriers of electrons could be arranged in a thermodynamically favorable sequence in the order of increasing oxidation-reduction potentials. This model, drawn as a "Z" as shown in Fig. 4.7, takes into account whether a given carrier is oxidized or reduced by each of the photosystems. It is also based on the assumption that two photosystems exist and *function in series.*

A detailed topographical arrangement of the electron carriers in the chloroplast membrane was devised by Trebst (1980) from available evidence. This arrangement, shown in Fig. 4.11, also appears as a Z scheme, like Hill and Bendall's scheme, across the membrane. The evolution of oxygen occurs on the inner surface of the thylakoid membrane, whereas NADPH is produced on the outer, stromal surface. But the electrons are thought to cross the membrane at least three times between the beginning and the end of the chain. A virtue of this proposed arrangement is its ability to explain the production of a proton gradient across the membrane and the stoichiometry of ATP synthesis (see Section X).

The two-photosystems-in-series arrangement, however, should not be considered a structural unit. Numerous experiments with a variety of different types of chloroplasts have revealed that there is not a 1:1 stoichiometry of the two photosystems. Within the same thylakoid membrane, the arrangements of the photosystems differ. Granal membranes contain most of the photosystem 2 reaction centers but are nearly depleted of photosystem 1 complexes. In contrast, the photosystem 1-rich stromal membranes contain only 10–20% of the photosystem 2 activity. Although it is not known with certainty, the photosystems may be functionally connected by the large, diffusible pools of plastoquinone and plastocyanin.

VII. Evidence for Two Photosystems

The primary evidence for the existence of two photochemical reactions in sequence is twofold. First, simply on the basis of energetics, it can be calculated

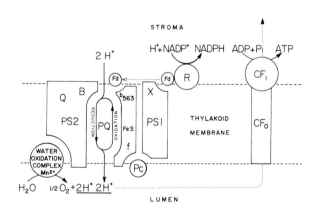

FIG. 4.11. The topographical distribution of electron carriers in thylakoid membranes. An electron carrier is considered to function on one side of the membrane or the other depending on its accessibility to chemical or immunological reagents. Q, electron acceptor of PS2; B, B protein; b_{563}, cytochrome b_{563} or b_6; PQ, plastoquinone; FeS, iron–sulfur protein; f, cytochrome f; Pc, plastocyanin; X, electron acceptor of PS1; Fd, ferredoxin; R, ferredoxin–NADP$^+$ reductase; PS1, photosystem 1; PS2, photosystem 2; CF_1, ATP synthetase (coupling factor 1); CF_0, membrane proteins associated with CF_1 that serve as a proton channel. (Adapted from Trebst, 1980.)

that reduction of $NADP^+$ by water requires a potential difference of at least 1.2 V but more likely 1.6 V. The larger value is the calculated difference between the oxidation–reduction potential of the photosystem 2 oxidant of water and that of the photosystem 1 reductant of ferredoxin. This potential difference, corresponding to about 154×10^3 J per Avogadro's number of electrons, is less than the energy available in an einstein (Avogadro's number of photons) of red light, which is 176×10^3 J at 680 nm. But considering that at least 31×10^3 J of energy must be supplied by the transport of Avogadro's number of electrons for the synthesis of ATP (see equation 4.5, p. 87) and that there are limitations to the overall efficiency of the system, it is clear that the energy required to functionally transfer one electron through the entire system is considerably greater than that available in a photon with a wavelength of 680 nm, which nevertheless is a wavelength of light that supports photosynthesis with maximal efficiency. The conclusion that emerges from these considerations is that the energy of at least

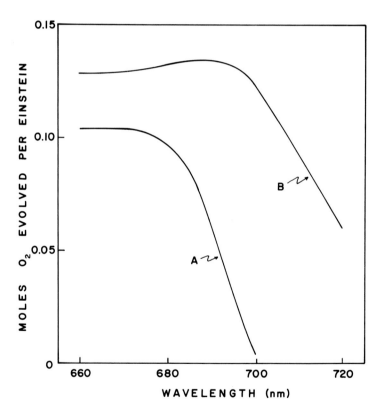

FIG. 4.12. Efficiency of photosynthesis at various wavelengths without (A) and with (B) a supplementary light source. Supplementary light was provided by a Hg–Cd lamp, which has several emission lines in the near ultraviolet and visible ranges, but the longest wavelength line is at 644 nm. The supplementary light greatly enhanced the efficiency of the primary monochromatic light input at wavelengths longer than 680 nm. This result indicated that activation of photosystem 2 by the 644-nm light was necessary so that electrons would be available for photosystem 1 activity at wavelengths above 680 nm. (Adapted from Emerson et al., 1957.)

two photons must be used to drive each electron through the system. Thus, two photochemical events, connected in series, are implied.

Second, direct experimental evidence for two photoreactions was provided by Emerson *et al.* (1957), who showed that although light of wavelengths longer than 680 nm is absorbed by chloroplasts, it is quite inefficient in promoting the production of oxygen without the simultaneous input of light of 640–650 nm (Fig. 4.12). This "enhancement" by shorter-wavelength light was interpreted as evidence for the involvement of two photoreactions. The photosystem that reduces $NADP^+$ requires long-wavelength light (photosystem 1) and is catalyzed by P700. The second photosystem, which oxidizes water, requires light of shorter wavelengths and is catalyzed by P680. For photosystem 1 to continue functioning, electrons must be fed to it by photosystem 2. Likewise, if photosystem 1 is not active, the electron carriers between the photosystems become reduced and can accept no more electrons from water through photosystem 2. In this situation, the energy absorbed by photosystem 2 is reemitted as fluorescence. This two-photoreaction scheme therefore predicts that at least 8 quanta of light are required to transfer four electrons from water to $NADP^+$ in the overall reaction.

Emerson (1958), Govindjee *et al.* (1968), and others made careful measurements to determine the minimal number of photons of light that must be absorbed for each O_2 molecule released. These studies, done mostly with the alga *Chlorella,* showed that between 8 and 12 quanta of light of 680 nm were required, depending on the physiological condition of the cells. The experimental values, when carefully measured, were never less than 8 quanta per O_2 molecule evolved. Since competing reactions occur, as well as cyclic electron flow around photosystem 1, which does not involve the oxidation of water (see Fig. 4.7), values somewhat greater than 8 quanta per O_2 molecule are expected. These results, therefore, are consistent with the two-photosystem scheme shown in Fig. 4.7. Most of the work on the mechanism of photosynthesis has supported this scheme, which remains the best model for current research on photosynthesis in green plants and algae.

VIII. An Alternate Proposal

Although the Z scheme of photosynthetic electron transport has played an important role in understanding this process, a number of alternatives to the Z scheme have been proposed over the years. An alternate mechanism has been developed by Daniel Arnon, who has long argued against the Z scheme of photophosphorylation. Arnon *et al.* (1981) have concluded from their experimental results that the two photosystems in thylakoid membranes do not work in series but rather act synchronously *in parallel.* Key features of this latter scheme are the roles of plastoquinone and the postulation of two photoreactions within photosystem 2. Arnon's group found that antagonists of plastoquinone, which blocked linear electron flow from photosystem 2 to photosystem 1, nevertheless did not block reduction of ferredoxin (and $NADP^+$) in the presence of uncoupling agents (Arnon *et al.,* 1981). The uncouplers in this case seem to be required for transport of protons away from the site of oxidation of water within the membrane. Plastoquinone normally would function in this role, as illustrated in Fig. 4.13, in addition to acting as a carrier of electrons between the two photosystems.

In Arnon's scheme, photosystem 2 contains two reaction centers. One of the centers may contain a few pheophytin molecules (chlorophyll lacking the central Mg^{2+} ion). Pheophytins exist in thylakoid membranes at about 0.02 the amount of chlorophyll and have a redox potential of around -0.61 V. This reducing potential would be sufficient to reduce ferredoxin, with electrons provided by the oxidation of water within the same photosystem 2 complex. The second reaction center, or the second photoreaction, is apparently required to reduce plastoquinone, by way of the primary acceptor Q, so it can serve to carry protons away from the site of the oxidation of water (Fig. 4.13). These protons are released from the membrane into the thylakoid lumen, while the electrons are returned to the reaction centers, thereby completing a short cyclic pathway.

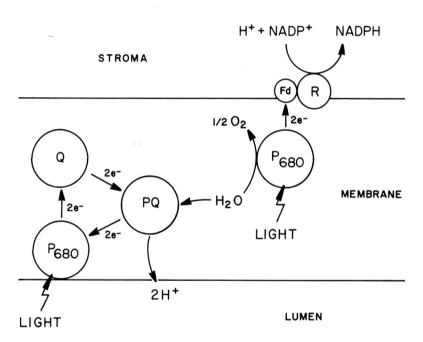

FIG. 4.13. Arnon's proposal for photosynthetic electron transport. In this "oxygenic" model, two forms of the photosystem 2 reaction center P680 occur. One form functions to reduce plastoquinone (PQ) through Q. Protons required for plastoquinone reduction are generated through oxidation of water by the second form of P680. This second form catalyzes electron flow from water to $NADP^+$ via ferredoxin (Fd). Reoxidation of plastoquinone releases protons into the thylakoid lumen to generate a proton gradient. Dissipation of the proton gradient through the coupling factor (CF_1) drives synthesis of ATP (as shown in Fig. 4.11). An "anoxygenic" system can function independently, as long as a supply of reduced plastoquinone (generated by the "oxygenic" system) is available. In this system, electrons are transferred from reduced plastoquinone to ferredoxin through P700. As in the Z scheme (Fig. 4.7), it is proposed that reduced ferredoxin can return electrons to the system at the level of plastoquinone. This cyclic electron flow, as the result of sequential oxidation and reduction of plastoquinone, functions essentially as a pump to transfer protons from the stroma into the thylakoid lumen. (Adapted from Arnon et al., 1981.)

Therefore, in this proposed mechanism photosystem 2 carries out *both* O_2 evolution *and* $NADP^+$ reduction. ATP synthesis also would occur, coupled to the flux of protons down a concentration gradient (see Section X). The stoichiometry of this system would be approximately one ATP molecule made per two NADPH molecules. Photosystem 1 in Arnon's scheme serves entirely as an additional ATP-generating system by **cyclic photophosphorylation.** Ferredoxin reduced by photosystem 1 would feed electrons back into an electron transport chain consisting of plastoquinone, cytochrome f, and plastocyanin. The flux of electrons from ferredoxin on the stromal side of the membrane of cytochrome f on the luminal side, through plastoquinone, would generate the proton gradient that drives ATP synthesis (see Figs. 4.7 and 4.11).

An important feature of this proposal is that, with the exception of a small feed of electrons from photosystem 2 to photosystem 1 to keep cyclic photophosphorylation functioning, the two photosystems can act essentially independently. The activity of each is modulated at the level of ferredoxin, which would be reduced by both photosystems 1 and 2. With an adequate supply of ATP, NADPH will be consumed by the CO_2-assimilation reactions (see Chapter 5), and a supply of $NADP^+$ will be available to accept electrons from ferredoxin. Consequently, ferredoxin will remain relatively oxidized and photosystem 2 will operate at a high rate. However, since insufficient ATP is generated by this system to maintain the stoichiometry required for CO_2 fixation, a depletion of ATP would cause a rise in the level of NADPH. The electronic backpressure will then permit electrons to flow through plastoquinone to photosystem 1 and thereby facilitate the activity of photosystem 1. It must be made clear that photosystem 1, by cyclic photophosphorylation, can operate independently for the generation of ATP. By the modulation of the rates of the two photosystems in this manner, the required balance of NADPH and ATP could be provided.

An attractive feature of Arnon's proposal is that, by functioning essentially independently, the two photosystems do not need to be present in equal amounts. In fact, the wide variability of the two photosystems among different plants, as well as their spatial separation between grana and stromal thylakoids, requires an awkward assumption to maintain the Z scheme. These aspects can easily be reconciled with the "parallel" scheme. Additional experimental work will determine the validity of this proposal.

Most observations made on photosynthetic electron transport, including the Emerson effect and the biochemical activities of the two photosystem reaction centers, support the Z scheme of Hill and Bendall, in which photosystem 2 oxidizes water and reduces plastoquinone while photosystem 1 oxidizes plastocyanin and reduces ferredoxin. Elegant confirmation of these conclusions was provided by Lam and Malkin (1982), who reconstituted the entire electron transport chain by mixing highly purified photosystem 2 particles, cytochrome b_6/f complexes, and photosystem 1 particles (Fig. 4.14). Active photoreduction of $NADP^+$ occurred only when all three of the complexes were present and supplemented with the soluble electron carriers plastocyanin, ferredoxin, and ferredoxin–$NADP^+$ reductase. Quinones present in the photosystem 2 particles transferred electrons to the cytochrome b_6/f complexes, whereas plastocyanin was required to transfer electrons into photosystem 1. The purified photosystem

2 particles, even when supplemented with the soluble factors, could not accomplish photoreduction of NADP$^+$

IX. Regulation of Energy Distribution from Light-Harvesting Complexes by Protein Phosphorylation

According to the noncyclic, Z scheme of photosynthetic electron transport, the two photosystems must operate at the same rate for maximal efficiency. However, depending on the wavelengths of incident light, this relation may become unbalanced as one of the photosystems is excited more than the other. Plants adapt to this type of situation by adjusting the distribution of energy delivered to the reaction centers from the light-harvesting complexes. In isolated thylakoid membranes, this regulation is expressed by the light-harvesting chlorophyll *a/b*–protein complex functioning almost entirely with photosystem 2 in the presence of monovalent or divalent cations but more with photosystem 1 in their absence.

Another factor that is part of this regulatory system and probably mediates this effect *in vivo* was found recently (Allen *et al.*, 1981). Thylakoid membranes contain a bound protein kinase that is activated by reduced plastoquinone. This protein kinase phosphorylates primarily the light-harvesting chlorophyll-binding protein on a threonine residue in a segment of the protein exposed on the surface of the membrane. In situations where the activity of photosystem 2 predominates over that of photosystem 1, reduction of plastoquinone activates the protein kinase. Phosphorylation of the light-harvesting complex results functionally in more excitation energy being transferred to photosystem 1, a process that apparently also involves cations as mentioned in the previous paragraph.

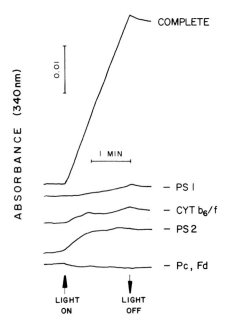

FIG. 4.14. Reconstruction of NADP$^+$ photoreduction. The amount of NADPH formed was measured by its absorbance at 340 nm. Purified particles containing the reaction centers of photosystem 1, photosystem 2, and the cytochrome b_6/f complex (see Fig. 4.11) were mixed with 5 mM MgCl$_2$, 15 mM NaCl, 2 mM NADP$^+$, 8 μM ferredoxin, 0.2 μM ferredoxin–NADP$^+$ reductase, and 1.2 μM plastocyanin. No NADPH was formed when either photosystem 1, the cytochrome b_6/f complex, photosystem 2, plastocyanin, or ferredoxin was omitted. (Adapted from Lam and Malkin, 1982.)

(Since the concentrations of the cations *in vivo* do not change significantly, their effects are mediated by the presence or absence of negatively charged phosphate groups on the proteins.) Consequently, as the activity of photosystem 1 increases, plastoquinone becomes more oxidized, and the light-harvesting chlorophyll protein is dephosphorylated by a membrane-bound protein phosphatase. This reversible phosphorylation, by altering the distribution of energy transfer from light-harvesting complexes to the reaction centers, thereby promotes a balance between the photosystems and maximizes the quantum efficiency of electron transfer between water and $NADP^+$.

Accompanying phosphorylation of the light-harvesting complex is a migration of 8-nm particles from grana into stromal thylakoids (Kyle *et al.*, 1983). At the same time, the content of chlorophyll *b* and chlorophyll *a/b*-binding polypeptides increases in stromal membranes. Upon dephosphorylation, the membrane structure returns to the original state. These results suggest that the regulation of energy distribution is determined by laterally mobile light-harvesting chlorophyll *a/b*-protein complexes contained within the 8-nm particles. These particles are observed on the P fracture face of the membrane and are distinct from the large particles on the E face, which contain photosystem 2 reaction centers.

X. Formation of ATP

Photosynthetic electron transport generates a high-energy state across the thylakoid membrane that can be converted into chemical energy in the form of ATP. Production of ATP, of course, is required for the fixation of CO_2. The process by which ATP is produced is becoming clear in outline form, although some details of the mechanism of photophosphorylation are still uncertain.

In an interesting experiment, reported in 1966, Jagendorf and Uribe demonstrated that ATP could be produced *in the dark* if a concentration gradient of H^+ ions existed across the thylakoid membrane. In their experiment, chloroplasts were suspended in a buffer at pH 4. After sufficient time had elasped to allow equilibration of protons across the membrane, the suspension was rapidly injected into a solution containing ADP, inorganic phosphate, and a buffer at pH 8. A rapid synthesis of ATP occurred, the rate of which decreased to zero as the pH inside the thylakoid vesicle became equal to that outside, i.e., as the proton gradient was spent. The pH transition, from acid to alkaline, was an absolute requirement in these experiments, as the same pH throughout or a gradient set up in the opposite direction did not lead to synthesis of ATP. Furthermore, if the thylakoid membranes were damaged and made permeable to H^+ ions by freezing and thawing, no ATP synthesis occurred. Thus, an intact, osmotically active thylakoid also was required in this phosphorylating system.

Hind and Jagendorf (1963) had noticed several years earlier that ATP could be synthesized, again in the dark, if thylakoid membranes were first illuminated in the absence of ADP and inorganic phosphate. After the light was turned off, resulting in a cessation of electron transport, the addition of ADP and inorganic phosphate resulted in the synthesis of substantial amounts of ATP (Fig. 4.15). A large amount of experimental evidence now exists that dem-

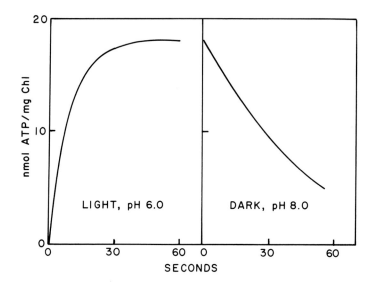

FIG. 4.15. The high-energy state during illumination of chloroplasts. Left: The high-energy state is formed during illumination at pH 6.0. After various times of illumination, the sample was rapidly mixed in the dark with ADP and inorganic phosphate, and the capacity to synthesize ATP was measured at pH 8.0. Right: Decay of the high-energy state begins immediately after cessation of illumination. In this experiment, chloroplasts were illuminated for 60 sec. ADP and inorganic phosphate then were added after various times in the dark. The decay in the capacity to synthesize ATP reflected decay of the proton gradient. The explanation of these results is that during illumination, in the absence of ADP and inorganic phosphate, protons accumulated within thylakoids. As the proton gradient developed, the capacity to synthesize ATP increased. After the light was turned off, the capacity to synthesize ATP, i.e., the proton gradient, was gradually lost as the result of leakage of protons through the membrane. (Adapted from Hind and Jagendorf, 1963.)

onstrates that a concentration gradient of H^+ ions is produced during photosynthetic electron transport, with the inside of the thylakoid more acidic than the outside. To explain the generation of this "high-energy" state, it was necessary to know the topography of the membrane components, as shown in Fig. 4.11, the site of each oxidation–reduction reaction within the membrane, and the directionality of release of the products.

The oxidation of two molecules of water to a molecule of O_2 yields four H^+ ions for each four electrons removed. These protons are released inside the thylakoid and are trapped by the membrane. Additional protons are transferred into the thylakoid vesicle by plastoquinone. Reduction of plastoquinone by the primary electron acceptor of photosystem 2 is thought to occur on the external surface of the membrane and requires the uptake of one H^+ from the chloroplast stroma for each electron accepted by plastoquinone (reaction 4.8):

[4.8]

Subsequently, reoxidation of plastoquinone by cytochrome *f*, located on the interior surface of the membrane, causes release of the protons inside the thylakoid (reaction 4.9):

$$+ 2 \text{ Cyt } f(\text{Fe}^{3+}) \quad \rightleftharpoons \quad + 2 \text{ Cyt } f(\text{Fe}^{2+}) + 2 \text{ H}^+ \qquad [4.9]$$

(To intra-
thylakoid
fluid)

Although a large pool of plastoquinone exists in the membrane, reaction 4.9 is the rate-limiting step in photosynthetic electron transport, since the internal concentration of H^+ ions influences the position of the equilibrium of this reaction. Chemicals that make the membrane permeable to H^+ ions, or act as H^+-transporting substances and thereby allow dissipation of the gradient (uncouplers), stimulate the rate of oxidation of plastoquinone.

The scheme shown in Fig. 4.11 illustrates that another H^+-consuming process, the reduction of the FAD coenzyme of ferredoxin–$NADP^+$ reductase to $FADH_2$, also occurs on the outside, stromal surface of the membrane. This reductive step consumes one H^+ ion for each electron transferred to the flavoprotein. In subsequent reactions, the equivalent of these H^+ ions and electrons are used to reduce 3-phosphoglyceroyl phosphate to glyceraldehyde 3-phosphate (see p. 114). The net result is that for each O_2 molecule evolved, at least four H^+ ions are generated from water on one side of the membrane, while an equal number are consumed on the other by subsequent reductive reactions in carbohydrate synthesis. Along with the H^+ ions transported across the membrane by plastoquinone, the passage of four electrons through the entire chain results in a net gain of eight H^+ ions by the interior of the thylakoid. Thus the H^+/e^- is 2. This value has been confirmed by experimental analyses by Witt (1975) and McCarty and Portis (1976).

The magnitude of the difference in the free H^+ ion concentration across the membrane during electron transfer, in the absence of ADP and inorganic phosphate, can be such that the interior of the thylakoids has a pH value nearly 4 units lower than that in the chloroplast stroma. Since pH is a logarithmic function, the concentration of H^+ ions inside the thylakoid therefore is nearly 10,000 times greater than that outside under these conditions. If the internal fluid has some buffering capacity, the concentration of H^+ ions actually available for driving ATP synthesis may be even greater than the amount indicated by the internal pH. However, when ADP and inorganic phosphate are available for the synthesis of ATP, the pH gradient is continuously expended and the steady-state level of the gradient is considerably less than 4 pH units. As indicated by experimental results, with respect to the synthesis of ATP, the actual pH values inside or outside the thylakoids are not important but only the difference in pH across the membrane. The magnitude of the H^+ ion concentration reached inside the thylakoid is proportional to the rate of electron flow, which in turn is related to the intensity of light (Fig. 4.16).

Thus, the "high-energy" state that drives the synthesis of ATP is presently considered to be the electrochemical gradient of H^+ ions across the thylakoid membrane. Yet, not only does potential energy exist in the *chemical* gradient of H$^+$ ions, but the imbalance in ions across the membrane also produces an *electrical* gradient, with the inside of the thylakoid positive with respect to the stroma. The magnitude of the voltage gradient within chloroplasts *in vivo* is not certain, but with isolated chloroplasts the value can be 100 mV or more. Therefore, a strong pull is exerted on the H$^+$ ions to pass through the membrane into the stroma. This flow of H$^+$ ions occurs at specific sites called **proton channels.** These channels are formed by specific membrane proteins.

Two approaches have been taken to determine the stoichiometry between proton flux and ATP synthesis. The relationship between the extent of phosphorylation of ADP and the H$^+$ concentration inside thylakoids was determined by McCarty and Portis (1976) (Fig. 4.16). The rate of electron flow in isolated thylakoid membranes was varied by increasing the intensity of incident light. The number of ATP molecules synthesized and the number of electrons passed from water to an electron acceptor were measured to obtain a P/e$^-$ ratio. The H$^+$ concentration inside the thylakoid was measured by the uptake of labeled hexylamine, which penetrated membranes in its uncharged form but became trapped inside when ionized by protonation. A plot of these values, as shown in Fig. 4.16, yielded a straight line, which when extrapolated to the ordinate, provided a maximal P/e$^-$ value of 0.66. Thus, since H$^+$/e$^-$ = 2, the minimal H$^+$/ATP = 3. That is, the energy released by three H$^+$ ions flowing down their electrochemical gradient is captured in the synthesis of one ATP molecule.

Schlodder *et al.* (1982) also obtained a value for H$^+$/ATP of 3 with experiments similar to those described in the previous paragraph. However, they obtained an elegant confirmation of this value by also examining the membrane potential as a driving force. In their experiments, thylakoid membranes were incubated in the presence of permeable buffers to prevent generation of a pH gradient. Short pulses of light or an external electric field provided a voltage gradient of about 150 mV across the membrane, as protons were transported into the thylakoid. As Fig. 4.17 shows, the ATP yield was relatively constant between a pH of 7–9 outside the thylakoids. Measurements of the concomitant ion flux gave a value of H$^+$/ATP = 3.

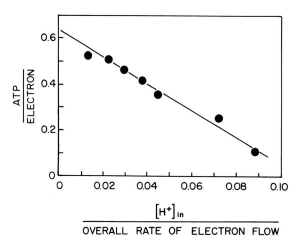

FIG. 4.16. The relationship between the observed phosphorylation efficiency (ATP/e$^-$) and the ratio [H$^+$]$_{in}$/(overall rate of electron flow). As the light intensity was increased, the rate of electron flow increased toward a maximal value, which caused the ratio [H$^+$]$_{in}$/(overall rate of electron flow) to decrease. Phosphorylation of ADP also rose along with the increase in the rate of electron flow. Extrapolation of the line to the ordinate, to the maximal value for ATP/e$^-$, provided a value near 0.66. Thus the value for ATP/2e$^-$ during noncyclic photosynthetic electron transport was about 1.3. (Adapted from McCarty and Portis, 1976.)

Therefore, the important factor in ATP synthesis is the flux of H^+ through the membrane. The functional proton-motive force, which drives the H^+ flux, may be composed of a pH difference across the membrane or a potential difference. Hangarter and Good (1982) carried out experiments with isolated spinach chloroplasts in which they simultaneously controlled both parameters. For the synthesis of ATP, an energy input of at least 3.1×10^4 joules per mole of ATP is necessary (see reaction 4.5). In the absence of a membrane potential, this threshold was provided by a H^+ concentration gradient of 2.7 pH units. Conversely, in the absence of a proton gradient, the threshold membrane potential was about 165 mV. Moreover, they found that the proton gradient and the membrane potential contributed to the proton-motive force in a precisely additive fashion, with a decrease of 1 unit in the pH difference counterbalanced by an increase in membrane potential of 60 mV. Both parameters are linked to the light-driven flux of electrons through the electron carriers. During steady-state photosynthesis in chloroplasts, the proton-motive force is provided primarily by the proton gradient.

The ratio $H^+/ATP = 3$ seems to be established. The theoretical value of 2 for the ratio H^+/e^- has experimental support but is less certain. However, if these values are used to calculate the overall stoichiometry, it then appears that, for each set of four electrons transferred through the system, one O_2 molecule is evolved, two $NADP^+$ molecules are reduced and 2.66 molecules of ATP are synthesized.

It should be apparent from this discussion that the magnitude of the $H^+/$ATP and H^+/e^- ratios are central to the interpretation of the energy yield during photosynthesis. If the values used in the previous paragraph are correct, they imply that insufficient ATP would be available, to provide a stoichiometric balance with NADPH, for synthesis of glucose during noncyclic photosynthesis

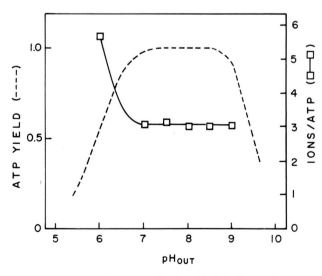

FIG. 4.17. The H^+/ATP ratio as a function of pH_{out}. Thylakoid membranes were incubated in permeable buffers to prevent generation of a pH gradient. A transmembrane voltage gradient was generated by external voltage or light pulses. Measurements of the ATP synthesized and the concomitant flux of ions across the membrane gave a value for H^+/ATP of 3 between pH 7 and 9. (Adapted from Schlodder *et al.*, 1982.)

(see p. 117). Several investigators suggested that additional protons may be transported per electron, and thus the ATP/4e$^-$ may be 4 rather than 2.66. If this ratio is 4, the ATP generated during noncyclic photophosphorylation would be adequate for CO_2 fixation. It is necessary to emphasize, however, that these conclusions are based upon the basic premise that photosynthetic electron transport proceeds according to the noncylic Z scheme shown in Fig. 4.7.

ATP can be synthesized, without evolution of O_2, by cyclic electron flow through photosystem 1. Since ATP is required for synthesis of compounds that subsequently are reduced by NADPH in the pathway for glucose synthesis, inadequate levels of ATP would cause a rise in the NADPH/NADP$^+$ ratio. Consequently, the rate of electron flow from ferredoxin to NADP$^+$ will decrease and reduced ferredoxin will accumulate. Reduced ferredoxin can donate electrons to plastoquinone on the stromal side of thylakoid membranes, thereby providing an additional flow of electrons to cytochrome *f*, an additional flux of protons into the lumen of the thylakoid, and a higher rate of ATP synthesis. This conservation of electrons maintains the balance of ATP and NADPH.

XI. The Chloroplast Coupling Factor (CF$_1$)

The terminal step in the synthesis of ATP is catalyzed by a large protein whose presence is required to transform or "couple" the electrochemical energy in the proton gradient into chemical energy as ATP. The mechanism of this reaction is not well understood, but the protein undergoes conformational changes during the reaction, apparently caused by protonation and deprotonation of sites on the enzyme. This **coupling factor** from higher plants has a molecular weight for the intact enzyme of 325,000 and consists of five different subunits with individual molecular weights of about 60,000 for α, 55,000 for β, 38,000 for γ, 20,000 for δ, and 15,000 for ϵ. The complete protein has a subunit stoichiometry of $\alpha_2\beta_2\gamma\delta\epsilon_2$. CF$_1$ from the alga *Chlamydomonas reinhardtii* recently was shown to have a molecular weight of about 400,000 and a subunit stiochiometry of $\alpha_3\beta_3\gamma\epsilon$. The δ subunit is required for association of CF$_1$ to the proton channel in the membrane, called CF$_0$. The available evidence suggests that the catalytic site for ATP synthesis resides on the β subunit, with the α, γ, and ϵ subunits playing regulatory roles. In addition, the γ subunit serves as a gate for protons conducted through CF$_0$ into CF$_1$.

Electron microscopy of thylakoid membranes revealed CF$_1$ as particles 9–10 nm in diameter on the surface of the membrane in contact with the chloroplast stroma. CF$_1$ is easily removed from the membrane by treatment with 1 mM ethylenediaminetetraacetate, a chelating agent that binds selectively Ca^{2+} and Mg^{2+}, and can be rebound to the membrane in the presence of Mg^{2+} ions. On release of the protein from the membrane, the thylakoid vesicles become more permeable to H$^+$ ions, and this permeability is largely lost when the protein is added back to the membrane. These results indicate that the CF$_1$ binds to specific sites on the membrane surface where the proton channels (CF$_0$) occur. Therefore, CF$_1$ apparently controls the flux of H$^+$ ions across the membrane by its ability to make ATP. The enzyme ordinarily does not hydrolyze ATP, but when freed from the membrane and treated under appropriate conditions, it exhibits an ATP hydrolase activity.

The evidence is compelling that the transfer of electrons through the membranes's electron transport system leads to a state of high potential energy, characterized both by the energy embodied in the proton gradient as well as by that in the electrical potential that accompanies a nonequilibrium state of ions across a membrane. The controlled flux of H^+ ions through the membrane, down the gradient into the coupling factor, leads to the synthesis of ATP.

The concept of ATP formation driven by an electrochemical gradient was first formulated by Peter Mitchell in 1961 and has been called the **chemiosmotic hypothesis.** The data that continue to accumulate concerning the process of photophosphorylation in chloroplasts increasingly tend to support Mitchell's hypothesis. The chemiosmotic model also reveals the essential *a priori* need for a membrane-enclosed compartment, of which many are present in the chloroplast as the thylakoids.

The overall light-driven processes of photosynthesis in chloroplasts are summarized in Fig. 4.18.

XII. Electron Transport in Photosynthetic Bacteria

Because of the greater simplicity of the photochemical reactions in chromatophores of photosynthetic bacteria, these organisms have been studied intensively as models for photosynthesis in plants. Chromatophores contain a single photosystem, similar to photosystem 1 in chloroplasts, which under normal conditions operates as a closed electron cycle. Photosynthetic bacteria do

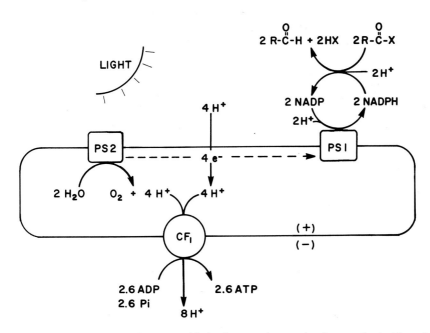

FIG. 4.18. Summary of the stoichiometry of light-driven processes in photosynthesis. The scheme assumes the Z arrangement of electron carriers and does not include cyclic photophosphorylation. The membrane is permeable to H_2O and O_2, but H^+ ions cross the membrane only at sites where CF_1 is bound to proton channels. (Because of its strong charge, an H^+ ion attracts a relatively large cloud of water. The hydrated radius is sufficiently large to prevent a significant rate of diffusion of protons across a membrane.)

not evolve oxygen. To keep the system primed, electron donors other than water are used. The principal donors in the green and purple sulfur bacteria are reduced forms of sulfur, such as S^{2-} (sulfide) and $S_2O_3^{2-}$ (thiosulfate), whereas the nonsulfur purple bacteria can use organic donors as succinate.

Reaction center complexes isolated from photosynthetic bacteria generally contain four molecules of bacteriochlorophyll, two molecules of bacteriopheophytin, two quinones, a nonheme iron center, and three polypeptides. In green bacteria the reaction centers contain a form of bacteriochlorophyll *a* that absorbs maximally at 840 nm (P840). Electron transport apparently also can proceed in a linear fashion in these organisms, as shown in Fig. 4.19B. Electrons

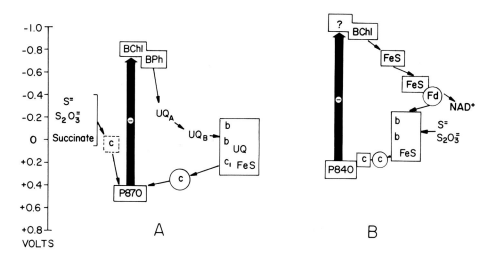

FIG. 4.19. Photosynthetic electron transport pathways in (A) purple (sulfur and nonsulfur) and (B) green (sulfur) photosynthetic bacteria. The components are arranged according to the midpoint of their oxidation–reduction potentials. The heavy vertical arrows indicate the change in reduction potential of the reaction centers when light is absorbed. The systems exhibit a cyclic electron flow, through a single reaction center, analogous to cyclic electron flow around photosystem 1 in green plants. A: In purple bacteria, the reaction center (P870) contains a bacteriochlorophyll dimer as the functional electron donor. The initial electron acceptor is a bacteriopheophytin (BPh) (a bacteriochlorophyll molecule minus the central Mg^{2+} ion). Electrons are transferred through several ubiquinone (UQ) species to a complex containing cytochrome *b*, cytochrome *c*, and an iron–sulfur protein. A *c*-type cytochrome, whose function is similar to that of cytochrome *f* in green plants, returns electrons to P870. Electrons can be donated to the system by sulfide, thiosulfate, or organic substrates such as succinate. This system functions to generate an electrochemical proton gradient that drives synthesis of ATP. Reducing compounds (as NADH) apparently are generated by reverse electron flow through the respiratory chain, which requires ATP. B: The reaction centers in green photosynthetic bacteria contain a form of bacteriochlorophyll that absorbs light at 840 nm (P840). The reaction centers donate electrons to iron–sulfur proteins, which in turn reduce ferredoxin (Fd). Sufficient reduction potential is retained by these electron acceptors to reduce NAD^+. Thus, these organisms are capable of a noncyclic electron flow from sulfide or thiosulfate to NAD^+. For cyclic flow, electrons are transferred from ferredoxin to a complex containing cytochrome *b* and iron–sulfur proteins. A *c*-type cytochrome returns electrons to P840. Green bacteria apparently do not synthesize ATP by photophosphorylation. Boxed components are part of integral membrane complexes; circled components are soluble or peripheral proteins. A *c*-type cytochrome present in some but not all purple bacteria is enclosed with a dashed-line box in (A). [Adapted from Wraight, 1982.]

are donated to the reaction centers by a *c*-type cytochrome, which is similar to cytochrome *f* in plant cells. Sufficient reducing potential is generated in illuminated cells to reduce ferredoxin and ultimately produce NADH (rather than NADPH).

More is known about the photosynthetic systems in purple than in green bacteria. The reaction center complex contains a dimer of bacteriochlorophyll *a*, which absorbs maximally at 870 nm (P870) and serves as the primary electron donor. When converted to its oxidized form in the light, P870 is completely bleached. The electrons are transferred through bacteriopheophytin to the primary acceptor. In contrast to green bacteria and photosystem 1 in plants, which reduce ferredoxin through iron–sulfur centers as the primary acceptors, in the purple bacteria the electron acceptor is a quinone compound, analogous to photosystem 2 in the plant cells. Two different quinone compounds, ubiquinone derivatives, provide the carriers of protons across the chromatophore membrane. A nonheme iron protein is associated with the quinones and facilitates electron flow between the quinone pools. Electrons then are transferred to a *b*-type cytochrome, through an iron–sulfur center, and on to two types of *c*-type cytochromes. Finally, electrons are fed back into the reaction center, P870 (Fig. 4.19A).

This cyclic electron flow in purple bacteria drives a flux of protons across the membrane, out of the cytoplasmic space into the extracellular space (or the interior of isolated, closed chromatophores). This flux of protons generates a membrane potential that is negative inside the cell. As protons cross the membrane into the cell in response to the electrochemical gradient, through a coupling factor on the cytoplasmic surface, ATP is synthesized. Since the reducing potential generated by the primary electron acceptor is too low ($E_{m,7} = -0.18$ V) to reduce NAD^+ ($E_{m,7} = -0.32$ V), NADH seems to be produced by an ATP-driven reversal of the respiratory electron transport chain on the nonchromatophore cell membrane or by oxidation of carbohydrates after phosphorylation by ATP.

Additional Reading

A list of sources for additional reading in photosynthesis is provided at the end of Chapter 5.

5

The Process of Photosynthesis

The Dark Reactions

I. Introduction

Thylakoid membranes of chloroplasts capture light energy and transform it into chemical energy in the form of ATP and NADPH. But cells do not, in fact cannot, accumulate a sizable store of energy reserves as ATP or as NADPH. Rather, these direct products of photosynthesis are used to synthesize more appropriate storage forms of energy. The final achievement of photosynthesis, therefore, is the biosynthesis of carbohydrate, as **glucose,** from CO_2 and H_2O. Glucose is the key carbohydrate around which the metabolism of most organisms revolves. The metabolism of glucose provides the compounds needed for the fabric of the cell, whereas the oxidation of glucose provides the energy needed by nonphotosynthetic organisms to live. However, more glucose can be synthesized than is needed for current life processes in plant cells. The excess is stored in the form of polymers, the most common of which is **starch.** Starch is an excellent storage form of carbohydrate, since it is stable, nearly inert, and exerts very little osmotic pressure, but is readily broken down enzymatically to provide glucose when needed.

The synthesis of glucose *per se* cannot be called "photosynthesis," since light is not required. However, the ATP and NADPH molecules generated by the light-driven reactions are required. The fixation of CO_2 into organic compounds is therefore called the "dark" reactions of photosynthesis. Whereas participants of the "light" reactions are confined to the thylakoid membrane, the "dark" reactions occur in the stroma of the chloroplast and are catalyzed by soluble enzymes.

II. Fixation of CO_2

A. The C_3 Pathway—The Reductive Pentose-Phosphate Cycle

1. Biochemistry of the Cycle

A major pathway for the assimilation of carbon from CO_2 into carbohydrate was elucidated by the work of Melvin Calvin and his colleagues in 1954, and is often referred to as the Calvin–Bassham–Benson cycle (Bassham *et al.*, 1954). The overall net reaction in this pathway is

$$CO_2 + 3\ ATP + 2\ NADPH + 2\ H^+ + 2\ H_2O \rightarrow \tfrac{1}{6}(C_6H_{12}O_6)$$
$$+ 3\ \text{inorganic phosphate } (P_i) + 3\ ADP + 2\ NADP^+ \quad [5.1]$$

By examining the kinetics of incorporation of the carbon into metabolic intermediates, Calvin and his co-workers were able to trace the path of carbon during CO_2 fixation. To do this, they used radioactive CO_2 containing the isotope carbon-14. Cultures of algal cells were incubated with $^{14}CO_2$ for short periods of time (0.4 to 15 sec). Incorporation was stopped, and the labeled products were extracted simultaneously, by dropping the cells into boiling methanol. The extracted material was then subjected to two-dimensional paper chromatography. To locate only those compounds that had incorporated the radioactive carbon, a sheet of photographic X-ray film was placed in contact with the chromatogram. The beta rays expelled as the ^{14}C decayed affected the silver halide crystals in the film in the same manner as light. After the film was developed, black spots indicated the positions of the labeled compounds.

The first product to become highly labeled from $^{14}CO_2$ in most higher plants and algae was the 3-carbon compound 3-phospho-D-glycerate, which is produced in a reaction of CO_2 with ribulose 1,5-bisphosphate. As Fig. 5.1 shows, 3-phosphoglycerate becomes labeled immediately and at a linear rate, which is expected for a primary product of CO_2 incorporation. Other compounds become labeled more slowly after a lag period. The mechanism of the CO_2-fixing reaction was established by Sue and Knowles in 1978. In this reaction

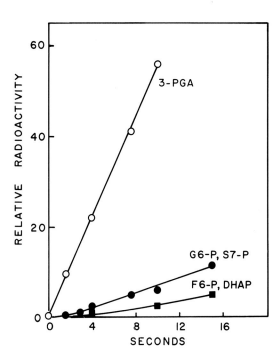

FIG. 5.1. Rates of incorporation of ^{14}C from $^{14}CO_2$ into intermediates in glucose synthesis. 3-PGA, 3-phosphoglyceric acid; G6-P, glucose 6-phosphate; S7-P, sedoheptulose 7-phosphate; F6-P, fructose 6-phosphate; DHAP, dihydroxyacetone phosphate. $^{14}CO_2$ was introduced into a culture of algal cells *(Scenedusmus obliquus)* cells, and samples were removed at the times indicated. (Adapted from Bassham *et al.*, 1954.)

(reaction 5.2), CO_2 condenses at carbon-2 of ribulose 1,5-bisphosphate to pro-duce a 6-carbon intermediate. HCO_3^- is not a substrate for this reaction. Several features of the reaction have provided clues to the mechanism. The oxygens at carbon-2 and -3 of the substrate are retained in the products, but the hydrogen on carbon-3 is lost to the solvent water as the reaction proceeds. Cleavage occurs between carbon-2 and -3 of ribulose 1,5-bisphosphate. Finally, a proton from water is added onto carbon-2 of the molecule of 3-phosphoglycerate that includes carbon-1 and -2 of the substrate. These data are consistent with the generation of a nucleophilic center, by the enzyme, at carbon-2 of ribulose 1,5-bisphosphate, which interacts with a molecule of CO_2.

Ribulose 1,5-
bisphosphate
(B=basic group on enzyme)

2-Carboxy-3-
ketoribitol 1,5-
bisphosphate

3-phospho-
glycerate

[5.2]

Reaction 5.2 provides a scheme that satisfactorily explains the electronic movements in the substrate during carboxylation. CO_2 is a linear, symmetrical molecule, and the attraction of the electrons toward the electronegative oxygen atoms renders the carbon atom particularly susceptible to nucleophilic attack. Although it is an apparent intermediate, 2-carboxy-3-ketoribitol 1,5-bisphos-phate has not been isolated from the reaction mixture and is thought to be a transient compound that remains bound to the enzyme. The carboxylation reac-tion is essentially irreversible. A large amount of energy, about 3.5×10^4 J/mole, is released as heat as the reaction proceeds.

The enzyme that catalyzes this incorporation of CO_2 into organic molecules binds ribulose 1,5-bisphosphate at a specific site. As this substrate interacts with functional groups in the enzyme, the nucleophilic center at carbon-2 is induced. The enzyme also has a binding site for CO_2 adjacent to the site for ribulose 1,5-bisphosphate. The combined action of bringing the two reactants close together and inducing shifts in the electronic distribution within the sub-strates allows the reaction to proceed with a high rate. This enzyme, **ribulose 1,5-bisphosphate carboxylase,** is an abundant protein in plant cells and very likely is the most abundant protein in nature. It is a large protein composed of 16 subunits of two types—8 subunits are 55,000 daltons in mass and 8 are 12,000–16,000 daltons in mass, depending on the type of plant cell from which they are obtained. The intact enzyme has a molecular weight of about 560,000. It is easily isolated from extracts of plant cells, and in some cases it accounts for up to one-half of the soluble protein in chloroplasts that fix CO_2 by the C_3 pathway. The catalytic sites reside on the large subunits of the enzyme.

Ribulose 1,5-bisphosphate, the substrate that must be present to achieve CO_2 fixation by the C_3 pathway, is synthesized from ribulose 5-phosphate and ATP by the enzyme **ribulose 5-phosphate kinase.** This is shown in reaction 5.3.

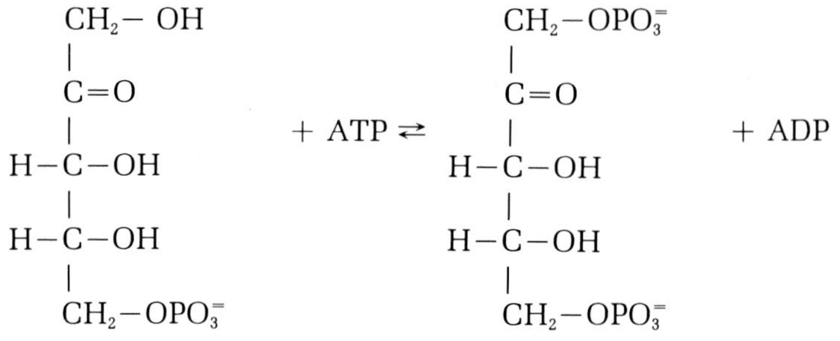

$$
\begin{array}{c}
CH_2-OH \\
| \\
C=O \\
| \\
H-C-OH \\
| \\
H-C-OH \\
| \\
CH_2-OPO_3^=
\end{array}
\quad + ATP \rightleftarrows \quad
\begin{array}{c}
CH_2-OPO_3^= \\
| \\
C=O \\
| \\
H-C-OH \\
| \\
H-C-OH \\
| \\
CH_2-OPO_3^=
\end{array}
\quad + ADP \qquad [5.3]
$$

Ribulose 5-phosphate   Ribulose 1,5-bisphosphate

The two molecules of 3-phosphoglycerate that are produced in the CO_2-fixing reaction are converted to 3-phosphoglyceroyl phosphate in a reaction that requires one molecule of ATP for each 3-phosphoglycerate molecule. This reaction, shown in reaction 5.4, is catalyzed by the enzyme **phosphoglycerate**

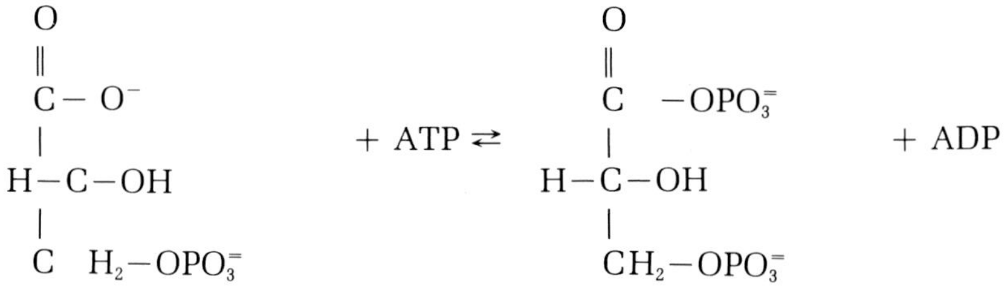

$$
\begin{array}{c}
O \\
\| \\
C-O^- \\
| \\
H-C-OH \\
| \\
C\ H_2-OPO_3^=
\end{array}
\quad + ATP \rightleftarrows \quad
\begin{array}{c}
O \\
\| \\
C\ -OPO_3^= \\
| \\
H-C-OH \\
| \\
CH_2-OPO_3^=
\end{array}
\quad + ADP \qquad [5.4]
$$

3-Phosphoglycerate   3-Phosphoglyceroyl phosphate

kinase. The two reactions shown by reactions 5.3 and 5.4 account for the three molecules of ATP needed for the fixation of each CO_2 molecule by this pathway.

In the next reaction, each 3-phosphoglyceroyl phosphate is reduced to glyceraldehyde 3-phosphate by NADPH (reaction 5.5). This step requires two mol-

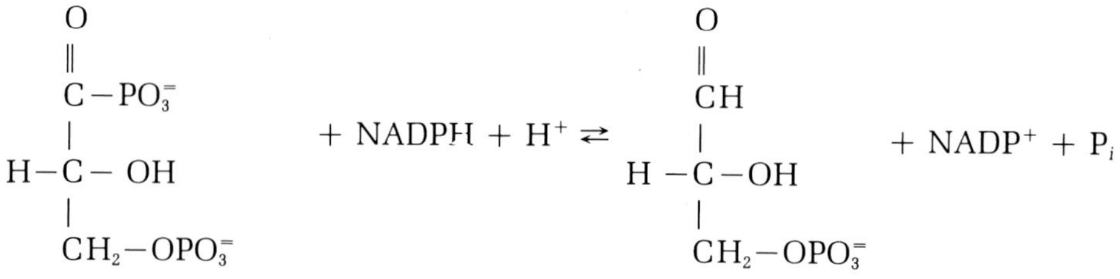

$$
\begin{array}{c}
O \\
\| \\
C-PO_3^= \\
| \\
H-C-\ OH \\
| \\
CH_2-OPO_3^=
\end{array}
\quad + NADPH + H^+ \rightleftarrows \quad
\begin{array}{c}
O \\
\| \\
CH \\
| \\
H\ -C-OH \\
| \\
CH_2-OPO_3^=
\end{array}
\quad + NADP^+ + P_i
$$

3-Phosphoglyceroyl phosphate   Glyceraldehyde 3-phosphate   [5.5]

ecules of NADPH to maintain stoichiometry with each CO_2 molecule that is assimilated. The enzyme that catalyzes this reaction, **glyceraldehyde 3-phosphate dehydrogenase (NADP)** is found only in chloroplasts. Although a similar reaction occurs in the glycolytic pathway in the cytosol of the cell, the next flux is in the opposite direction, and the cytosolic enzyme involved in glycolysis is specific for the coenzyme NAD^+. The two enzymes also are structurally different proteins.

D-Glyceraldehyde 3-phosphate, by a reversal of the analogous reactions in glycolysis, is converted to glucose 6-phosphate. These reactions are shown in Table 5.1.

On paper, these reactions can be arranged in a neat pathway, with glucose

TABLE 5.1. Reactions Involved in CO_2 Fixation

1. Synthesis of glucose

$$\text{Glyceraldehyde 3-phosphate} \xrightarrow{\text{Triose phosphate isomerase}} \text{dihydroxyacetone phosphate}$$

$$\begin{array}{c}\text{Glyceraldehyde 3-phosphate}\\ \text{+ dihydroxyacetone phosphate}\end{array} \xrightarrow[\text{aldolase}]{\text{Fructose bisphosphate}} \text{fructose 1,6-bisphosphate}$$

$$\text{Fructose 1,6-bisphosphate} \xrightarrow{\text{Fructose bisphosphatase}} \text{fructose 6-phosphate} + 3P_i$$

$$\text{Fructose 6-phosphate} \xrightarrow[\text{isomerase}]{\text{Glucose phosphate}} \text{glucose 6-phosphate}$$

$$\text{Glucose 6-phosphate} \xrightarrow{\text{Glucose 6-phosphatase}} \text{glucose} + P_i$$

2. Regeneration of ribulose 1,5-bisphosphate

$$\text{Fructose 6-phosphate + glyceraldehyde 3-phosphate} \xrightarrow{\text{Transketolase}}$$
$$\text{xylulose 5-phosphate + erythrose 4-phosphate}$$

$$\begin{array}{c}\text{Erythrose 4-phosphate}\\ \text{+ dihydroxyacetone phosphate}\end{array} \xrightarrow[]{\begin{array}{c}\text{Fructose bisphosphate}\\ \text{aldolase}\end{array}} \text{sedoheptulose 1,7-bisphosphate}$$

$$\text{Sedoheptulose 1,7-bisphosphate} \xrightarrow[\text{bisphosphatase}]{\text{Sedoheptulose}} \text{sedoheptulose 7-phosphate} + P_i$$

$$\text{Sedoheptulose 7-phosphate + glyceraldehyde 3-phosphate} \xrightarrow{\text{Transketolase}}$$
$$\text{ribose 5-phosphate + xylulose 5-phosphate}$$

$$\text{Ribose 5-phosphate} \xrightarrow{\text{Ribose phosphate isomerase}} \text{ribulose 5-phosphate}$$

$$\text{Xylulose 5-phosphate} \xrightarrow[\text{3-epimerase}]{\text{Ribulose phosphate-}} \text{Ribulose 5-phosphate}$$

$$\text{Ribulose 5-phosphate + ATP} \xrightarrow[\text{kinase}]{\text{Ribulose 5-phosphate}} \text{ribulose 1,5-bisphosphate + ADP}$$

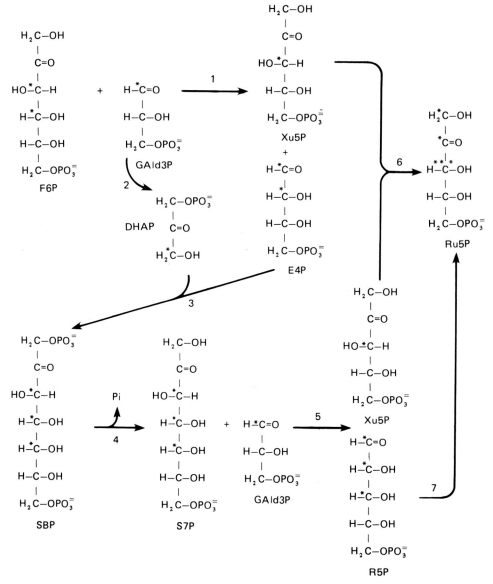

Reaction:

1. Fructose 6-phosphate + glyceraldehyde 3-phosphate $\xrightarrow{\text{transketolase}}$
   xylulose 5-phosphate + erythrose 4-phosphate

2. Glyceraldehyde 3-phosphate $\xrightarrow{\text{triose phosphate isomerase}}$ dihydroxyacetone phosphate

3. Erythrose 4-phosphate + dihydroxyacetone phosphate $\xrightarrow{\text{aldolase}}$ sedoheptulose 1,7-bisphosphate

4. Sedoheptulose 1,7-bisphosphate $\xrightarrow{\text{sedoheptulose bisphosphatase}}$
   sedoheptulose 7-phosphate + inorganic phosphate

5. Sedoheptulose 7-phosphate + glyceraldehyde 3-phosphate $\xrightarrow{\text{transketolase}}$
   xylulose 5-phosphate + ribose 5-phosphate

6. Xylulose 5-phosphate $\xrightarrow{\text{ribulose phosphate-3-epimerase}}$ ribulose 5-phosphate

7. Ribose 5-phosphate $\xrightarrow{\text{ribose phosphate isomerase}}$ ribulose 5-phosphate

6-phosphate as the end product. But an important question requires attention. How is the initial substrate, ribulose 5-phosphate, produced? If this pentose phosphate were made from glucose, by the well-known reaction catalyzed by glucose 6-phosphate dehydrogenase, a loss of one carbon atom would occur. Consequently, there would be no net uptake of carbon by the system. A significant achievement of Calvin and his colleagues was the recognition in 1954 that the reaction can operate as a cycle. Although we can, again on paper, write reactions that could operate as a cycle, it is important to examine the experimental evidence and deduce from it the correct pathway.

Calvin's group purified the compounds that became labeled rapidly from $^{14}CO_2$ and carefully degraded each to determine which carbons contained radioactivity. They could in this way trace the path of carbon from CO_2 into and through the sugar phosphates within the cell. Any pathway that was devised would have to be compatible with the observed labeling patterns. For example, the phosphoglycerate was labeled in carbon-1, as expected from reaction 5.2. Hexoses were initially labeled only in carbon-3 and -4, which was again predicted by the condensation of two triose phosphate molecules, labeled at carbon-1. These latter reactions are analogous to a reversal of glycolysis. The critical determination, however, was the labeling pattern in ribulose 5-phosphate. The observed pattern, along with the labeling pattern found in sedoheptulose 7-phosphate, suggested the reactions shown in Fig. 5.2. By these reactions, a reshuffling of carbon fragments allows formation of ribulose 5-phosphate without a loss of carbon.

A simple conversion of a 6-carbon sugar to a 5-carbon sugar would not provide the labeling pattern found in ribulose 5-phosphate. However, as shown in Fig. 5.2, reactions involving a **transketolase,** an **aldolase, isomerases,** and **epimerases** would provide this pattern. These enzymes were discovered and actively investigated during the 1950s, and these studies gave support to the reactions shown in the scheme. Thus the appearance of radioactive carbon in ribulose 5-phosphate within seconds after exposure of cells to $^{14}CO_2$ indicated that a cyclic pathway existed. The labeling pattern provided evidence for the interconversions that occur within the cycle. The cycle deduced by Calvin is shown in Fig. 5.3. Six turns of the cycle are required for the synthesis of one glucose molecule in the sense of net synthesis from CO_2. The overall stoichiometry of the C_3 pathway for the assimilation of CO_2 is therefore

$$6 \ CO_2 + 18 \ ATP + 12 \ NADPH + 12 \ H^+ + 12 \ H_2O \rightarrow$$
$$Glucose + 18 \ P_i + 18 \ ADP + 12 \ NADP^+ \qquad [5.6]$$

·

←

FIG. 5.2. Labeling patterns determined by chemical degradation of the intermediates in the synthesis of glucose and the reactions proposed to account for the labeling patterns. The (*) indicates the relative amount of radioactivity at the various positions in the sugar phosphates as the result of the incorporation of $^{14}CO_2$. To obtain these labeling patterns, algal cells *(Scenedesmus obliquus)* were exposed to $^{14}CO_2$ for less than 10 sec, and then each of the sugar phosphates was isolated and stepwise, carbon-by-carbon, degraded by chemical means. F6P, fructose 6-phosphate; GAld3P, glyceraldehyde 3-phosphate; DHAP, dihydroxyacetone phosphate; Xu5P, xylulose 5-phosphate; E4P, erythrose 4-phosphate; SBP, sedoheptulose 1,7-bisphosphate; S7P, sedoheptulose 7-phosphate; P_i, inorganic orthophosphate; Ru5P, ribulose 5-phosphate; R5P, ribose 5-phosphate. (Adapted from Bassham *et al.,* 1954.)

The most commonly studied higher plant tissue that exhibits the C_3 pathway is spinach. Intact spinach leaves are capable of fixing CO_2, at saturating light intensity, at rates of about 110 μmoles of CO_2 per milligram of chlorophyll per hour in normal air (330 ppm CO_2) and at about 260 μmoles of CO_2 per milligram of chlorophyll per hour at saturating CO_2 concentrations (1000 ppm CO_2).

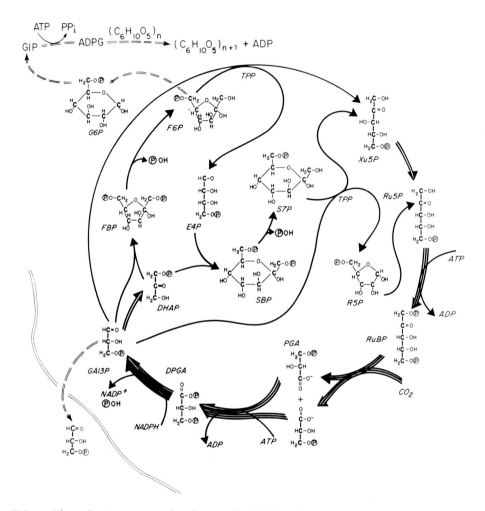

FIG. 5.3. The reductive pentose–phosphate cycle. The heavy lines indicate reactions of the cycle. Light, broken double lines indicate removal of intermediate compounds from the cycle for other biosynthetic processes. The number of heavy lines in each arrow indicates the stoichiometry of each step. For example, three molecules of CO_2 and three molecules of ribulose 1,5-bisphosphate are used for synthesis of six molecules of glyceraldehyde 3-phosphate. In one complete turn of the cycle, the **net** reaction is three molecules of CO_2 converted to one molecule of glyceraldehyde 3-phosphate. RuBP, ribulose 1,5-bisphosphate; PGA, 3-phosphoglycerate; DPGA, 1,3-diphosphoglycerate; GAl3P, glyceraldehyde 3-phosphate; DHAP, dihydroxyacetone phosphate; FBP, fructose 1,6-bisphosphate; G6P, glucose 6-phosphate; E4P, erythrose 4-phosphate; SBP, sedoheptulose 1,7-bisphosphate; S7P, sedoheptulose 7-phosphate; Xu5P, xylulose 5-phosphate; R5P, ribose 5-phosphate; Ru5P, ribulose 5-phosphate; ADPG, adenosine diphosphoglucose; G1P, glucose 1-phosphate; TPP, thiamine pyrophosphate (an enzyme cofactor). (Adapted from Bassham, 1977. Figure courtesy of J. A. Bassham.)

The concentration of CO_2 that allows a half-maximal velocity of fixation is a parameter defined as the K_m. For intact leaves and isolated chloroplasts, this concentration is about 10 μM CO_2. (At equilibrium, the concentration of dissolved CO_2 in water exposed to normal air is about 10 μM at 25°C.) Until recently, a disconcerting finding was that the K_m of purified ribulose 1,5-bisphosphate carboxylase was over 10 times higher than the apparent K_m for the CO_2-fixing activity in intact leaves or isolated chloroplasts. An enzyme with a K_m this much greater than the actual substrate concentration could not support a significant rate of CO_2 fixation. However, this problem was solved by Bahr and Jensen, who found in 1974 that a low K_m form of the enzyme exists *in vivo*. When the activity of the enzyme was measured immediately after lysis of chloroplasts, the enzyme exhibited a K_m for CO_2 that was the same as that in isolated chloroplasts or intact leaves. During purification of the enzyme, it changed into the high K_m form.

Jensen's laboratory succeeded in isolating intact chloroplasts capable of fixing CO_2 at rates approaching the *in vivo* rate (Stumpf and Jensen, 1982). However, after about 10 min of incubation the rate gradually falls to zero. Studies with these chloroplasts now are directed toward determining the regulatory aspects of CO_2 fixation. It is known that the activity of ribulose 1,5-bisphosphate carboxylase is controlled by factors in its environment. The major activating agents are CO_2 and Mg^{2+}, which stabilize the activity of the enzyme and must be bound to the protein before it can functionally bind ribulose 1,5-bisphosphate. The catalytically competent form of the enzyme is generated by the following equilibrium reactions.

$$\underset{\text{(Inactive)}}{\text{Enzyme}} + {}^A CO_2 \overset{\text{Slow}}{\rightleftharpoons} \text{enzyme-}{}^A CO_2 + Mg^{2+} \overset{\text{Fast}}{\rightleftharpoons} \underset{\text{(Active)}}{\text{enzyme-}{}^A CO_2 - Mg^{2+}} \quad [5.7]$$

The **activator** CO_2 (${}^A CO_2$) is different from the **substrate** CO_2 and binds to the ϵ-amino group of a lysine residue at position 201 in the large subunit of the enzyme to form a carbamate derivative.

$$\overset{201}{\frown}\text{Lys-NH}_3^+ \underset{+H^+}{\overset{-H^+}{\rightleftharpoons}} \overset{201}{\frown}\text{Lys-NH}_2 + {}^A CO_2 \underset{+H^+}{\overset{-H^+}{\rightleftharpoons}} \overset{201}{\frown}\text{Lys-NH-COO}^- \quad [5.8]$$

The equilibria of these reactions are pH sensitive, and the activated species is favored at high pH values. In intact chloroplasts, illumination results in activation, in part by causing an increase in pH of the stroma as protons are transported into the thylakoids.

2. Dynamics of the Cycle

The flow of carbon in the forward direction through the cycle, toward glucose, is maintained by the energy provided by the light reactions in the thylakoid membrane. Although most reactions in the cycle are reversible to a degree, the carboxylation of ribulose 1,5-bisphosphate to form 3-phosphoglycerate is irreversible. Therefore, the phosphorylation of ribulose 5-phosphate by ATP maintains the substrate for a unidirectional fixation of CO_2. The equilibrium for the subsequent two reactions (reactions 5.4 and 5.5), shown in the coupled form as

$$\text{3-Phosphoglycerate} + \text{ATP} + \text{NADPH} + \text{H}^+ \rightleftharpoons$$
$$\text{glyceraldehyde 3-phosphate} + \text{ADP} + \text{P}_i + \text{NADP}^+ \quad [5.9]$$

lies in the direction of 3-phosphoglycerate. However, with the dynamic situation that exists in the cell, the products of the reactions, ADP and NADP^+, are continuously recycled into the substrates by the "light" reactions. Consequently, by association of the "light" reactions with the "dark" reactions of the CO_2 assimilation pathway, the photosynthesis of carbohydrates is an efficient, forward-moving process.

3. A Regulatory Role for Light

A fascinating phenomenon has recently been uncovered that involves light-dependent activation of the reductive pentose-phosphate pathway. Although the synthesis of glucose by this pathway is driven by substrates provided by the "light" reactions in thylakoid membranes, it has become clear that light also plays an additional, regulatory role. Several of the enzymes in the pathway have little or no activity in the dark but are rapidly activated upon illumination in intact chloroplasts. However, light has no direct effect on these enzymes, since illumination of extracted or purified enzymes does not cause activation.

A striking example of activation by light occurs with ribulose 5-phosphate kinase, which exhibits nearly a 100-fold increase in activity after several minutes of illumination in intact chloroplasts. A 5-fold activation of glyceraldehyde 3-phosphate dehydrogenase (NADP) occurs over several minutes of illumination. Similar activation also was found with other enzymes in the pathway, such as fructose 1,6-bisphosphatase and sedoheptulose 1,7-bisphosphatase. Interestingly, enzymes involved in the breakdown of starch and glucose, such as phosphofructokinase, phosphoglucomutase, and phosphoglucoisomerase, are inactivated by illumination.

Activation of the enzymes can be studied in extracts of chloroplasts. These enzymes exist in two forms. At pH 6.8–7.0, the form that is relatively inactive contains a disulfide link between two cysteine residues in the protein. This form can be activated by addition of a sulfhydryl-containing reducing agent (dithiothreitol) to reduce the disulfide to free sulfhydryl groups. However, these enzymes are also activated by an increase in the pH to 7.8–8.0. At the higher pH, a reducing agent has no further effect on activity.

In the chloroplast several pathways exist to reduce the disulfide group. As shown in Fig. 5.4, the initial step is the reduction of a membrane electron acceptor by photosystem 1. According to one mechanism, a protein called **light effect mediator** (LEM) is reduced by the initial electron acceptor to the sulfhydryl form, which in turn reduces the enzyme. The LEM system seems to be associated with thylakoid membranes. In the second mechanism of activation, ferredoxin, which is reduced by photosystem 1, is used to reduce the small protein **thioredoxin** in a reaction catalyzed by **ferredoxin–thioredoxin reductase.**

Reduced thioredoxin then reduces a disulfide group on the enzyme. This latter system is composed of soluble components of the stroma.

Illumination, then, by generating reducing agents and by raising the pH of the stroma as the result of transfer of protons into the thylakoids, provides the proper environment for optimal activity of chloroplast enzymes. The activation of synthetic enzymes, and the inactivation of degradative enzymes, is indispensable to the function of the C_3 cycle in biosynthesis. Consequently, this is an important *regulatory* role for light in photosynthesis.

4. The Synthesis of Starch

The final storage form of the carbon fixed by photosynthesis in most plants is starch, a linear polymer of glucose units. Starch is synthesized within chloroplasts from glucose 6-phosphate by the following reactions.

$$\text{D-Glucose 6-phosphate} \xrightarrow{\text{Phosphoglucomutase}} \alpha\text{-D-glucose 1-phosphate} \qquad [5.10]$$

$$\alpha\text{-D-Glucose 1-phosphate} + \text{ATP} \xrightarrow{\text{ADP-glucose pyrophosphorylase}} \text{ADP-glucose} + \text{pyrophosphate} \qquad [5.11]$$

$$\text{ADP-glucose} + (\text{glucose})_n \xrightarrow{\text{Starch synthase}} (\text{glucose})_{n+1} + \text{ADP} \qquad [5.12]$$

These three reactions occur in higher plants exclusively within the chloroplast. Although starch is primarily a linear polymer of glucose, with few branch points, a starch branching enzyme also occurs in chloroplasts.

A system for the degradation of starch occurs as well. The initial step in starch breakdown is an endolytic cleavage by **amylase,** which yields maltose and a complex mixture of oligosaccharides. Branch chains are removed by a **debranching enzyme** (R-enzyme), whereas a **transglycosylase** (D-enzyme) cat-

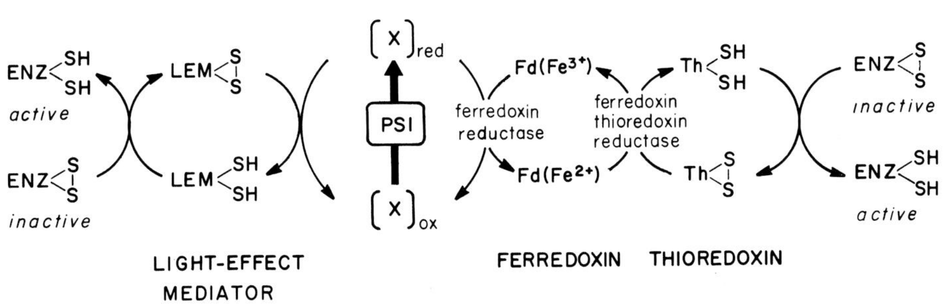

FIG. 5.4. A scheme illustrating proposed reactions by which light causes activation of chloroplast enzymes. In one pathway, a light-effect mediator (LEM) is reduced by photosystem 1 and directly activates the enzyme by reduction of a disulfide bond. This reaction occurs in association with thylakoid membranes. In the other pathway, ferredoxin (Fd) is reduced by photosystem 1. Reduced ferredoxin is used to reduce thioredoxin (Th), which then reduces a disulfide bond in the inactive enzyme. Thioredoxin is a small, soluble protein in the chloroplast stroma. (Adapted from Buchanan, 1980.)

alyzes the condensation of short oligosaccharides to longer linear chains. The linear oligosaccharides then are degraded to glucose 1-phosphate by the enzyme **phosphorylase.** During periods of active photosynthesis, the degradative system is nearly inactive. At night, when photosynthesis is inactive, starch is mobilized to support the metabolic requirements of the cell.

Isolated chloroplasts suspended in a suitable buffer, containing a substance (such as sucrose) to maintain osmotic balance across the envelope, bicarbonate (a source of CO_2), and phosphate, can fix CO_2 as rapidly as the rate of photosynthesis in a leaf. This activity is quickly lost if the chloroplast envelope is broken. These experiments have convincingly demonstrated that complete photosynthesis, from the fixation of CO_2 to the formation of storage forms of carbohydrate, is compartmentalized within the chloroplast.

5. The Synthesis of Sucrose

Sucrose, a major product of photosynthesis in many plants, is not made by chloroplasts, however. Chloroplasts lack the enzymes for sucrose synthesis, and the chloroplast envelope has a low permeability to sucrose. During photosynthesis the major products that leave the chloroplast are 3-phosphoglycerate, the predominant product, and dihydroxyacetone phosphate. These triose phosphate esters leave the chloroplast in exchange for incoming inorganic phosphate. This exchange is facilitated by a specific transport system called the **phosphate translocator,** which is a major protein of the inner membrane of the chloroplast envelope. In the cytosol, the triose phosphates are converted to glucose 6-phosphate and then to sucrose. Two pathways exist for the synthesis of sucrose. In some plants, such as sugar cane, the following reactions occur.

$$\text{Glucose 6-phosphate} \rightarrow \text{glucose 1-phosphate} \qquad [5.13]$$

$$\text{Glucose 1-phosphate} + \text{UTP} \rightarrow \text{UDP–glucose} + \text{pyrophosphate} \quad [5.14]$$

$$\text{UDP-glucose} + \text{fructose 6-phosphate} \rightarrow \text{sucrose 6'-phosphate} + \text{UDP} \qquad [5.15]$$

$$\text{Sucrose 6'-phosphate} + H_2O \rightarrow \text{sucrose} + P_i \qquad [5.16]$$

This pathway is strongly exergonic and essentially irreversible. As a result the concentration of sucrose can reach relatively high concentrations from very dilute monosaccharide precursors. Plants bearing this pathway are therefore good commercial sources of sucrose.

In other plants, the synthesis of sucrose occurs by a simpler process, in which fructose rather than fructose 6-phosphate acts as the glucosyl acceptor.

$$\text{UDP-glucose} + \text{fructose} \rightarrow \text{sucrose} + \text{UDP} \qquad [5.17]$$

6. Regulation of Synthesis of Storage Carbohydrates

The flux of phosphate esters through the phosphate translocator plays an important regulatory role in cellular carbohydrate metabolism. The trigger for this regulatory mechanism is the concentration of inorganic phosphate in the cytosol. A low concentration of phosphate would result from a diminished flux

123

PHOTOSYNTHESIS:
THE DARK
REACTIONS

of metabolites through cytosolic pathways, an indication of low demand. Consequently, less phosphate would be available under these conditions for exchange across the chloroplast envelope, and thus export of triose phosphates from the chloroplast would decrease. This situation leads to a high ratio of 3-phosphoglycerate to phosphate in the chloroplast, which is the major factor in activation of ADP-glucose pyrophosphorylase (shown in reaction 5.11). These conditions, therefore, favor a flux of carbon into starch as the triose phosphates are converted to glucose 1-phosphate. Of course, the high level of 3-phosphoglycerate would only occur in the light during periods of active photosynthesis.

Conversely, a high phosphate concentration outside the chloroplast would promote a flux of triose phosphates into the cytosol, through the envelop translocator. A low 3-phosphoglycerate-phosphate ratio in the chloroplast stroma inactivates ADP-glucose pyrophosphorylase and lowers the rate of starch synthesis. Starch breakdown proceeds under these conditions, catalyzed by the enzyme starch phosphorlyase in the reaction

$$(\text{Glucose})_n + \text{P}_i \rightarrow (\text{glucose})_{n-1} + \text{glucose 1-phosphate} \qquad [5.18]$$

The name *phosphorylase* for the enzyme represents the fact that, in the reaction of starch breakdown, phosphate rather than water is used to break the glycosidic bonds between the glucose units. The reaction, therefore, is a phosphorolysis rather than a hydrolysis. The metabolism of starch also illustrates the important concept that *synthesis* of many biological compounds and macromolecules is not simply the reversal of the *breakdown* pathway. The two are catalyzed by different enzymes and are independently regulated.

7. Quantum Yield

As shown on p. 107, the minimum of 8 quanta of red light theoretically yields, under optimal conditions, about 2.6 molecules of ATP and 2 molecules of NADPH for each O_2 molecule evolved. But as reaction 5.1 shows, and as can be deduced by examining the C_3 cycle shown in Fig. 5.3, at least 3 molecules of ATP are required for the fixation of one molecule of CO_2, along with the 2 molecules of NADPH. Therefore, on paper a net deficiency of ATP is inferred (see also p. 105). Consequently, additional quanta of light presumably are required to overcome this deficit.

Experimental measurements of the quantum yield have shown that at least 10 to 12 quanta of light are required for each O_2 molecule produced (Govindjee, et al., 1968). However, more energy is needed for the net fixation of one molecule of CO_2. Ehleringer and Pearcy (1983) found that in C_3 plants 18 to 20 quanta of light were required for uptake of one CO_2 molecule. In C_4 plants only 15 to 16 quanta were absorbed per one molecule of CO_2 fixed. When the additional demand for ATP in the initial fixation of CO_2 in mesophyll cells in C_4 plants is considered (see p. 124), these measurements show that photosynthesis in C_3 plants is considerably less efficient than in C_4 plants. Nevertheless, in both types of plants, a mechanism must exist to provide more ATP per O_2 molecule than the noncyclic photophosphorylation scheme would predict.

Two possible mechanisms have been proposed by which the required balance of ATP and NADPH could be attained. One scheme proposes a *cyclic* photophosphorylation pathway to allow production of ATP without the reduction of $NADP^+$ (see Fig. 4.7). The extent of cyclic photophosphorylation would be regulated by the level of NADPH within the chloroplast. In this scheme, the amount of CO_2 fixation that is driven by the *noncyclic* electron flow is limited by the production of ATP. Therefore, the limited synthesis of 3-phosphoglyceroyl phosphate would result in an accumulation of NADPH, or an elevated $NADPH/NADP^+$ ratio. This situation would then cause a backup of electrons, with an accumulation of reduced ferredoxin. Ferredoxin, as the endogenous catalyst for cyclic photophosphorylation, would return electrons to the electron transport chain between photosystems 1 and 2. The cyclic flow of electrons, through photosystem 1, could then lead to the synthesis of ATP without the concomitant production of NADPH, thereby restoring the balance between these cofactors needed for the operation of the C_3 cycle.

An alternate hypothesis suggests that under conditions when the rate of CO_2 fixation is limiting, that is, when the $NADPH/NADP^+$ ratio is high, a significant amount of the reductant of photosystem 1 is oxidized by oxygen. As shown in Fig. 5.5 oxygen uptake as well as evolution occurs in the chloroplast, and both processes require light. The generation of H_2O_2, the product of oxygen uptake, has been demonstrated during photosynthetic electron transport, and this process is thought to be an essential part of photosynthesis. Chloroplasts are rich in catalase, the enzyme that catalyzes the breakdown of H_2O_2 by the reaction

$$2 H_2O_2 \rightarrow 2 H_2O + O_2. \qquad [5.19]$$

Hydrogen peroxide does not seem to be produced when the supply of ATP does not limit the rate of NADPH utilization. Therefore, whether electrons flow from photosystem 1 to either $NADP^+$ or to O_2 apparently depends, according to this scheme, indirectly on the concentration of ATP in the chloroplasts.

The consequence of these mechanisms is that the quantum yield for O_2 evolution should reflect the additional absorption of light for cyclic photophosphorylation, or the utilization of part of the liberated O_2 for the production of H_2O_2. Thus, rather than the theoretical value of 8 quanta based only on electron transport, a minimum of 10 to 12 quanta/O_2 evolved or CO_2 fixed should be observed for the whole photosynthetic process. Experimentally determined values for the quantum yield are within this range.

B. The C_4 Pathway—The Hatch–Slack Pathway

A number of hot-weather plants (e.g., corn, sugar cane, and crabgrass) fix CO_2 at rates roughly three times greater than rates measured for a typical C_3 plant (e.g., spinach). These plants are among the minor species of Angiosperms (flowering plants) that have relatively recently evolved a modification of the CO_2-fixing pathway. Investigations into the mechanism of CO_2 fixation in these plants have shown that, rather than an early appearance of labeled carbon from radioactive $^{14}CO_2$ in 3-phosphoglycerate, as occurred with the C_3 (Calvin) cycle, the earliest products are predominantly four-carbon compounds. The current concept of the process of photosynthesis in these plants reflects the anatomy of their leaf tissue. Arranged concentrically around bundles of phloem tubules are layers of cells with differing chloroplast morphology (Fig. 5.6A). Adjacent to the vascular bundle are the **bundle sheath** cells, whose chloroplasts generally, although not universally, contain thylakoids that are not differentiated into granal and stromal regions but rather exist primarily as single units. These "paucigranal" chloroplasts carry out photosynthesis and make glucose by the C_3 pathway. Surrounding the bundle sheath cells is a layer of **mesophyll** cells, whose chloroplasts contain amounts of grana similar to typical C_3 plants (see Fig. 5.6A). Yet these mesophyll cells do not carry out CO_2 fixation by the C_3 pathway. A different pathway, called in short the **C_4 pathway,** occurs

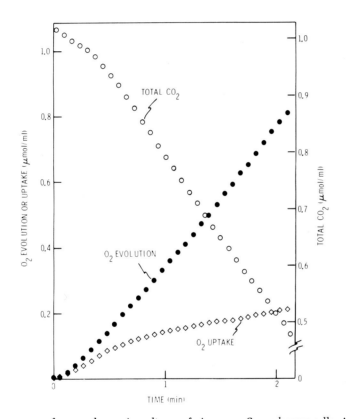

FIG. 5.5. Time course of gas exchange in cultures of air-grown *Scenedesmus* cells. At time zero, the culture was exposed to light, and the exchange of O_2 and CO_2 between the algal cells and the liquid medium was determined by measuring concentrations of dissolved gases with a mass spectrograph. O_2 evolution was measured as the production of $^{16}O_2$ from $H_2^{16}O$. O_2 uptake was measured by loss of $^{18}O_2$ present at the beginning of the experiment. Uptake of O_2 was dependent on the availability of CO_2 and the activity of photosystem 1. In CO_2-depleted cells, the rate of O_2 uptake was nearly equal to O_2 evolution. No appreciable O_2 uptake was observed with a mutant strain lacking photosystem 1 activity. These results indicate that O_2 can be an electron acceptor at the reducing side of photosystem 1. H_2O_2 is the product of reduction of O_2; the superoxide anion (O_2^-) is an intermediate in this reaction. Chloroplasts contain relatively high amounts of the enzyme superoxide dismutase, which catalyzes the reaction $O_2^- + O_2^- + 2H^+ \rightarrow H_2O_2 + O_2$.

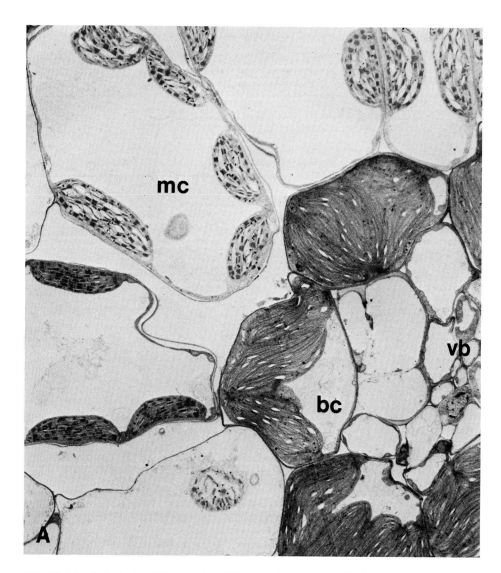

FIG. 5.6. Morphological and biochemical differences between bundle sheath and mesophyll cells in crabgrass (*Digitaria sanguinalis* (L.) Scop.). Crabgrass is a common plant that synthesizes carbohydrates by the C_4 pathway. A: Essentially a monolayer of bundle sheath cells (bc) surrounds the vascular bundle (vb). Concentrically arranged around the bundle sheath cells is a layer of mesophyll cells (mc). The dimorphism between the bundle sheath cells, which contain chloroplasts with very little differentiation of thylakoids into grana, and the outer mesophyll cells, whose chloroplasts contain distinct grana, is quite apparent. This arrangement has been referred to as "Kranz" (literally, *garland*) morphology. ($\times 4000$. From Black and Mollenhauer, 1971. Micrograph courtesy of C. C. Black, Jr.) B: The major proteins in bundle sheath and mesophyll cells are different. Samples were analyzed on a polyacrylamide gradient (10–17%) gel in the presence of sodium dodecyl sulfate and stained with Coomassie blue. Lanes 1 and 5 show the polypeptides recovered in the soluble and particulate fractions, respectively, from whole crabgrass leaves. Bundle sheath (lane 2) and mesophyll (lane 4) cells were separated and analyzed individually. Prominent polypeptides in lane 2 are the large (molecular weight 56,000) and the small (molecular weight 14,000) subunits of ribulose 1,5-bisphosphate carboxylase. This enzyme is not present in mesophyll cells (lane 4). A prominent polypeptide, 94,000 in molecular weight, in lane 4 is phosphoenol pyruvate carboxylase

in these cells, and the evidence indicates that these mesophyll cells provide the very high rates of CO_2 fixation found in these hot-weather plants.

In 1966 the Australian biochemists Hatch and Slack proposed a mechanism by which the mesophyll cells fix CO_2. A major protein *in the cytosol* of these cells is **phosphoenol pyruvate carboxylase** (see Fig. 5.6B), which catalyzes the essentially irreversible reaction shown in reaction 5.20.

$$H_2C = \overset{\overset{OPO_3^=}{|}}{C} - O^- + CO_2 \longrightarrow {}^-O - \overset{\overset{O}{\|}}{C} - CH_2 - \overset{\overset{O}{\|}}{C} - \overset{\overset{O}{\|}}{C} - O - + P_i \qquad [5.20]$$

Phosphoenol   Oxaloacetate
pyruvate

The oxaloacetate synthesized by this reaction is then converted to malate by reduction or to aspartate by transamination (reactions 5.21 and 5.22).

Some of the plants have a preference for making one or the other. Either or both of the products of these reactions then apparently diffuse into adjacent

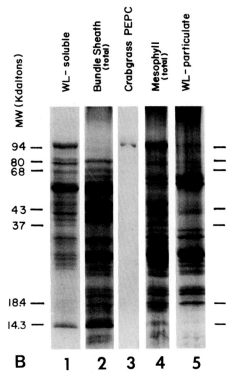

(PEPC), which comigrates with the purified enzyme (lane 3). This enzyme is not present in bundle sheath cells (lane 2). Identifiable particulate (membrane) components, which are more abundant in mesophyll cells, are the α and β subunits, 62,000 and 58,000 in molecular weight, of the coupling factor 1 and the major polypeptide of the light-harvesting chlorophyll–protein complex, with a molecular weight of 26,000. (From Potter and Black, 1982. Used with permission.)

$$\overset{O}{\overset{\|}{C}}-CH_2-\overset{O}{\overset{\|}{C}}-C\overset{O}{\underset{O^-}{}} + NADH + H^+ \rightleftharpoons \overset{O}{\overset{\|}{C}}-CH_2-\overset{OH}{\underset{H}{C}}-C\overset{O}{\underset{O^-}{}} + NAD^+ \qquad [5.21]$$

Oxaloacetate   Malate

$$\overset{O}{\overset{\|}{C}}-CH_2-\overset{O}{\overset{\|}{C}}-C\overset{O}{\underset{O^-}{}} + \overset{O}{\overset{\|}{C}}-CH_2CH_2-\overset{+NH_3}{\underset{H}{C}}-C\overset{O}{\underset{O^-}{}} \rightleftharpoons \overset{O}{\overset{\|}{C}}-CH_2-\overset{+NH_3}{\underset{H}{C}}-C\overset{O}{\underset{O^-}{}} + \overset{O}{\overset{\|}{C}}-CH_2CH_2-C\overset{O}{\underset{O^-}{}}$$

Oxaloacetate   Glutamate   Aspartate   α-Ketoglutarate

$$[5.22]$$

bundle sheath cells where the four-carbon compounds are decarboxylated (Fig. 5.7). This CO_2 released in the bundle sheath cells then is fixed again but now into 3-phosphoglycerate by the ribulose 1,5-bisphosphate carboxylase *in the chloroplast* of these cells. The remaining steps of glucose synthesis are those of the C_3 pathway.

Although seemingly futile, there apparently are definite advantages for the initial fixation of CO_2 into the four-carbon compounds. Phosphoenol pyruvate carboxylase uses HCO_3^- as substrate rather than CO_2. At the pH of the cellular cytoplasm, the concentration of HCO_3^- is much higher than that of dissolved CO_2. But perhaps more importantly, the delivery of "CO_2-carriers" to the bundle sheath cells, such as the four-carbon compounds malate and aspartate, and their subsequent enzymatic decarboxylation provide a supply of CO_2 at concentrations up to five times what ambient conditions would permit. Therefore, ribulose 1,5-bisphosphate carboxylase functions more nearly at its maximal velocity in the chloroplast of the bundle sheath cells. The cycle is completed by return of the three-carbon compounds pyruvate or alanine to the mesophyll cells. It should be noted that the C_4 pathway is only an "appendage" to the C_3 pathway.

The locations of enzymes involved in the C_4 pathway are consistent with the operation of this system. Analyses for the sites of activity of key enzymes in the pathway have shown a highly selective distribution, with some of the enzymes located only in bundle sheath cells or only in mesophyll cells, as expected for the operation of the C_4 pathway. Table 5.2 shows data for some of these enzymes. Bundle sheath cells rapidly decarboxylate added malate or aspartate and refix the CO_2 by the C_3 pathway. Ribulose 1,5-bisphosphate carboxylase cannot be detected in mesophyll cells of C_4 plants, but phosphoenol pyruvate carboxylase is a major protein in these cells (see Fig. 5.6B). The reverse is true in bundle sheath cells. The mesophyll cells contain the enzymes needed to reduce oxaloacetate to malate or to make aspartate by transamination, whereas the bundle sheath cells have high activities of the enzymes needed for the decarboxylation of these products. The selective localization of these enzymes is dramatically illustrated by Fig. 5-8, which shows the presence of ribulose 1,5-bisphosphate carboxylase in bundle sheath but not in mesophyll cells by immunocytochemistry.

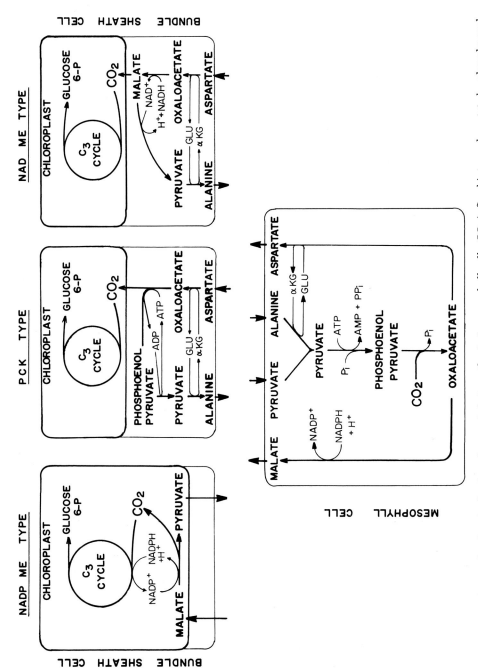

FIG. 5.7. The fluxes of carbon in the C$_4$ pathway of photosynthesis. In mesophyll cells, CO$_2$ is fixed into oxaloacetate by phosphoenol pyruvate carboxylase. The oxaloacetate is either reduced to malate or transaminated to aspartate. These intermediates are transported to adjacent bundle sheath cells, where a reversal of these reactions provides the source of CO$_2$. This CO$_2$ is again fixed in the bundle sheath cells, but now by the reductive pentose–phosphate (C$_3$) pathway. (Adapted from Hatch and Kagawa, 1976.)

TABLE 5.2. Enzyme Activities in Extracts of Whole Leaves, Mesophyll Cells, and Bundle Sheath Strands of Nutsedge, a C_4 Plant[a]

Enzymes of photosynthesis	Whole leaf	Mesophyll cells	Bundle sheath cells
Phosphoenol pyruvate carboxylase	1350	2220	27
Pyruvate P_i dikinase	496	230	7
Malate dehydrogenase ($NADP^+$)	280	600	(0)
Ribulose 1,5-bisP carboxylase	253	5	523
Phosphoribulose kinase	1500	20	3820
Fructose 1,6-bisP aldolase	200	44	440
Malic enzyme	504	14	1040

[a]Values in μmoles/mg chlorophyll per hour. Data taken from Chen *et al.*, 1974.

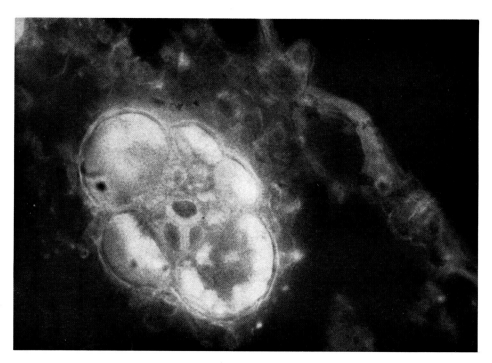

FIG. 5.8. Immunocytochemical localization of ribulose 1,5-bisphosphate carboxylase in a cross section of a leaf of *Zea mays* (corn). The enzyme was purified from the leaf, and antibodies were produced by injecting the protein into rabbits. The section was treated with the antibody in solution to allow the antigen–antibody complex to form. The tissue was washed free of unbound antibody and then treated with antiserum against rabbit antibodies produced in sheep. The sheep antibodies were labeled with fluorescein, a fluorescent dye. After unbound sheep antibodies were washed away, the section was photographed through a microscope that transmitted fluorescent light emitted by the dye. Specific fluorescence was localized only to chloroplasts in bundle sheath cells. (From Sayre *et al.*, 1979. Reprinted with permission.)

A conceptually difficult aspect of the C_4 pathway is the apparent requirement for a flux of four-carbon compounds by simple diffusion from mesophyll
to bundle sheath cells and the return of three-carbon compounds, pyruvate or
alanine. In most C_4 plants there are structural features at cell junctions (plas-
modesmata) that may facilitate these fluxes. However, in the nutsedge, a C_4
plant, a curious situation exists in that the two types of cells, rather than being
juxtaposed as in most C_4 plants, are separated by an apparently nonphotosynthetic layer of cells. Whether this layer of cells presents a significant barrier to
the flux of metabolites or in fact aids the flux between the two types of photosynthetic cells is not known.

Although the mechanistic advantages of the C_4 pathway are not entirely
obvious, it is quite clear that in some plants, cells are present that do not have
the enzymatic activities to carry out the reductive pentose phosphate cycle, yet
very actively fix CO_2 into organic compounds. It is not enough to assume that
aspartate and malate are metabolized to other cellular compounds by the normal pathways of metabolism, since any subsequent pathway involves at some
stage a decarboxylation back to a three- or even two-carbon compound. In this
situation no net incorporation of carbon into organic compounds could be
achieved. Cooperation between two types of cells provides a way out of this
quandary.

A consequence of the C_4 pathway is that the demand for ATP is greater
than for the C_3 pathway. Mesophyll cells must generate sufficient phosphoenol
pyruvate for the subsequent carboxylation reaction. This substrate is made in
the reaction catalyzed by a chloroplast enzyme, **pyruvate phosphate dikinase**
(reaction 5.23). This interesting enzyme is present in cells of C_4 plants but cannot be detected in C_3 plants. In the reaction catalyzed by this enzyme, the
equivalent of two "high-energy" pyrophosphate bonds is spent in the conversion of pyruvate to phosphoenol pyruvate. The other product, pyrophosphate,
is hydrolyzed to inorganic phosphate by an active inorganic pyrophosphatase
in the cell (reaction 5.22), thereby pulling the reaction toward products. In fact,
the equilibrium of the dikinase reaction lies in the direction of substrates.
Because of the hydrolytic removal of the product pyrophosphate, the conversion of substrates to products is favored. This is a common mechanism in cells
to drive reactions essentially to completion.

$$\text{ATP} + CH_3-C-C \quad + P_i \quad \rightleftharpoons \quad \text{AMP} + H_2C=C-C \quad + PP_i \qquad [5.23]$$

$$PP_i + H_2O \longrightarrow 2\,P_i \qquad [5.24]$$

$$\text{Net: ATP} + CH_3-C-C \longrightarrow \text{AMP} + H_2C=C-C \quad + P_i \qquad [5.25]$$

Pyruvate   Phosphoenol
                          pyruvate

Pyruvate phosphate dikinase is among the group of enzymes that is activated by a light-dependent mechanism. Inactivation in the dark requires ADP, a trace amount of ATP, and a regulatory protein to catalyze the process. Dark-inactivated enzyme can be reactivated by inorganic phosphate in a reaction apparently catalyzed by the same regulatory protein (Fig. 5.9). Greater than tenfold changes in activity have been observed between the enzyme obtained from dark-adapted and illuminated leaves. Malate dehydrogenase (NADP) in the chloroplasts of mesophyll cells (see Fig. 5.7) also has a near-absolute requirement for light for activity *in vivo*. Thus, these observations demonstrate that light plays a regulatory role in controlling the C_4 as well as the C_3 pathway of CO_2 fixation (see Fig. 5.9).

A curious aspect of the C_4 system is that the substrate for CO_2 fixation, phosphoenol pyruvate, is made in the chloroplast, whereas the enzyme that catalyzes the carboxylation reaction is located in the cytosol of the cell. A transport system apparently exists to transfer phosphoenol pyruvate across the chloroplast envelope. This process, and the exchange of four-carbon intermediates between cells of C_4 plants, demonstrate the importance of transport processes in photosynthesis by the C_4 pathway.

The fixation of one CO_2 molecule by the C_4 pathway requires two molecules of NADPH and five or six molecules of ATP, depending on the pathway of decarboxylation (Fig. 5.7). This is about twice the amount of ATP required

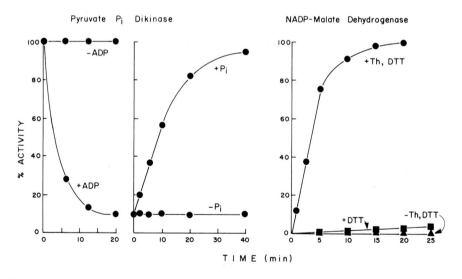

FIG. 5.9. *In vitro* activation of enzymes that are activated by light *in vivo*. Left: Inactivation of pyruvate phosphate dikinase in the dark is dependent on ADP; a trace amount of ATP is also required. Activation of the enzyme is dependent on the presence of inorganic phosphate. Inactivation and activation of the enzyme are catalyzed by the same regulatory protein. These results suggest that pyruvate phosphate dikinase is modulated by light through changes in the concentrations of ATP, ADP, and inorganic phosphate in the chloroplast stroma. Therefore, this regulatory mechanism is different from the types diagramed in Fig. 5.4. (Adapted from Burnell and Hatch, 1983.) Right: Activation of NADP$^+$-linked malate dehydrogenase is dependent on reduced thioredoxin (Th). In this *in vitro* system, dithiothreitol (DTT) replaced the light-generated reductant. Dithiothreitol itself provided essentially no activation. Activation of this enzyme, therefore, occurs by the thioredoxin-mediated pathway diagramed in Fig. 5.4. (Adapted from Ferte *et al.*, 1982.)

for the C_3 pathway. Considering the fact that the C_3 pathway operates at a deficit of ATP, without cyclic photophosphorylation, this problem becomes even more serious for the C_4 pathway. Again, however, the deficiency may be made up by cyclic photophosphorylation. It has been found that bundle sheath cells, in which carbohydrate synthesis occurs from CO_2 by the C_3 pathway, have a considerably lower photosystem 2 activity and higher photosystem 1 activity than the mesophyll cells from the same leaf tissue. Plants of the C_4 type also require a higher light intensity for optimal growth than do C_3 plants. Nevertheless, C_4 plants synthesize hexoses more rapidly per unit leaf area, grow faster, and function more efficiently at high light intensities and low CO_2 concentrations than C_3 plants.

C. The CAM Pathway

Yet a third pathway exists in higher plants to fix CO_2. This pathway, referred to as **crassulacean acid metabolism** (CAM), is similar to the C_4 scheme but is temporally different. In the relatively few species of plants that have this pathway (the *Crassulaceae* family, which grow in hot, dry exposed places, chiefly in South Africa), significant amounts of CO_2 are fixed by carboxylation of phosphoenol pyruvate *in the dark* (Fig. 5.10). The oxaloacetate produced then is reduced to malate, which is stored in the large vacuole of the cell. The substrates and energy required for this process apparently are derived from the metabolism of glucose, stored as starch during the day, and mitochondrial oxidative phosphorylation. During the subsequent daytime, the malate is decarboxylated by malate dehydrogenases, an NAD-linked enzyme in the mitochondria, and an NADP-linked form of the enzyme in the cytosol (Fig. 5.10). The rates of CO_2 fixation by this system typically are lower than for the C_3 or C_4 pathway.

The interesting aspect of the metabolism of these plants is the temporal separation of gas exchange with the atmosphere and of photosynthetic CO_2 fixation. Gas exchange, when the stomata are open, is limited to the hours of darkness, which minimizes the loss of water. CO_2 taken up at night is stored as malate and then released by decarboxylation and refixed by the photosynthetic reactions during the day. Thus gas exchange is unnecessary during the day when the stomata are closed (Fig. 5.11).

D. The Reductive Carboxylic Acid Cycle

Another cyclic pathway for the assimilation of CO_2, which so far is known to exist only in anaerobic photosynthetic bacteria, was discovered by Evans et al. in 1966. This pathway, shown in Fig. 5.12, resembles a reversal of the normal respiratory tricarboxylic acid cycle. In the direction of photosynthesis, four reactions incorporate carbon from CO_2 into organic products. The net result is that for each turn of the cycle one molecule of oxaloacetate is produced from four equivalents of CO_2.

Two of the reactions by which CO_2 is incorporated in this cycle are practically irreversible in the direction of decarboxylation during respiration. However, these reactions—the decarboxylation of pyruvate to acetyl-coenzyme A and of α-ketoglutarate to succinyl-coenzyme A—are driven in the reverse direction during photosynthesis by the interaction of a strong reductant, reduced ferredoxin. The two other equivalents of CO_2 are incorporated by normally reversible reactions catalyzed by phosphoenol pyruvate carboxylase and isocitrate dehydrogenase. The unique feature of this cycle, which permits it to operate in the direction of CO_2 fixation, is the involvement of ferredoxin. In these photosynthetic bacteria, ferredoxin is reduced by light-driven reactions as described on p. 108.

During the operation of the reverse tricarboxylic acid cycle, intermediates can be removed for the synthesis of other compounds. Thus, oxaloacetate can be converted to aspartate or to malate, two products normally seen also during C_4 photosynthesis. Pyruvate can be transaminated to alanine, and glutamate can be produced from α-ketoglutarate either by transamination or by reductive amination with NH_4^+ and NADH in a reaction catalyzed by glutamate dehydrogenase. These products are also found in significant quantities during photosynthesis in C_4-type plants. However, the existence of this cycle in higher plants has not been established.

The reverse tricarboxylic acid cycle is possibly the most efficient mechanism in nature for fixing CO_2. For each turn of the cycle, ten electrons are

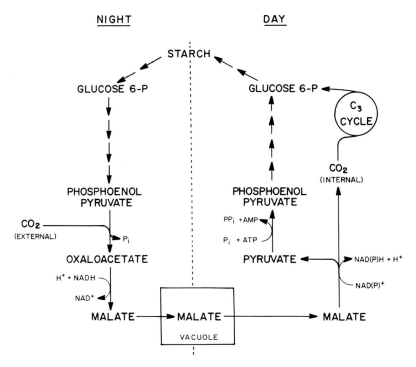

FIG. 5.10. Proposed scheme for the flux of carbon in crassulacean acid metabolism (CAM). During the night, starch is degraded through a glycolytic pathway to phosphoenol pyruvate. Exogenous CO_2, entering leaves from the environment through open stomata, is fixed into oxaloacetate by phosphoenol pyruvate carboxylase. Oxaloacetate becomes reduced by cytosolic NAD^+–malate dehydrogenase to malate, which is transported into and accumulates in the vacuole in the cell. During daytime hours, malate is released from the vacuole and is converted to pyruvate through a decarboxylation reaction catalyzed by either mitochondrial or cytosolic malic enzymes. Pyruvate is converted by pyruvate phosphate dikinase to phosphoenol pyruvate, which provides the substrate for glucose and starch synthesis by a reversal of glycolysis (gluconeogenesis). The CO_2 released by decarboxylation of malate is refixed photosynthetically by the reductive pentose–phosphate cycle. Loss of endogenously produced CO_2 is prevented by closure of the stomata during the day. (Adapted from Spalding *et al.*, 1979.)

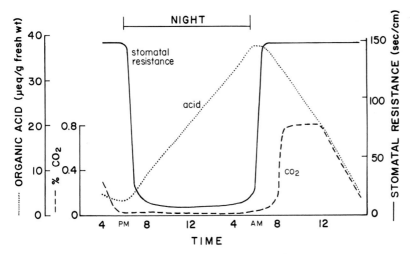

FIG. 5.11. Changes in the internal gas phase CO_2 concentration (- - - -), the acid content (primarily malate) (......), and the stomatal resistance to diffusion of water vapor (———). The drop in stomatal resistance at nightfall marked the opening of these leaf structures. CO_2 entering through the stomata at night was incorporated into malate by the CAM pathway (see Fig. 5.10). At dawn, the stomata closed and subsequent decarboxylation of malate markedly elevated CO_2 levels within the plant during the day. Over the course of the day, the released CO_2 was fixed into carbohydrate (starch) by the reductive pentose–phosphate cycle. (Adapted from Cockburn *et al.*, 1979.)

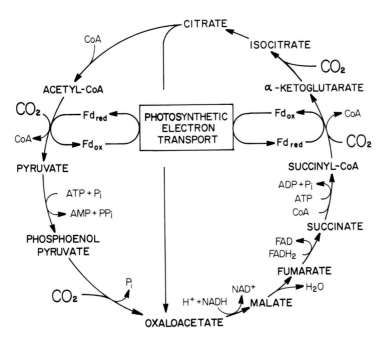

FIG. 5.12. The reductive carboxylic acid cycle in *Clostridium thiosulfatophilum*, an anaerobic organism that uses H_2S as a source of electrons for photosynthesis rather than H_2O. One turn of the cycle results in the incorporation of four molecules of CO_2. The cycle is in essence the reverse of the oxidative tricarboxylic acid cycle in aerobic organisms. In this photosynthetic bacterium, reduced ferredoxin, generated by photosynthetic electron transport, provides a strong reductant to drive the operation of the cycle in the direction of CO_2 incorporation. Intermediates of the cycle can be removed by transamination reactions with pyruvate to alanine, oxaloacetate to aspartate, and α-ketoglutarate to glutamate for other biosynthetic processes. (Adapted from Evans *et al.*, 1966.)

required, or only 2.5 electrons per CO_2. If acetyl coenzyme A must be made from acetate, and phosphoenol pyruvate is made from pyruvate, two reactions that each require the expenditure of the equivalent of two ATP molecules, then a total of four ATP molecules is needed per turn of the cycle, or one ATP molecule per CO_2. However, to make a proper comparison with the C_3 and C_4 pathways, the end product must be the same. In both the C_3 and C_4 pathways, a product on the way to glucose is glyceraldehyde 3-phosphate. Oxaloacetate can be converted to this product as well, but the conversion requires the expenditure of two ATP molecules and one NADH per oxaloacetate molecule and results in the loss of one of the CO_2 molecules that were fixed by the reverse tricarboxylic acid cycle. The corrected requirement, therefore, for this cycle and the subsequent synthesis of glyceraldehyde 3-phosphate (or glucose) is four electrons (as NADPH, NADH, and ferredoxin) and two ATP molecules per CO_2 assimilated. Thus, the requirement for ATP still is only two thirds that needed by the C_3 pathway.

A number of questions remain to be answered concerning the process of photosynthesis. The existence of the C_3 pathway has been firmly established. The existence in higher plants of another system, in which four-carbon rather than three-carbon products are initially formed by CO_2 fixation, is also established. But some of the critical details of these pathways, in particular how all the necessary ATP is generated and how adjacent cells cooperate in the assimilation of CO_2, are still in the process of being defined.

III. Requirement of CO_2 for O_2 Evolution

An interesting relationship exists between the presence of CO_2, or HCO_3^-, and the ability of thylakoid membranes to produce oxygen. The German biochemist Otto Warburg (1964) made the observation that illuminated membranes were relatively inactive in producing oxygen unless HCO_3^- was added. His work led him to support the concept, held since the reports of Ingen-Housz and

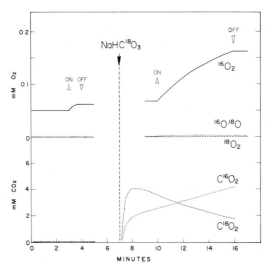

FIG. 5.13. Analysis of the isotopic composition of O_2 produced photosynthetically by thylakoid membranes before and after introduction of $NaHC^{18}O_3$. Concentrations of dissolved gases were determined with a mass spectrograph. $K_3Fe(CN)_6$ was present as an electron acceptor from photosystem 2. Without added HCO_3^-, O_2 was evolved at a slow rate (upper left). Addition of HCO_3^- caused a marked increase in the rate of O_2 evolution (upper right). Although ^{18}O exchanged rapidly between $HC^{18}O_3^-$ and H_2O (lower right) by the reactions

$$H^+ + HCO_3^- \rightleftharpoons H_2CO_3 \rightleftharpoons H_2O + CO_2,$$

the analysis showed that initially only $^{16}O_2$ was evolved (upper right). Thus, the O_2 could not have come from CO_2 but rather came from H_2O. (Redrawn from Stemler and Radmer, 1975.)

Senebier in the 18th century, that oxygen is formed by the photolysis of CO_2 or its hydrated form H_2CO_3. However, this source of O_2 was questioned by a number of other investigators. In 1940 Hill found that purified chloroplast membranes would produce oxygen in the light in the absence of CO_2 if they were incubated with ferric salts as electron acceptors of photosystem 2. Later, Ruben and Kamen and their colleagues (1963) added ^{18}O-labeled water to cultures of algal cells *(Chlorella)* and found that the evolved O_2 was labeled. But in these experiments with whole cells, exchange of oxygen between water and CO_2 could not be controlled, preventing definite conclusions. Results of a critical experiment were reported by Stemler and Radmer in 1975. In the presence of ferricyanide as the electron acceptor, isolated chloroplasts evolved very little O_2 in the absence of HCO_3^- (CO_2) but were quite active in its presence (Fig. 5.13). The isotopic content of the evolved O_2 was measured after introduction of $HC^{18}O_3^-$. If O_2 was derived from CO_2, it should contain ^{18}O. In the experiment, only $^{16}O_2$ was produced initially, proving that the O_2 came instead from the unlabeled water. Oxygen atoms in water and CO_2 are exchanged by the reversible hydration of CO_2 to H_2CO_3, which is catalyzed by carbonic anhydrase, an enzyme present at relatively high amounts in chloroplasts of C_3 plants. Therefore, the conclusion of the source of the O_2 had to be drawn from the initial labeling pattern. The ability to identify H_2O as the source of O_2 in this situation is a dramatic demonstration of the utility of isotopes.

Govindjee and his colleagues (Vermas and Govindjee, 1982) recently demonstrated that the addition of CO_2 or HCO_3^- to thylakoid membranes enhances electron flow from the primary electron acceptor (Q) of photosystem 2 to the plastoquinone pool. This enhancement of electron flow can be as high as tenfold. As described on p. 89, "B protein," located on the stromal side of thylakoid membranes, requires bound CO_2 or HCO_3^- for electron transfer activity. Therefore, this protein appears to be primarily responsible for the need of CO_2 for photosynthetic oxygen evolution.

IV. Inhibition of CO_2 Fixation by O_2: Photorespiration

Although CO_2 is required for the production of O_2, high concentrations of oxygen competitively inhibit the assimilation of CO_2. The mechanism of this inhibition of CO_2 fixation appears to involve a recently discovered second reaction catalyzed by ribulose 1,5-bisphosphate carboxylase. This enzyme has the ability to act also as a ribulose 1,5-bisphosphate oxygenase. Oxygen competes with CO_2 in the active site on the enzyme and results in an *oxygenation* rather than a *carboxylation* of the substrate, as shown in reaction 5.26.

$$
\begin{array}{ccc}
\text{H}_2\text{C}-\text{OPO}_3^= & & \text{H}_2\text{C}-\text{OPO}_3^= \\
| & & | \\
\text{C}=\text{O} & & \text{C}=\text{O} \\
| & & \quad\searrow\text{O}^- \\
\text{H}-\text{C}-\text{OH} & +\text{O}_2 \longrightarrow & \qquad\qquad\text{Phosphoglycolate} \\
| & & + \\
\text{H}-\text{C}-\text{OH} & & \quad\nearrow^{\text{O}} \\
| & & \text{C}-\text{O}^- \\
\text{H}_2\text{C}-\text{OPO}_3^= & & | \\
& & \text{H}-\text{C}-\text{OH} \qquad \text{3-Phosphoglycerate} \\
\text{Ribulose 1,5-bisphosphate} & & | \\
& & \text{H}_2\text{C}-\text{OPO}_3^=
\end{array}
\qquad [5.26]
$$

Since the early work on $^{14}CO_2$ incorporation during photosynthesis, glycolate was known to be an early product of this process. However, only after the activity of the carboxylase as an oxygenase was discovered by Bowes *et al.* in 1971 did the source of glycolate become apparent. Formation of phosphoglycolate (reaction 5.26), followed by the release of inorganic phosphate by a specific phosphoglycolate phosphatase in the chloroplast, accounts for the experimental observations that (1) a large portion of fixed CO_2 can appear as glycolate in C_3 plants, (2) glycolate formation is favored by high pO_2/pCO_2 ratios, (3) glycolate produced during photosynthesis is equally labeled in both carbons, and (4) only one atom of ^{18}O from $^{18}O_2$ is incorporated into the carboxyl group of glycolate. The enzyme that carries out these reactions, therefore, should properly be named **ribulose 1,5-bisphosphate carboxylase/oxygenase.**

Associated with the metabolism of glycolate is another phenomenon that occurs in most plants but particularly in those that fix CO_2 entirely by the C_3 pathway. During photosynthesis, these plants also *release* previously fixed CO_2 in an oxygen- and light-dependent process known as **photorespiration.** The metabolic pathway shown in Fig. 5.14 has been proposed to account for the release of CO_2. Glycolate diffuses from the chloroplast to the peroxisome, where it is oxidized to glyoxylate and subsequently converted to glycine by a transamination reaction. The glycine that is formed by this sequence of reactions then diffuses into the mitochondria, where serine, NH_3, and CO_2 are produced from two molecules of glycine by the reaction

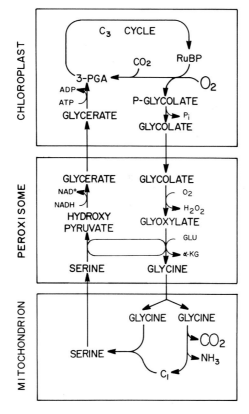

FIG. 5.14. The major reactions proposed for photorespiration. Phosphoglycolate is produced from ribulose 1,5-bisphosphate (RuBP) by the oxidative reaction catalyzed by ribulose 1,5-bisphosphate carboxylase/oxygenase. A phosphatase allows glycolate to leave the chloroplast and enter peroxisomes, where it is metabolized to glycine. Oxidation of glycine in mitochondria produces CO_2 and a one-carbon fragment, which is transferred to a second molecule of glycine to form serine. Tetrahydrofolate is required as a carrier of the one-carbon fragment in this reaction, catalyzed by the pyridoxal phosphate-containing enzyme serine hydroxymethyltransferase. Serine is converted in a transamination reaction with glyoxylate to hydroxypyruvate in peroxisomes. Glycerate, formed by reduction of hydroxypyruvate in peroxisomes, enters the chloroplast, where it feeds into the reductive pentose phosphate (C_3) cycle. (Adapted from Sommerville and Ogren, 1982.)

$$2\ H_3\overset{+}{N}-CH_2-\overset{\overset{\textstyle O}{\|}}{C}-O^- \longrightarrow HO-CH_2-\overset{\overset{\textstyle \overset{+}{N}H_4}{|}}{\underset{\textstyle H}{C}}-C\overset{\nearrow O}{\diagdown O^-} + \overset{+}{N}H_4 + CO_2 \qquad [5.27]$$

Glycine   Serine

This metabolism of glycine is considered to be the major source of photorespiratory CO_2. Serine can be further metabolized to 3-phosphoglycerate according to the reactions shown in the scheme in Fig. 5.14.

All major crop plants that fix CO_2 primarily by the C_3 (Calvin) pathway exhibit appreciable photorespiration. Obviously, the extent of photorespiration will limit the efficiency of the conversion of CO_2 to carbohydrate in such plants. In contrast, maize, sorghum, sugar cane, and other C_4 plants do not show a significant amount of photorespiration. In C_4 plants phosphoenol pyruvate carboxylation is not inhibited by O_2, and since decarboxylation of 4-carbon compounds in the bundle sheath cells provides an elevated concentration of CO_2 for refixation by the C_3 cycle, CO_2 incorporation into carbohydrate in these plants is relatively unaffected by the concentration of O_2.

The concentration of CO_2 that provides a rate of fixation equal to that lost by photorespiration is called the **compensation point.** Whereas in C_3 plants loss of fixed carbon by photorespiration can be substantial and an ambient concentration of CO_2 above about 60 ppm (the compensation point) must be available to promote net photosynthesis, in C_4 plants the compensation point is at very low CO_2 concentrations (about 0.1 ppm). Thus the efficiency of photosynthetic carbohydrate production in C_4 plants is correspondingly greater than that in C_3 plants.

Although most algae are considered to be C_3 plants, until recently attempts to determine compensation points in these organisms had failed. Birmingham et al. (1982) found that freshwater algae, in slightly alkaline media optimal for photosynthesis, exhibited quite low compensation points in the range of 0.2–5 ppm. The feature of these cells that apparently allows maintenance of photosynthesis at 'low concentrations of CO_2 is their ability to actively take up HCO_3^- from the surrounding aqueous environment. The equilibrium between HCO_3^- and CO_2, maintained by the enzyme carbonic anhydrase, then provides relatively high CO_2 concentrations inside the cells for photosynthesis.

The studies on photosynthesis, particularly those with the isotopes ^{14}C and ^{18}O, clearly demonstrated the dynamic aspects of cellular metabolism. The flux of material through metabolic pathways is extraordinarily rapid, yet during "steady-state" photosynthesis the concentrations of intermediates do not change significantly. Metabolic pathways are interconnected but always finely balanced with each other by regulatory mechanisms. In a sense, an equilibrium exists that can be termed a "dynamic equilibrium," in which everything is rapidly changing yet stays the same.

V. Assimilation of Nitrogen and Sulfur

Higher plants assimilate the elements carbon, nitrogen, and sulfur primarily as the oxides CO_2, NO_3^- (nitrate), and SO_4^{2-} (sulfate). The chloroplast plays an indispensable role in the reduction of these oxides into forms that can be used by the cell for synthesis of proteins and nucleic acids.

Nitrate is produced in the atmosphere by electrical discharges and falls to earth dissolved in rain. It is also supplied to soil as fertilizer. Nitrate is reduced to ammonia in two steps. In eukaryotic plants, **nitrate reductase** catalyzes the two-electron reduction of nitrate to nitrite by the electron donor NADH (or NADPH).

$$\text{NADH} + \text{H}^+ + \text{NO}_3^- \xrightarrow[\text{(Cytoplasm)}]{\text{Nitrate reductase}} \text{NO}_2^- + \text{NAD}^+ + \text{H}_2\text{O} \qquad [5.28]$$

This nitrate reductase is an oligomeric protein of about 220,000 in molecular weight and contains flavin (as flavin adenine dinucleotide, FAD), several iron–sulfur (Fe_4S_4) centers, and molybdenum ions as cofactors. A b-type cytochrome (b_{557}) is also included within the enzyme. The probable path of electrons during enzyme action is

$$\text{NAD(P)H} \rightarrow (\text{FAD} \rightarrow \text{cyt } b_{557} \rightarrow \text{Mo}) \rightarrow \text{NO}_3^- \qquad [5.29]$$

Reduction of nitrate apparently occurs in the cytoplasmic compartment of the cell with NADH generated by glycolysis. Thus, the reaction can proceed in the light or dark. Synthesis of the reductase is induced by nitrate or nitrite and is repressed by ammonia and amino acids, the end products of the assimilation of nitrate.

Nitrate reductase in prokaryotic cyanobacteria is dependent on ferredoxin as the electron donor rather than NADH. This enzyme also contains molybdenum within a single polypeptide chain of 75,000 in molecular weight.

Reduction of nitrite, produced by reaction 5.28, occurs within the chloroplast, utilizing reduced ferredoxin generated by photosynthetic electron transport. Nitrite is reduced to ammonia, without the release of intermediates, in a six-electron transfer reaction catalyzed by **nitrite reductase:**

$$\text{NO}_2^- + 6 \text{ Fd}(red) + 8 \text{ H}^+ \xrightarrow[\text{(Chloroplast)}]{\text{Nitrite reductase}} \text{NH}_4^+ + 6 \text{ Fd}(ox) + 2 \text{ H}_2\text{O} \qquad [5.30]$$

Nitrite reductase from plant cells is reddish-brown in color and contains an iron–sulfur (Fe_4S_4) center and a heme group. The molecular weight of the enzyme is about 65,000.

Fixation of atmospheric nitrogen (N_2) into ammonia occurs in root nodules of leguminous plants. These nodules form when the host plant surrounds invading nitrogen-fixing bacteria—particularly the *Rhizobium* bacteria—within a cyst. The bacterial cells provide the **nitrogenase complex,** which is composed of a reductase containing an iron–sulfur (Fe_4S_4) center and the nitrogenase, an enzyme that contains molybdenum in addition to iron–sulfur (Fe_4S_4) groups. The reductase is the smaller protein, with a molecular weight of about 60,000, and is composed of two identical subunits. The nitrogenase has a molecular weight of about 200,000 and has an $\alpha_2\beta_2$ subunit composition. The stoichiometry of the reaction catalyzed by the complex is as follows:

$$N_2 + 6 \text{ Fd(red)} + 12 \text{ ATP} + 12 \text{ H}_2\text{O} \xrightarrow{\text{Nitrogenase}} 2 \text{ NH}_4^+$$
$$+ 6 \text{ Fd(ox)} + 12 \text{ ADP} + 12 \text{ P}_i + 4 \text{ H}^+ \quad [5.31]$$

Metabolism within the bacterial cells of carbohydrate, provided by the plant host, generates ATP and a strong reductant in the form of a protein similar to ferredoxin. Electrons are transferred first to the nitrogenase, in a reaction that requires ATP. The nitrogenase then reduces N_2 to ammonia. In addition to the energy supply, the plant host also provides the appropriate environment for fixation of N_2. This process is very sensitive to oxygen, and the root cells in the nodule synthesize an oxygen-binding protein, **leghemoglobin** (legume hemoglobin), whose function is to maintain a very low concentration of free oxygen. This protein is similar to the oxygen-transporting protein hemoglobin in red blood cells of animals.

As reactions 5.28, 5.30, and 5.31 show, nitrogen assimilation is energetically an expensive process. Roughly a fourth of the electrons photosynthetically abstracted from water are used by the plant to sustain these processes. At least 0.6 mole of glucose, generated in chloroplasts within leaves, is required for fixation of 1 mole of N_2 to form 2 moles of ammonia in roots of plants. Also, the minimal energy expenditure required to produce 2 moles of ammonia from nitrate is equivalent to 0.66 mole of glucose. Thus, the cost to the plant, for the production of ammonia, is approximately the same regardless of whether N_2 or NO_3^- is the starting point. However, the rate of 2×10^4 megatons of nitrogen assimilated annually by nitrate-reducing systems is two orders of magnitude greater than that recovered by nitrogen fixation.

Plant cells also assimilate sulfur by the reduction of SO_4^{2-}. The initial step in this process (Fig. 5.15) is the formation of adenosine 5'-phosphosulfate (APS) in a reaction catalyzed by **ATP sulfurylase,** a chloroplast enzyme. The activated sulfate then is transferred to the sulfhydryl group of a carrier molecule by the second enzyme, **APS sulfotransferase.** The carrier molecule is bound to the

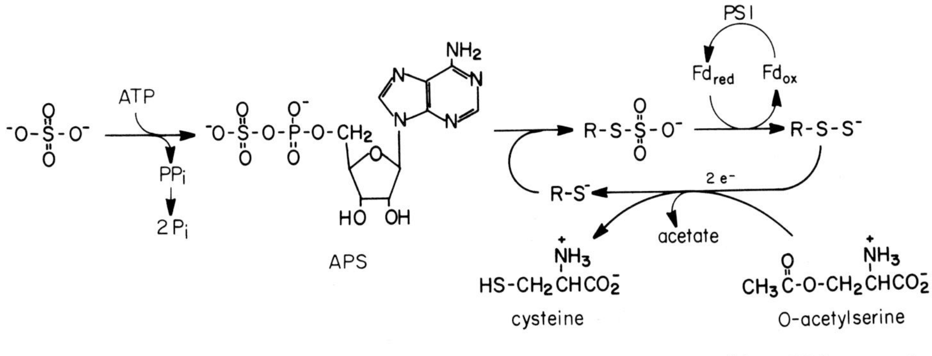

FIG. 5.15. The pathway for sulfate assimilation in plants. Sulfate is activated by ATP in a reaction catalyzed by ATP sulfurylase to form adenosine 5'-phosphosulfate (APS). The sulfate group then is transferred to a carrier molecule (designated R-S⁻) by APS sulfotransferase. The sulfonate group is reduced by ferredoxin in a reaction catalyzed by thiosulfonate reductase. The thiol group then is used to form cysteine by O-acetyl-L-serine sulfhydrylase. (Adapted from Schmidt et al., 1974.)

next enzyme, **thiosulfonate reductase,** which utilizes the reducing power of ferredoxin to reduce the bound sulfonate group. Finally, the carrier–S-S⁻ group reacts with O-acetylserine, in a reaction catalyzed by **O-acetylserine sulfhydrylase,** to produce the amino acid cysteine.

Reduction of the —S-SO_3^- group to —S-S⁻ requires six electrons in the form of reduced ferredoxin, generated by photosynthetic electron transport. An additional pair of electrons is required in the formation of cysteine. This process, consequently, also is energetically costly. However, sulfur is not an abundant element in plant cells, occurring in a ratio of carbon to sulfur of 700–1000:1.

VI. Summary for Chapters 4 and 5

Photosynthesis is possible because of the existence of chlorophyll. The properties of chlorophyll permit light-driven oxidation–reduction reactions within thylakoid membranes that ultimately produce NADPH and ATP. Electrons are transferred from water to $NADP^+$ through a sequence of electron carriers. The evidence suggests that this electron flow requires two photosystems connected in a linear series. Electrons also can act catalytically around photosystem 1 in cyclic photophosphorylation, in which ATP is produced but not O_2 or NADPH. Linked to the electron transport is the generation of a concentration gradient of H^+ ions across the membrane, with the difference in concentration as high as a factor of 10^3 to 10^4. The flux of H^+ ions down this concentration gradient and through the coupling factor drives the synthesis of ATP.

The products of the "light" reactions in the membranes, NADPH and ATP, are used for the fixation of CO_2 in "dark" reactions within the chloroplast stroma. Two primary enzymes are involved in incorporating CO_2 into organic compounds. The major enzyme, found in nearly all photosynthetic organisms, is ribulose 1,5-bisphosphate carboxylase of the C_3 cycle. The major products of this cycle and its associated reactions are glucose and polymeric starch. In the C_4 pathway, CO_2 is fixed by phosphoenol pyruvate carboxylase, to produce four-carbon compounds within the mesophyll cells of certain plants. In adjacent bundle sheath cells, the four-carbon compounds are decarboxylated, and the released CO_2 is refixed by the C_3 pathway. Minor pathways of CO_2 fixation are found in the *Crassulaceae* family of plants, which differs temporally from the C_4 pathway, and in anaerobic photosynthetic bacteria.

In C_3 plants, but not in C_4 plants, an appreciable amount of photosynthetically fixed carbon is lost by the process of photorespiration.

Literature Cited

Allen, J. F., Bennett, J., Steinback, K. E., and Arntzen, C. J. (1981) Chloroplast protein phosphorylation couples plastoquinone redox state to distribution of excitation energy between photosystems, *Nature* (London) **291**:25–29.

Arnon, D. I., Tsujimoto, H. Y., and Tang, G. M.-S. (1981) Proton transport in photooxidation of water: A new perspective on photosynthesis, *Proc. Natl. Acad. Sci. USA* **78**:2942–2946.

Bahr, J. T., and Jensen, R. G. (1974) Ribulose diphosphate carboxylase from freshly ruptured spinach chloroplasts having an *in vivo* Km[CO_2], *Plant Physiol.* **53**:39–44.

Bassham, J. A. (1977) Increasing crop production through more controlled photosynthesis, *Science* **197**:630–638.

Bassham, J. A., Benson, A. A., Kay, L. D., Harris, A. Z., Wilson, A. T., and Calvin, M. (1954) The path of carbon in photosynthesis, XXI. The cyclic regeneration of carbon dioxide acceptor, *J. Am. Chem. Soc.* **76**:1760–1770.

Bendall, D. S., Davenport, H. E., and Hill, R. (1971) Cytochrome components in chloroplasts of the higher plants, *Meth. Enzymol.* **23**:327–344.

Birmingham, B. C., Coleman, J. R., and Colman, B. (1982) Measurement of photorespiration in algae, *Plant Physiol.* **69**:259–262.

Black, C. C., Jr., and Mollenhauer, H. H. (1971) Structure and distribution of chloroplasts and other organelles in leaves with various rates of synthesis, *Plant Physiol.* **47**:15–23.

Bowes, G., Ogren, W. L., and Hagman, R. H. (1971) Phosphoglycolate production catalyzed by ribulose diphosphate carboxylase, *Biochem. Biophys. Res. Commun.* **45**:716–722.

Buchanan, B. B. (1980) Role of light in regulation of chloroplast enzymes, *Annu. Rev. Plant Physiol.* **33**:341–374.

Burnell, J. N., and Hatch, M. D. (1983) Dark-light regulation of pyruvate, Pi dikinase in C4 plants: Evidence that the same protein catalyses activation and inactivation, *Biochem. Biophys. Res. Commun.* **111**:288–293.

Chen, T. M., Dittrich, P., Campbell, W. H., and Black, C. C. (1974) Metabolism of epidermal tissues, mesophyll cells, and bundle sheath strands resolved from mature nutsedge leaves, *Arch. Biochem. Biophys.* **163**:246–262.

Cockburn, W., Ting, I. P., and Sternberg, L. O. (1979) Relationships between stomatal behavior and internal carbon dioxide concentration in crassulacean acid metabolism plants, *Plant Physiol.* **63**:1029–1032.

Ehleringer, J., and Pearcy, R. W. (1983) Variation in quantum yield for CO_2 uptake among C3 and C4 plants, *Plant Physiol.* **73**:555–559.

Emerson, R. (1958) The quantum yield of photosynthesis, *Annu. Rev. Plant Physiol.* **9**:1–24.

Emerson, R., Chalmers, R., and Cederstrand, C. (1957) Some factors influencing the long-wave limit of photosynthesis, *Proc. Natl. Acad. Sci. USA* **43**:133–143.

Evans, M. C. W., Buchanan, B. B., and Arnon, D. I. (1966) A new ferredoxin-dependent carbon reduction cycle in a photosynthetic bacterium, *Proc. Natl. Acad. Sci. USA* **55**:928–934.

Ferte, N., Meunier, J.-C., Ricard, J., Buc, J., and Sauve, P. (1982) Molecular properties and thioredoxin-mediated activation of spinach chloroplastic NADP-malate dehydrogenase. *FEBS Lett.* **146**:133–138.

Govindjee and Govindjee, R. (1975) Introduction to photosynthesis, in *Bioenergetics of Photosynthesis* (Govindjee, ed.), Academic Press, New York, pp. 1–50.

Govindjee, R., Rabinowitch, E., and Govindjee (1968) Maximum quantum yield and action spectrum of photosynthesis and fluorescence in *Chlorella*, *Biochim. Biophys. Acta* **162**:539–544.

Hall, D. O. (1976) Photobiological energy conversion, *FEBS Lett.* **64**:6–16.

Hangarter, R. P., and Good, N. E. (1982) Energy thresholds for ATP synthesis in chloroplasts, *Biochim. Biophys. Acta* **681**:397–404.

Hatch, M. D., and Kagawa, T. (1976) Photosynthetic activities of isolated bundle sheath cells in relation to differing mechanisms of C4 pathway photosynthesis, *Arch. Biochem. Biophys.* **175**:39–53.

Hatch, M. D., and Slack, C. R. (1966) Photosynthesis by sugar-cane leaves. A new carboxylation reaction and the pathway of sugar formation, *Biochem. J.* **101**:103–111.

Hill, R. (1951) Oxidoreduction in chloroplasts, *Adv. Enzymol. Rel. Sub. Biochem.* **12**:1–39.

Hill, R., and Bendall, F. (1960) Function of the two cytochrome components in chloroplasts: A working hypothesis, *Nature* (London) **186**:136–137.

Hind, G., and Jagendorf, A. T. (1963) Separation of the light and dark stages in photophosphorylation, *Proc. Natl. Acad. Sci. USA* **49**:715–722.

Jagendorf, A. T., and Uribe, E. (1966) ATP formation caused by acid-base transition of spinach chloroplasts, *Proc. Natl. Acad. Sci. USA* **55**:170–177.

Joliot, P. (1972) Modulated light source use with the oxygen electrode, *Meth. Enzymol.* **24**:123–134.

Kyle, D. J., Staehelin, L. A., and Arntzen, C. J. (1983) Lateral mobility of the light-harvesting complex in chloroplast membranes controls excitation energy distribution in higher plants. *Arch. Biochem. Biophys.* **222**:527–541.

Lam, E., and Malkin, R. (1982) Reconstruction of the chloroplast noncyclic electron transport pathway from water to NADP with three integral protein complexes, *Proc. Natl. Acad. Sci. USA* **79:**5494–5498.

McCarty, R. E., and Portis, A. R., Jr. (1976) A simple, quantitative approach to the coupling of photophosphorylation to electron flow in terms of proton fluxes, *Biochemistry* **15:**5110–5114.

Mitchell, P. (1961) Coupling of phosphorylation to electron and hydrogen transfer by a chemiosmotic type of mechanism, *Nature* (London) **191:**144–148.

Potter, J. W., and Black, C. C., Jr. (1982) Differential protein composition and gene expression in leaf mesophyll cells and bundle sheath cells of the C4 plant *Digitaria sanguinalis*, *Plant Physiol.* **70:**590–597.

Radmer, R., and Ollinger, O. (1980) Light-driven uptake of oxygen, carbon dioxide, and bicarbonate by the green alga *Scenedesmus*, *Plant Physiol.* **65:**723–729.

Ruben, S., Randall, M., Kamen, M., and Hyde, J. L. (1963) Heavy oxygen (O^{18}) as a tracer in the study of photosynthesis, *J. Am. Chem. Soc.* **63:**877–879.

Sauer, K. (1975) Primary events and the trapping of energy, in *Bioenergetics of Photosynthesis* (Govindjee, ed.), Academic Press, New York, pp. 115–181.

Sayre, R. T., Kennedy, R. A., and Pringnitz, D. J. (1979) Photosynthetic enzyme activities and localization in *Mollugo verticillata* populations differing in the levels of C3 and C4 cycle operation, *Plant Physiol.* **64:**293–299.

Schlodder, E., Rögner, M., and Witt, H. T. (1982) ATP synthesis in chloroplasts induced by a transmembrane electric potential difference as a function of the proton concentration, *FEBS Lett.* **138:**13–18.

Schmidt, A., Abrams, W. R., and Schiff, J. A. (1974) Reduction of adenosine 5′-phosphosulfate to cysteine in extracts from *Chlorella* and mutants blocked for sulfate reduction, *Eur. J. Biochem.* **47:**423–434.

Sommerville, C. R., and Ogren, W. L. (1982) Genetic modification of photorespiration, *Trends Biochem. Sci.* **7:**171–174.

Spalding, M. H., Schmitt, M. R., Ku, S. B., and Edwards, G. E. (1979) Intracellular localization of some key enzymes of crassulacean acid metabolism in *Sedum praealtum*, *Plant Physiol.* **63:**738–743.

Stemler, A., and Radmer, R. (1975) Source of photosynthetic oxygen in bicarbonate-stimulated Hill reaction, *Science* **190:**457–458.

Stumpf, D. K., and Jensen, R. G. (1982) Photosynthetic CO_2 fixation at air levels of CO_2 by isolated spinach chloroplasts, *Plant Physiol.* **69:**1263–1267.

Sue, J. M., and Knowles, J. R. (1978) Retention of the oxygens at C-2 and C-3 of D-ribose 1,5-bisphosphate in the reaction catalyzed by ribulose-1,5-bisphosphate carboxylase, *Biochemistry* **17:**4041–4044.

Trebst, A. (1980) Inhibitors in electron flow: Tools for the functional and structural localization of carriers and energy conservation sites, *Meth. Enzymol.* **69:**675–715.

Vermaas, W. F. J., and Govindjee (1982) Bicarbonate or carbon dioxide as a requirement for efficient electron transport on the acceptor side of photosystem II, in *Photosynthesis: Development, Carbon Metabolism and Plant Productivity*, Vol. 2 (Govindjee, ed.), Academic Press, New York, pp. 541–558.

Warburg, O. (1964) Prefatory chapter, *Annu. Rev. Biochem.* **33:**1–14.

Witt, H. T. (1975) Primary acts of energy conservation in the functional membrane of photosynthesis, in *Bioenergetics of Photosynthesis* (Govindjee, ed.), Academic Press, New York, pp. 493–554.

Wraight, C. A. (1982) Current attitudes in photosynthesis research, in *Photosynthesis: Energy Conversion by Plants and Bacteria*, Vol. 1 (Govindjee, ed.), Academic Press, New York, pp. 17–61.

Additional Reading

A recently published series of excellent, detailed reviews is contained in:

Govindjee (ed.) (1982) *Photosynthesis:* Vol. 1, *Energy Conversion by Plants and Bacteria.* Vol. 2, *Development, Carbon Metabolism and Plant Productivity*, Academic Press, New York.

Further sources of information on various aspects of photosynthesis are the following:

Andrew, T. J., and Ballment, B. (1983) The function of the small subunits of ribulose bisphosphate carboxylase-oxygenase, *J. Biol. Chem.* **258**:7514–7518.

Bendall, D. S. (1982) Photosynthetic cytochromes of oxygenic organisms, *Biochim. Biophys. Acta* **683**:119–151.

Bahr, J. T., and Jensen, R. G. (1978) Activation of ribulose bisphosphate carboxylase in intact chloroplasts by CO_2 and light, *Arch. Biochem. Biophys.* **185**:39–48.

Bassham, J. A. (1971) The control of photosynthetic carbon metabolism, *Science* **172**:526–534.

Blankenship, R. E., and Parson, W. W. (1978) The photochemical electron transfer reactions of photosynthetic bacteria and plants. *Annu. Rev. Biochem.* **47**:635–653.

Buchanan, B. B., and Arnon, D. I. (1970) Ferredoxins: Chemistry and function in photosynthesis, nitrogen fixation and fermentative metabolism, *Adv. Enzymol. Rel. Areas Mol. Biol.* **33**:119–176.

Clark, R. D., and Hind, G. (1983) Isolation of a five-polypeptide cytochrome b-f complex from spinach chloroplasts, *J. Biol. Chem.* **258**:10348–10354.

Clayton, R. K. (1980) *Photosynthesis: Physical Mechanisms and Chemical Patterns,* Cambridge University Press, Cambridge.

Clayton, R. K., and Sistrom, W. R., (eds.) (1978) *The Photosynthetic Bacteria,* Plenum, New York.

Cogdell, R. J. (1983) Photosynthetic reaction centers. *Annu. Rev. Plant Physiol.* **34**:21–45.

Ellis, R. J. (1979) The most abundant protein in the world, *Trends Biochem. Sci.* **4**:241–244.

Govindjee (ed.) (1975) *Bioenergetics of Photosynthesis,* Academic Press, New York.

Guerrero, M. G., Vega, J. M., and Losada, M. (1981) The assimilatory nitrate-reducing system and its regulation, *Annu. Rev. Plant Physiol.* **32**:169–204.

Hinkle, P. C., and McCarty, R. E. (1978) How cells make ATP, *Sci. Am.* **238**(3):104–123.

Kohlbrenner, W. E., and Boyer, P. D. (1983) Probes of catalytic site cooperativitiy during catalysis by the chloroplast adenosine triphosphatase and the adenosine triphosphate synthase, *J. Biol. Chem.* **258**:10881–10886.

Lorimer, G. H. (1981) The carboxylation and oxygenation of ribulose 1,5-bisphosphate: The primary events in photosynthesis and photorespiration, *Annu. Rev. Plant Physiol.* **32**:349–383.

Miller, K. R. (1979) The photosynthetic membrane, *Sci. Am.* **241**(4):102–113.

Miziorko, H. M., and Lorimer, G. H. (1983) Ribulose-1,5-bisphosphate carboxylase-oxygenase, *Annu. Rev. Biochem.* **52**:507–535.

San Pietro, A. (ed.) (1980) *Photosynthesis and Nitrogen Fixation* (Part C), Meth. Enzymol., Vol. 69, Academic Press, New York. (This volume contains chapters on nearly all thylakoid components and describes methods for analysis of photosynthetic activities.)

Saver, B. G., and Knowles, J. R. (1982) Ribulose-1,5-bisphosphate carboxylase: Enzyme-catalyzed appearance of solvent tritium at carbon 3 of ribulose 1,5-bisphosphate reisolated after partial reaction, *Biochemistry* **21**:5398–5403.

Shavit, N. (1980) Energy transduction in chloroplasts: Structure and function of the ATPase complex, *Annu. Rev. Biochem.* **49**:111–138.

Staehelin, L. A. (1983) Control and regulation of the spatial organization of membrane components by membrane-membrane interactions, *Modern Cell Biol.* **2**:73–92.

Trebst, A. (1974) Energy conservation in photosynthetic electron transport of chloroplasts, *Annu. Rev. Plant Physiol.* **25**:423–458.

Zelitch, I. (1975) Pathways of carbon fixation in green plants, *Annu. Rev. Biochem.* **44**:123–145.

The Chloroplast Genome and Its Expression

I. Introduction

The studies of Schimper, Strasburger, and others in the 1880s demonstrated that chloroplasts in a sense have an existence of their own. Chloroplasts were found to proliferate by division of existing plastids and were passed on to daughter cells at the time of cell division. But more recent findings have indicated that the chloroplasts' independence is quite limited. Plastids have their own DNA, which is quite distinct from nuclear DNA, and also have the ability to express the genetic information in their DNA. A question with much current interest, however, is the critical one of how much information chloroplast DNA actually contains. Although the *total* mass of DNA in a chloroplast is generally somewhat more than that in a bacterial cell, the genetic capacity of chloroplast DNA is much less than that of the bacterial chromosome. The reason for this paradox is that chloroplasts contain multiple copies of a relatively small molecule of DNA. As with mitochondria, the information in chloroplast DNA apparently is only that which is required, separate from the nucleus, to synthesize a few necessary functional proteins. However, even to accomplish the synthesis of these relatively *few* proteins, the chloroplast must contain its own ribosomes and complete machinery for protein synthesis. As work on the biosynthetic capabilities of the chloroplast proceeds, it is becoming quite clear that *most* of the plastid's proteins and properties are determined by the nuclear genome and that *most* of its proteins are synthesized on cytoplasmic ribosomes.

II. Purification of Chloroplast DNA

The amount of DNA in a mature chloroplast varies somewhat from organism to organism but generally lies within the range of 5–50×10^{-15} g (or, expressed in another way, 3–30×10^9 daltons*). Fig. 6.1 shows the structures of

---

*1 dalton = $\frac{1}{12}$ the mass of a carbon atom or 1.67×10^{-24} g. The *mass* of a single molecule, expressed in units of daltons, has the same number as molecular weight. However, the molecular weight is a relative number and has no units.

components of nucleic acids. A useful index for describing the composition of DNA molecules is their guanine + cytosine (G + C) content. Chloroplast DNA from higher plants and most algae have a characteristic composition of 37 ± 1% G + C (Table 6.1). Since the density of DNA is directly related to its composition, or its G + C content, chloroplast DNA from these cells has a density

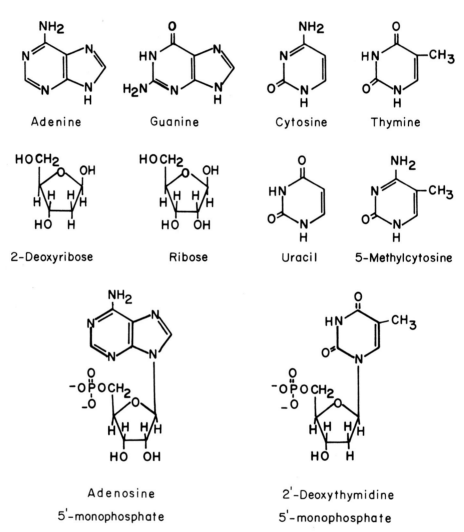

FIG. 6.1. The building blocks of chloroplast nucleic acids. RNA contains uracil, cytosine, adenine, and guanine as the four nitrogenous bases and the pentose sugar ribose. DNA contains thymine, cytosine, adenine, and guanine as the bases and the pentose sugar 2-deoxyribose. Both nucleic acids contain phosphate as the linking agent between nucleosides (the combination of a base and a sugar molecule). In plants, nuclear DNA contains a small amount of 5-methyl cytosine. Although chloroplast DNA usually lacks 5-methyl cytosine, during mating of the alga *Chlamydomonas reinhardtii* the chloroplast DNA of the (+) mating type becomes heavily methylated on cytosine residues. Adenosine 5′-monophosphate (AMP) is an example of a nucleotide unit of RNA. 2′-Deoxythymidine 5′-monophosphate (dTMP) is an example of a nucleotide unit of DNA. ("Primed" numbers refer to positions on the sugar moiety.)

TABLE 6.1. Compositions of Chloroplast DNA

Source	Nucleoside (mole %)				
	dT	dG	dA	dC	(G + C)
Spinach	31.4	18.0	32.1	18.5	36.5[a]
Euglena	35.5	14.2	36.3	14.0	28.2

[a]Most higher plants and green algae contain chloroplast DNA with a (G + C) content of 37 ± 1%. (From Crouse *et al.*, 1978.)

of 1.695–1.697 g cm^{-3}. A notable exception is the chloroplast DNA from the alga *Euglena*, which has a G + C content of 28.2% (Table 6.1) and a density of 1.685 g cm^{-3}. The nuclear DNA of plant cells is more variable in composition, with densities ranging from 1.69 g cm^{-3} in many higher plants to 1.72 g cm^{-3} in some algae. (The density of DNA is determined by subjecting it to a centrifugal field in a density gradient of CsCl with an ultracentrifuge. The gradient is centrifuged for a sufficient time to allow the DNA to reach an equilibrium position in the gradient where the density of the medium is equal to its own. The rate of sedimentation of the DNA then becomes zero. The measured densities are referred to as **buoyant** densities.)

Chloroplast DNA can be purified on the basis of either its location within the organelle or its distinct composition. If the density of nuclear DNA is sufficiently different from that of chloroplast DNA, these types of DNA can be separated directly by centrifugation of a cell extract in a concentration gradient of CsCl (Fig. 6.2). But with some higher plants, chloroplast and nuclear DNA molecules have similar overall compositions, and for the purification of chloroplast DNA a prior isolation of intact chloroplasts is necessary. These intact chloroplasts are then treated with the enzyme deoxyribonuclease to digest any

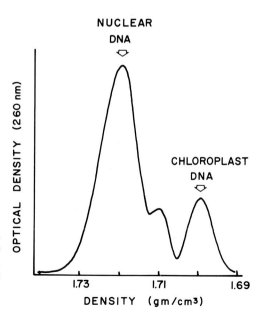

FIG. 6.2. CsCl density gradient profile of DNA extracted from vegetative cells of *Chlamydomonas reinhardtii*. Chloroplast DNA has a density of 1.695 g/cm³ in this gradient and is separated from other species of cellular DNA. (Adapted from Howell and Walker, 1976.)

nuclear DNA adhering to their outer surface and are subsequently lysed in the presence of detergents to solubilize membranes. Final purification of the DNA is accomplished by sedimentation through a concentration gradient of CsCl.

Carefully prepared, highly purified chloroplast DNA appears as circular structures when viewed with the electron microscope (Fig. 6.3). The contour lengths of these structures from **higher plants** lie within the range of 43–46 μm. These sizes correspond to molecular weights of 9–10 × 10⁷. DNA molecules of this size contain about 140,000 base pairs. The sizes of **algal** chloroplast DNA molecules are more variable. That from a Chrysophyte, *Sphaerocarpus donnellii*, is only 36 μm in contour length, whereas DNA from the chloroplast of

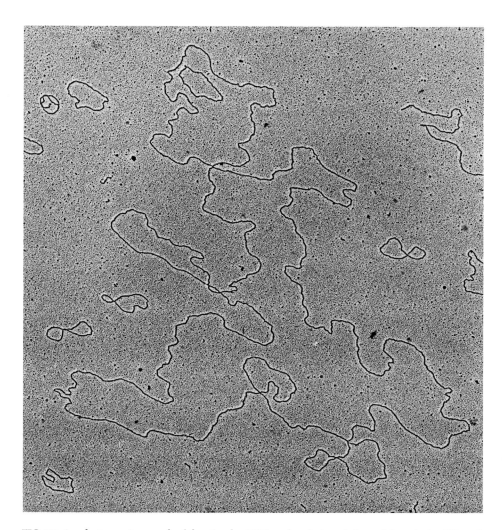

FIG. 6.3. An electron micrograph of the circular DNA molecule from lettuce chloroplasts. Although this micrograph shows a "relaxed" molecule, most of the DNA molecules exist as tight, highly twisted ("supercoiled") structures. The DNA has a contour length of about 43 μm and a molecular weight of 9.6 × 10⁷. The small circular molecules are φX RF II bacteriophage DNA, added as calibration markers. (From Kolodner and Tewari, 1975. Micrograph courtesy of R. Kolodner.)

the green alga *Chlamydomonas reinhardtii* is 62 μm. The molecular weight of the latter DNA is 12.6 × 10⁷. Considering the amount of DNA in a chloroplast, simple calculations suggest that on the order of 50 to more than 1000 copies of these molecules are present per plastid in plant cells (Table 6.2).

III. Chloroplast DNA Molecules Are Identical

The current knowledge of chloroplast DNA indicates that the multiple copies in each chloroplast from a given cell type are identical. This evidence for homogeneity came from two types of experimental procedures. The first was the study of the kinetics of renaturation. In this approach, strands of fragmented DNA molecules were separated either by heating the solution or by making the solution alkaline to pH 12. Upon a rapid drop in the temperature or shift in the pH to neutrality, complementary polynucleotide sequences again aligned themselves by base-pairing, reforming double-stranded DNA. The denaturation and renaturation can be conveniently followed by measuring first the increase and then the decrease in the absorbancy of the solution (Fig. 6.4). The increase in the absorbance of the solution as the DNA strands come apart (denaturation of the DNA) is referred to as **hyperchromicity,** and is a consequence of the relaxation of the restraint on electronic resonance in the bases within the DNA helix. Subsequently, as the single-stranded DNA base-pairs and reforms the double helix, the absorbancy of the solution will decrease.

The probability that complementary strands find each other depends on their concentration in solution. The smaller the molecular size of the DNA, the more frequently a specific nucleotide sequence will occur in a given amount of DNA. From an analysis of the rate of renaturation for a given amount of DNA, according to second-order kinetic equations, it is possible to calculate how often repeated sequences occur. One repeat length is considered as the genome size. Measurements of this type revealed that a sample of chloroplast DNA

TABLE 6.2. DNA Content in Cells and Chloroplasts in (1) the Alga *Chlamydomonas reinhardtii* and (2) the Higher Plant *Triticum aestivum* (Wheat)

Growth stage	DNA/cell (g × 10⁻¹⁴)	Plastids/cell	DNA/plastid (g × 10⁻¹⁴)	Genome copies/plastid (approximate)
1. *Chlamydomonas reinhardtii*				
Vegetative cells	12.4	1	1.74	82
Gametes	12.3	1	0.86	40
2. *Triticum aestivum*				
cm from base[a]: 1–2	3600	<50	12.1	810
2–3	3900	>50	15.3	1020
4–5	4500	155	4.8	320
6–7	4500	155	4.7	310

[a]The distance of the cells from the base of the leaf is an indication of their age (see Fig. 7.4). The DNA level in the plastid decreases with development, since chloroplast division continues for a time after plastid DNA replication stops. (Data on *Chlamydomonas* was obtained from Chiang, 1971. Data for *Triticum* was obtained from Boffey and Leech, 1982.)

from a given organism behaved as a uniform population of molecules. The genome size calculated by this technique agreed closely with the size determined directly by electron microscopy.

A second approach that has become a powerful tool for the analysis of DNA structure is the use of endodeoxyribonucleases called **restriction enzymes.** These enzymes, isolated from bacterial cells, recognize specific sequences in DNA and cleave both strands of the duplex. A large number of these enzymes, with different sequence specificities, have been purified. For example, an enzyme from *Escherichia coli,* called EcoRI (*E. coli* restriction enzyme 1), cleaves DNA only between guanine and adenine nucleotides in the sequence G-A-A-T-T-C (5′ → 3′). The same sequence, of course, occurs on the opposite strand as the complement and is likewise hydrolyzed. The sequence G-G-A-T-C-C (5′ → 3′) is cleaved between the quanine nucleotides by the enzyme BamHI isolated from *Bacillus amyloliquefaciens* strain H. When a sample of DNA is digested with these enzymes, a discrete set of fragments is produced. The number of fragments depends only on the frequency with which the specific sequence occurs in the DNA molecule. Restriction enzymes with

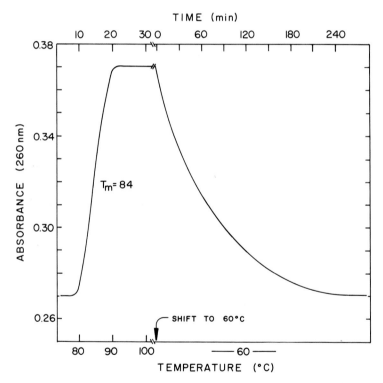

FIG. 6.4. The changes in absorbance that accompany melting and renaturation of chloroplast DNA from pea leaves. DNA strands separate when thermal energy breaks hydrogen bonds between complementary bases. For chloroplast DNA, the T_m (the temperature at the midpoint in the melting curve) is 84°C. (The T_m is influenced by base composition of the DNA and the ionic strength of the medium.) When the temperature is rapidly dropped to 60°C, the DNA reanneals or renatures. The genome size can be calculated with a second-order rate equation from the kinetics of renaturation. (Adapted from Kolodner and Tewari, 1972.)

different specificities produce different sets of fragments. These fragments can be separated on the basis of their size by electrophoresis through a porous gel, usually made from agarose (Fig. 6.5). Within the electrophoretic pattern of fragments, the relative frequency of each fragment can be determined from an absorbance scan of the gel and by relating the total absorbance of a fragment to its size. A summation of the masses of the fragments, taking into account any multiple sequences, allows determination of the minimal molecular size. This procedure has fewer technical and interpretative difficulties than measurements of renaturation kinetics, since the rate of renaturation is influenced by the base composition of the DNA and the conditions of the analysis. The mass of chloroplast DNA molecules calculated from the fragment sizes produced by restriction enzymes again agreed well with the value derived from the lengths of the molecules obtained by electron microscopy.

By using several restriction enzymes, each with a different specificity, and by analyzing the sizes of the fragments produced by these nucleases alone and in series, it is possible to construct a physical (in contrast to a genetic) map of the chloroplast chromosome. Construction of the physical map involves fitting fragments, produced by two or more restriction enzymes, according to length, into overlapping fragments obtained with only one of the enzymes. The procedure is analogous to fitting together pieces of a puzzle. Each fragment generated by EcoR1, shown in Fig. 6.5, then can be arranged in sequence around the circular DNA molecule.

Several genes have been localized to specific restriction fragments either by hybridization of the RNA product or by alignment of nucleotide sequences in the DNA with corresponding amino acid sequences in proteins. Since the position of a restriction fragment within the physical map of the DNA is known, these procedures allow development of a genetic map without doing a genetic experiment. This ability becomes particularly important in those cases where genetic experiments are not feasible. Particularly with chloroplasts, in which DNA exists in multiple copies, genetic experiments become complicated. Fig. 6.6 shows the positions, determined by physical mapping, of several genes in the chloroplast genome of four different plant cells. Determination of positions on the physical map and nucleotide sequences of genes on the chloroplast genome is an area currently receiving intense effort. Positions of the ribosomal RNA genes, most of the transfer RNA genes, and the structural genes for nearly ten proteins already are known. Within a few years, the complete genome most likely will be elucidated. Aspects of the physical maps and the techniques that allow determination of gene location and expression will be described in the following sections.

It is interesting to note that gene arrangements in the chloroplast DNA of C_3 plants, of which *Triticum aestivum* (wheat) is an example, are similar to those of C_4 plants, such as *Zea mays* (maize). Variations do occur in gene order and position among different chloroplast DNA molecules, but these differences are not along major metabolic divisions. Therefore, the development of C_4 metabolism, including repression of the gene for the large subunit of ribulose 1,5-bisphosphate carboxylase in mesophyll cells, must be determined by nuclear genes.

With several exceptions, the differences in gene arrangements within chlo-

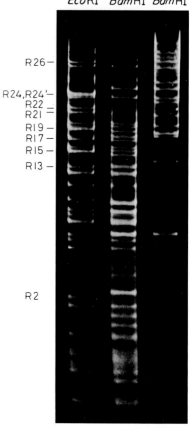

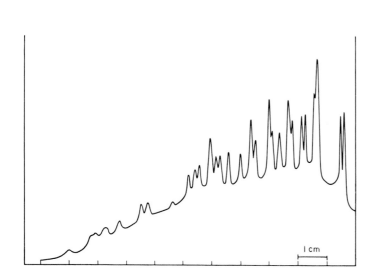

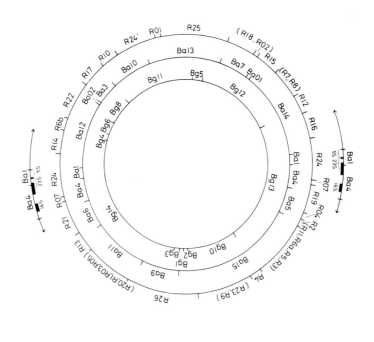

roplast DNAs suggest simply evolutionary variation of a somewhat plastic structure. However, within an individual gene the nucleotide sequences are highly conserved.

IV. Synthesis of Chloroplast DNA

Chloroplasts contain a **DNA polymerase** that catalyzes the synthesis of DNA from deoxyribonucleoside triphosphates by the usual semiconservative process (Fig. 6.7). The regulation of this enzyme is distinct from that of the replicative enzyme in the nucleus. As shown in Fig. 6.8, in cells of the alga *Chlamydomonas reinhardtii*, induced to grow synchronously by cycles of 12 hr of light/12 hr of dark, chloroplast DNA is replicated during the cell cycle at a different time (between hours 2 and 5 in the light phase) than the nuclear DNA (between hours 14 and 20 in the dark phase). Shortly after nuclear DNA was replicated, cell division occurred. In these cells chloroplast DNA accounts for about 14% of the total cellular DNA (Table 6.2).

Several unusual features of the chloroplast have permitted examination of DNA synthesis in the organelle without significant interference by nuclear DNA synthesis. When *Chlamydomonas reinhardtii* cells were incubated with radioactive thymidine, only chloroplast DNA became labeled (Fig. 6.9). Since thymidine must be phosphorylated before it is incorporated into DNA, these results have been interpreted to indicate that thymidine kinase, the enzyme that catalyzes addition of the first phosphoryl group to the form deoxyTMP, occurs only within the chloroplast. However, this compartmentation of the enzyme has not been established.

Chlamydomonas cells, when treated briefly with a low concentration of toluene, become permeable to nucleotide precursors of DNA and RNA. A curious feature of this system is that the incorporation of the precursors is nearly all into chloroplast nucleic acids in these treated cells. These results clearly demonstrate that chloroplasts have an independent system for synthesis of their own DNA.

A detailed analysis of DNA synthesis in chloroplasts has not been accomplished.

V. Inheritance of Chloroplast DNA

From the work of Gregor Mendel (1822–1884), the pattern of inheritance of genetic traits was known to follow a statistical distribution, with equal proba-

---

FIG. 6.5. Patterns of fragments obtained by electrophoresis in an agarose gel of restriction enzyme digests of chloroplast DNA from *Chlamydomonas reinhardtii*. At the top left of the figure is a densitometric scan of the lane containing the EcoRI digest, shown at the top right. The genome size can be determined by summation of the sizes of the individual fragments. Also shown are patterns produced by digestion with BamHI and with both restriction nucleases. In the lower portion of the figure is the physical map, which shows the positions of these fragments within the DNA molecule. (From Rochaix, 1978. Reprinted with permission.)

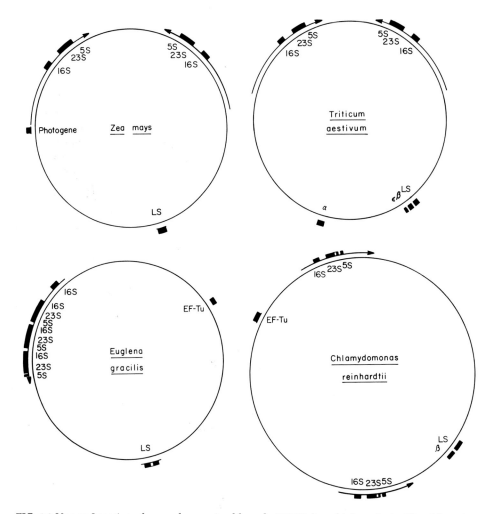

FIG. 6.6 Upper: Location of several genes in chloroplast DNA from higher plants. The chloroplast genomes of *Zea mays* (maize) and *Triticum aestivum* (wheat) are representative of DNA from most higher plants. The physical maps are characterized by two inverted repeats containing genes for the 16 S, 23 S, and 5 S ribosomal RNA. The arrowheads indicate the direction of synthesis of the RNA product (5′ → 3′). Examples of other species that contain similar inverted repeats are tobacco, spinach, soybean, mung bean, mustard, and petunia. However, pea, chick-pea, and broad-bean chloroplast DNA contain only one set of ribosomal RNA genes and no inverted repeat regions. The map for *Zea mays* shows the position of the "photogene," which codes for a 32,000-molecular-weight polypeptide in thylakoid membranes. Synthesis of this polypeptide is induced by light. The position of the gene for the large subunit of ribulose 1,5-bisphosphate carboxylase (LS) is indicated, which is similar in chloroplast DNA of most higher plants. The positions of the genes for the α, β, and ϵ subunits of coupling factor CF_1 in the *Triticum* genome are also shown. Genes for transfer RNA are dispersed throughout the DNA molecules. The DNA molecule from *Zea mays* contains 140,000 base-pairs, whereas that from *Triticum* contains 135,000 base-pairs. Lower: Representations of the chloroplast genome in two algae, *Euglena gracilis* and *Chlamydomonas reinhardtii*. *Euglena* DNA contains three sets of ribosomal genes arranged in tandem repeats, along with a supplemental gene for 16 S ribosomal RNA at the 5′ end of the region. The gene for the large subunit of the carboxylase (LS) is transcribed in the same direction as the ribosomal RNA genes and apparently contains an intron that interrupts the coding sequence of the gene. Chloroplast

bility of the trait from either parent appearing in the progeny. Correns and Baur separately published in 1909 their discovery of traits that did not follow Mendel's rules of inheritance. These traits were unusual by virtue of the fact that they were inherited by progeny only from the maternal (egg) cells. Traits determined by chloroplast DNA of the paternal (pollen) cells rarely were passed on to the next generation.

Chloroplast inheritance has been extensively studied in the alga *Chlamydomonas reinhardtii*, a system developed for this purpose by Ruth Sager (1977). Fig. 6.10 illustrates the pattern of inheritance in this organism during its sexual life cycle. Haploid vegetative cells, which contain a single chloroplast, differentiate into **gametes** upon transfer to N-free medium. Gametes of opposite mating type pair and completely fuse to form a **zygote.** Within the zygote, the two nuclei and the two chloroplasts fuse to form a diploid cell with a single chloroplast. Maturation of the zygote, which occurs over a period of a week, results in germination and release of four haploid progeny, products of meiosis. Inheritance of nuclear markers, such as the mating types (mt^+ and mt^-), is characterized by a 2:2 segregation, following the Mendelian pattern. In contrast, inheritance of markers in chloroplast DNA yields a 4:0 pattern, or a **uniparental** pattern of inheritance, with only the trait from the mt^+ (maternal) cell appearing in the progeny. Only if the cells are perturbed, for example, by irradiation of mt^+ gametes with ultraviolet light prior to zygote formation, is biparental or paternal inheritance observed to a significant extent.

A striking demonstration of uniparental inheritance of chloroplast DNA in *Chlamydomonas* was provided by Grant, Gillham, and Boynton (1980). These investigators isolated a mutant strain that contained deletions of about 100 base-pairs within the inverted repeat region, near the 16 S RNA gene end (refer to the physical map for *Chlamydomonas* chloroplast DNA in Fig. 6.6). Inheritance of the deletion was examined by crossing this mutant strain with other strains carrying uniparental antibiotic-resistance markers. As shown in Fig. 6.11, the deletion, which was clearly observed in a restriction enzyme digest of the purified DNA, was inherited only from the maternal cell.

A subject of considerable interest has been the mechanism that determines uniparental transmission of chloroplast genes to progeny. Early studies on this question indicated that chloroplast DNA from the paternal (mt^-) gametes was degraded within the zygote and thus made unavailable for recombination. Sager and her colleagues (Burton *et al.*, 1979; Sano *et al.*, 1981) obtained evidence for a possible mechanism that controls selective degradation of chloro-

DNA from *Chlamydomonas*, in contrast, contains inverted repeats that are more distant from each other than in higher-plant DNA. The 23 S ribosomal RNA gene in *Chlamydomonas* contains an intron near the 3' end of the gene (also see Fig. 6.19). The positions of the genes for elongation factor Tu (EF-Tu), a protein involved in protein synthesis, have also been located on the algal genomes. The *Euglena* chloroplast DNA molecule contains 145,000 base-pairs, whereas the DNA in *Chlamydomonas* is 190,000 base-pairs in size. (Maps redrawn from: Bogorad, 1981, *Zea mays*; Howe *et al.*, 1982, 1983, *Triticum aestivum*; Passavant *et al.*, 1983, *Euglena gracilis*; Rochaix and Malnoe, 1978, *Chlamydomonas reinhardtii*.)

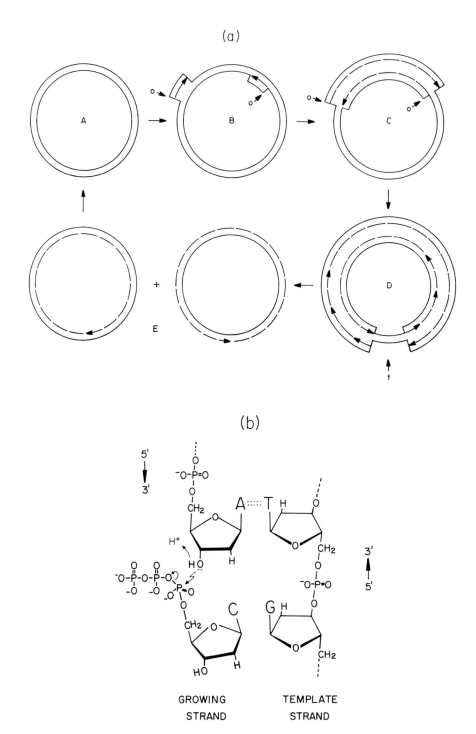

FIG. 6.7. (a) A model for the semiconservative, bidirectional replication of chloroplast DNA. A represents the closed, circular parent molecule. B indicates the formation of displacement loops (D loops) at the origin points (o) of replication for each strand. As in other systems, initiation of

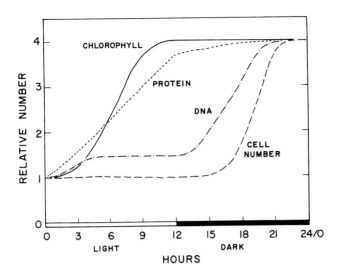

FIG. 6.8. Synthesis of DNA during the division cycle of vegetative cells of *Chlamydomonas rein-hardtii*. Cells were induced to grow in synchrony by 12-hr-light–12-hour-dark cycles. After the cell density had reached about 1×10^6 cells/ml, samples were removed periodically and the cellular DNA was measured. The increase in DNA early in the light phase of the cycle was caused by replication of chloroplast DNA. Nuclear DNA was replicated during the dark phase, just before cell division. In this organism, each cell produces four daughter cells. The amount of chlorophyll in the culture, which increases only during the light phase, is also shown. (Adapted from Chiang and Sueoka, 1967.)

replication probably involves synthesis of a short RNA "primer" by RNA polymerase. The primer is extended in the $5' \rightarrow 3'$ direction with deoxyribonucleotides by DNA polymerase. The RNA primer then is removed by a ribonuclease. As shown in C, each strand may be synthesized in a continuous fashion until the origin of the opposite strand is reached. Subsequently, as the parental strands separate to form a replication fork, synthesis of the "leading" strand will proceed in a continuous fashion, but synthesis of the "lagging" strand will require additional initiation events followed by retrograde synthesis in the $5' \rightarrow 3'$ direction. These latter events, which generate rel- atively short strands, are indicated by the multiple arrowheads in D. The replication forks move toward the terminus (t) opposite the origin sites. After the primers are removed, the resulting "gaps" are filled in by DNA polymerase. The ends of the newly synthesized strands finally are linked together by the enzyme polynucleotide ligase, to form the covalently closed, circular progeny mol- ecules indicated by E. (Adapted from Tewari *et al.*, 1976). (b) Mechanism of addition of 2'-deoxy- ribonucleotides to the 3' end of the growing strand by DNA polymerase. The incoming nucleotide, which contains the base cytosine (C), is selected by its ability to form a base-pair by hydrogen- bonding with guanine (G) on the template (right) strand. The oxygen of the 3'-hydroxyl group on the terminal nucleotide, which contains the base adenine (A), of the growing (left) strand attacks the α-phosphate of the incoming deoxycytidine triphosphate in a S_N2 reaction, liberating the β and γ phosphates as inorganic pyrophosphate. The hydrogen atom on the sugar hydroxyl group of deoxy- ribose is lost into the medium as a H^+. The strand grows in the $5' \rightarrow 3'$ direction and the template strand is read in the $3' \rightarrow 5'$ direction. The double-stranded product contains strands oriented with opposite polarity. The strands are held together by hydrogen bonds between adenine- and thy- mine-containing nucleotides (A-T pairs) and guanine- and cytosine-containing nucleotides (G-C pairs) and by hydrophobic association of bases-pairs along the central axis of the molecule.

plast DNA in *Chlamydomonas*. Chloroplast DNA usually is not methylated during vegetative growth of these cells or differentiation into gametes. However, shortly after fusion of gametes to form the zygote, chloroplast DNA from the mt^+ gamete becomes extensively methylated on cytosine residues, forming 5-methylcytosine (Fig. 6.1). Methylation is essentially completed by 4–6 hr, the time at which chloroplasts fuse, after the initial formation of the zygote. However, DNA in the chloroplast from the mt^- gamete does not become methylated (Fig. 6.12), and is destroyed within 2 hr after formation of the zygote, before chloroplast fusion occurs. These results suggest that methylation may be involved in protection of maternal chloroplast DNA by blocking the cleavage of DNA by an endonuclease similar to restriction enzymes. The increase in methyltransferase activity, as shown in Table 6.3, is compatible with this suggestion. Presumably, the endonuclease activity appears after modification of the DNA has occurred, but the degradative enzyme has not been identified.

Methylation as the mechanism for preservation of maternal chloroplast DNA was challenged by results obtained in Gillham's laboratory with a strain of *Chlamydomonas* containing a nuclear mutation (Bolen *et al.*, 1982). More than 35% of the cytosine residues in chloroplast DNA were methylated constitutively in vegetative mutant cells. The methylated bases were conserved in mt^- gametes. But chloroplast genes still were not transmitted to progeny from mt^- parental cells during mating.

Sager and Grabowy (1983) recently found that this extensive methylation of chloroplast DNA in the mutant strain apparently is unrelated to inheritance. Even in the mutant, additional methyl groups were added to DNA bases during

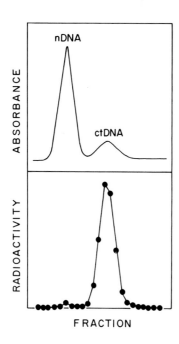

FIG. **6.9.** Incorporation of thymidine into DNA of *Chlamydomonas reinhardtii*. [^{3}H]thymidine was added to cultures of the algal cells, and after a period of incubation, the DNA was isolated and subjected to CsCl density centrifugation. Determination of the radioactivity in each fraction of the gradient indicated incorporation only into chloroplast DNA (ctDNA). In contrast, labeled thymidine 5′-monophosphate was incorporated into both nuclear (nDNA) and chloroplast DNA. These results suggest that thymidine kinase, which catalyzes the reaction

$$\text{TdR (thymidine)} + \text{ATP} \rightleftharpoons \text{dTMP} + \text{ADP}$$

is located only within the chloroplast of his organism. (Adapted from Swinton and Hanawalt, 1972.)

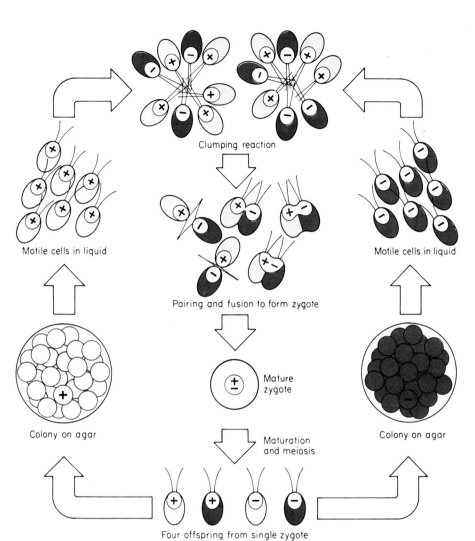

FIG. 6.10. The life cycle of *Chlamydomonas reinhardtii.* The alga exists as two mating types, designated + and −, which can be grown indefinitely as vegetative cells. These cells differentiate into gametes when transferred into nitrogen-free medium and incubated for 12 hr. In a mixture of mating-competent cells, gametes of opposite mating types pair by flagellar adhesion and subsequently fuse to form zygotes (for more details see Martin and Goodenough, 1975; and Goodenough and Weiss, 1975). The zygote is the only diploid stage in the normal cell cycle. Maturation of the zygote usually requires several days, after which four zoospores are released, the four products of meiosis. Mating type, determined by a nuclear gene, and another unlinked nuclear gene, indicated by light and dark shading of the cells, segregate 2:2 in a Mendelian pattern in the offspring. However, segregation of chloroplast DNA markers is 4:0, with only markers from the + parent appearing in the progeny. (From Sager, 1972. Reprinted with permission.)

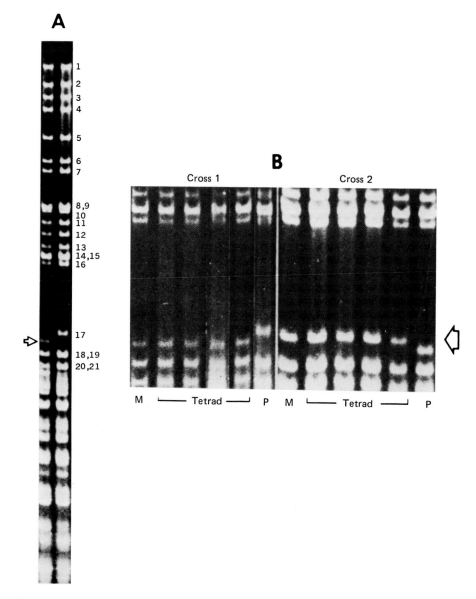

FIG. 6.11. Direct evidence for maternal inheritance of chloroplast DNA in *Chlamydomonas rein-hardtii*. A mutant strain was obtained that contained a deletion of about 100 base-pairs near the 16 S end in each of the inverted repeat regions. A: After electrophoresis through an agarose gel of chloroplast DNA digested ,with the restriction enzyme MspI (from *Moraxella* sp., hydrolyzes at 5′ ... C↓C G G ...3′), the deletion is indicated by the shortening of fragment 17 and consequently a more rapid migration (arrow). B: The mutant was crossed with another uniparental mutant strain conferring resistance to the antibiotic erythromycin in cross 1 or streptomycin (*sr*) in cross 2. The mating-type cells were allowed to fuse into zygotes, which matured for six days before germination. The four tetrad meiotic progeny were isolated and cloned. DNA from each was analyzed as in A. M is the pattern of fragments 13–21 from the maternal (mt^+) chloroplast DNA, and P is the pattern for the paternal (mt^-) DNA. Cross 1 was mt^+, deletion × mt^-, erythromycin resistant. All four progeny of this cross contained the deletion in fragment 17 (arrow) but were erythromycin sensitive

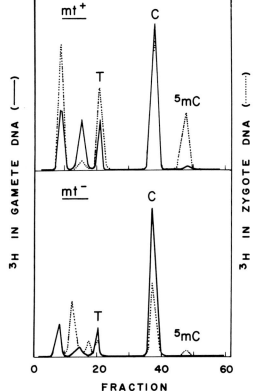

FIG. 6.12. Occurrence of 5-methyl cytosine in chloroplast DNA of *Chlamydomonas reinhardtii* zygotes. Vegetative cells of one mating type were labeled during growth with [³H]deoxycytidine, transferred to N-free medium for differentiation into gametes, and then allowed to fuse with unlabeled gametes of the other mating type. The presence of 5-methyl cytosine (⁵mC) in chloroplast DNA of gametes (———) and in zygotes 6 hr after fusion (......) was determined by hydrolysis of the DNA and chromatography of the bases. (Adapted from Burton *et al.*, 1979.)

gametogenesis of *mt*⁺ (maternal) but not *mt*⁻ (paternal) cells. These data further support the hypothesis that uniparental inheritance involves degradation of *mt*⁻ chloroplast DNA during zygote formation and that DNA of *mt*⁺ cells is protected from degradation by specific methylation.

Whether a process as described for the alga *Chlamydomonas* also operates in higher plants has not been determined. In some species of plants, maternal inheritance is determined by the exclusion of plastids from the paternal (pollen) cells from entering the egg. As a result, only chloroplast genes in the maternal cell appear in the progeny. However, this is not true for all higher plants, and occasionally biparental inheritance of chloroplast markers is observed in mutant strains, which suggest that plastids from both parents enter the initial zygote. Very little biochemical information is available in these systems to explain the genetic results.

(inheritance pattern 4:0). Cross 2 was *mt*⁺, streptomycin resistant × *mt*⁻, deletion. The deletion, which in this case was in the *mt*⁻ strain, did not appear in any of the progeny. However, all progeny were streptomycin resistant (inheritance pattern 4:0). Mating type of the progeny of these crosses, determined by a nuclear gene, segregated 2:2. (From Grant *et al.*, 1980. Reprinted with permission.)

It is interesting to note that inheritance of chloroplast genes, possibly through the involvement of the DNA modification and restriction process, is determined by nuclear (mating type) genes.

VI. Expression of Chloroplast Genetic Information

The genetic information in chloroplast DNA is expressed by the synthesis of appropriate RNA molecules. Chloroplasts contain at least one enzyme that catalyzes synthesis of RNA, a **RNA polymerase.** This enzyme is distinct in its physical properties from the RNA polymerases in the nucleus of the cell. Rifampicin, an antibiotic that inhibits RNA polymerase in bacterial cells, inhibits the enzyme located in the chloroplast but not those in the nucleus of *Chlamydomonas.* A **sigma factor** is present in *Chlamydomonas* that controls the activity of the chloroplast RNA polymerase. This factor also functions with the bacterial RNA polymerase. These results indicate that transcription in the chloroplast has a decidedly prokaryotic nature. The chloroplast enzyme functions in the synthesis of the usual types of RNA—ribosomal, transfer, and messenger RNA—for expression of the chloroplast genome. More is known about the products made by the RNA polymerase than about the enzyme itself.

A. Chloroplast Ribosomes

1. Characteristics

Chloroplast ribosomes can be obtained, after lysis of the purified organelle, by pelleting first the membranes and starch particles in low gravitational fields

TABLE 6.3. DNA Methyltransferase Activities in *Chlamydomonas reinhardtii*[a]

| | Enzyme, units per 10^9 cells | |
Stage in life cycle	Total	200,000-mol.-wt. form
mt⁺ vegetative	3.4	0.6
mt⁺ gametes	3.1	1.4
mt⁻ gametes	1.8	0
mt⁻*mat*-1 gametes	6.7	1.6
Zygote	22.8	22.8

[a]Two forms of the enzyme were observed in these cells, one 60,000 and one 200,000 in molecular weight. The larger form appeared to be the functional form. This activity greatly increased after fusion of gametes to form the zygote; the bulk of the activity would be expected to occur in the chloroplast derived from the *mt*⁺ gamete. Only if the *mt*⁻ cells contained the *mat*-1 mutation, which confers biparental transmission of chloroplast genes to progeny, did significant methyltransferase activity occur in gametes of this strain. Chloroplast DNA also was methylated in this mutant strain. Methyltransferase activity was measured by the transfer of methyl groups from S-adenosyl[*methyl*-³H]methionine to DNA. The enzyme primarily methylated a cytosine residue next to a guanine nucleotide. (Adapted from Sano *et al.*, 1981.)

(40,000g for 30 min) and then the ribosomes at high fields (150,000g for several hours). Alternatively, they can be separated from cytoplasmic ribosomes by sedimenting all ribosomes in a whole-cell extract through a concentration gradient of sucrose (Fig. 6.13). Chloroplast ribosomes exhibit a sedimentation coefficient of 70 S, similar to that of bacterial ribosomes. Because chloroplast ribosomes are smaller than those in the cytoplasm, which have a sedimentation coefficient of 82 S, they sediment more slowly. In this type of analysis, the gradient of sucrose simply prevents mixing of the contents of the tube, which could result from disturbances during acceleration and deceleration of the centrifuge rotor or from thermal gradients generated by high rotor speeds.

The ribosomes, which are specific complexes of RNA and proteins, are composed of two subunits. The association of the subunits of chloroplast ribosomes is more sensitive to the concentration of Mg^{2+} than that of cytoplasmic ribosomes. As a result, chloroplast ribosomes dissociate into subunits when the Mg^{2+} concentration is lowered below 10 mM, whereas cytoplasmic ribosomes dissociate only below 2 mM Mg^{2+}. The subunits of chloroplast ribosomes have sedimentation coefficients of about 50 S and 30–35 S (Fig. 6.14), which correspond to aggregate masses of about 1.7×10^6 and 1.0×10^6 daltons, respectively. The larger subunit from all chloroplasts contains a 23 S and a 5 S RNA. In addition, those from higher plants contain a 4.5 S RNA. The large subunit from the alga *Chlamydomonas reinhardtii*, however, does not contain the 4.5 S RNA but rather 3 S and 7 S RNA molecules, that are tightly associated with the 23 S RNA. A 5.8 S RNA molecule, present in the large subunit of cytoplasmic ribosomes, is not found in the chloroplast counterpart.

The smaller chloroplast ribosomal subunit contains only a 16 S RNA, which is similar in size and extensively homologous to bacterial 16 S ribosomal RNA (see p. 174).

The mature ribosome is roughly 50% protein by weight. The number of proteins in each subunit may vary slightly, depending on the source, but usually

FIG. 6.13. Analysis of ribosomes from *Chlamydomonas reinhardtii* by sucrose gradient sedimentation. The algal cells were broken with a pressure cell and centrifuged at 10,000 g o remove particulate material and membranes. A sample of the supernatant, containing the ribosomes, was layered over an isokinetic sucrose gradient. After centrifugation for 5 hr at 38,000 rpm, the gradient was passed through a monitor with which the absorbance by the RNA at 254 nm was measured. The slow sedimenting species, with a sedimentation coefficient of 70 S, are the chloroplast ribosomes. Cytoplasmic ribosomes have a sedimentation coefficient of 82 S. Material sedimentating more rapidly than 82 S was present as polysomes of cytoplasmic ribosomes. (Redrawn from Hoober and Blobel, 1969.)

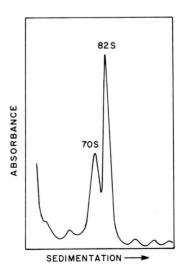

is in the range of 33–35 for the large subunit and 25–31 for the small subunit. These proteins are distinct from those in cytoplasmic ribosomes. Fig. 6.15 shows an analysis of ribosomal proteins in subunits of *Chlamydomonas*. Of the 33 proteins in the large subunit, at least five are made in the chloroplast and the remainder on cytoplasmic ribosomes. Fourteen of the thirty-one small subunit proteins are made within the chloroplast. These proteins made on chloroplast ribosomes are probably coded by chloroplast DNA and make up about one-half of the total proteins shown to be synthesized within the organelle.

2. Ribosomal RNA

The RNA molecules in chloroplast ribosomes have been analyzed by sedimentation in sucrose gradients and, more preferably, by electrophoretic migration through polyacrylamide gels (Fig. 6.14). The sedimentation coefficients of the two major RNA molecules, 23 S and 16 S, correspond to molecular weights of 1.05×10^6 and 0.56×10^6, respectively.

These ribosomal RNA species were the first to be identified as products of transcription on chloroplast DNA, through the technique of **hybridization.** For this procedure, the DNA and the ribosomal RNA were obtained in highly purified form. The RNA was also made radioactive by culturing cells, before isolating the RNA, in a medium containing inorganic ^{32}P-phosphate salts. The DNA was denatured, fixed onto nitrocellulose filters (which bind single-

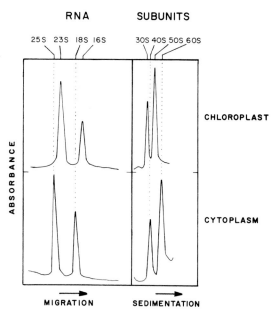

FIG. 6.14. Left: Analysis of the major RNA species of chloroplast and cytoplasmic ribosomes by electrophoresis through a polyacrylamide gel. The rate of migration through the porous gel in response to the electric field is proportional to the size of the RNA, with the smaller molecules able to permeate the gel more rapidly than larger molecules. Chloroplast ribosomes contain 23 S and 16 S RNA molecules, whose size is similar to the RNA in bacterial ribosomes. Plant cytoplasmic ribosomes contain 25 S and 18 S RNA species. The size of the 18 S RNA is similar to that in animal cytoplasmic ribosomes, but the larger RNA in the animal ribosomes is 28 S. Right: Analysis of the subunits of chloroplast and cytoplasmic ribosomes by sedimentation through a gradient of sucrose during centrifugation in an ultracentrifuge. The sedimentation coefficient of the larger subunit is 50 S, and that of the smaller subunit is 30–34 S, depending on conditions. Subunits of cytoplasmic ribosomes have sedimentation coefficients of 60 S and 40 S. Chloroplast and cytoplasmic ribosomal subunits are similar in size to bacterial and animal counterparts, respectively. (Adapted from Brügger and Boschetti, 1975.)

stranded DNA), and then incubated with the single-stranded RNA. If the RNA was a complement of a sequence on one of the strands of the DNA, and base-paired perfectly to give a hybrid RNA:DNA duplex, the RNA became resistant to ribonucleases specific for single-stranded RNA. The extent of hybridization was, therefore, measured by determining how much ribonuclease-resistant, radioactive RNA was retained on the filter.

The hybridization procedure also can be done in solution, in which case the RNA and DNA are mixed at elevated temperatures and slowly cooled. Hybrid molecules then are isolated by chromatography on a column of hydroxylapatite. Truly complementary RNA:DNA hybrids will show a sharp increase in absorbance over a narrow range of temperature as the structures denature or "melt," in a manner similar to the denaturation of double-stranded DNA (see Fig. 6.4).

Fig. 6.16 shows an example of the hybridization of ribosomal RNA with chloroplast DNA. Both the 23 S and 16 S RNA species hybridized to filter-bound DNA in an additive manner, and together they were complementary to 3.8–4.0% of the DNA at the saturation level. From the molecular weight values for the RNA and DNA, a simple calculation based on this extent of hybridization, which corresponded to about 8% of the coding capacity of the DNA, indicated that each chloroplast genome contains two genes for each of the ribosomal RNA molecules (see also Fig. 6.6).

Hybridization of ribosomal RNA to fragments of DNA, produced by digestion with restriction enzymes, was the first step toward converting the physical map of chloroplast DNA (Fig. 6.5) into a genetic map. After separation by electrophoresis through an agarose gel, the fragments of DNA were transferred to nitrocellulose or derivatized paper filters by transverse elution. Then by hybridization of ^{32}P-labeled RNA to the fragments of DNA, it was determined which of the fragments contain nucleotide sequences complementary to a given species of RNA. Since the linear order of the restriction fragments of chloroplast DNA from several plants is known, this technique permitted determination of the position on the DNA molecule of the gene for that RNA. This type of experiment is illustrated in Fig. 6.17.

The genes for ribosomal RNA molecules lie on two separate coding regions in chloroplast DNA from most higher plants and green algae. The positions of these regions are particularly interesting. As shown in Fig. 6.6, the two regions are nearly opposite from each other on the physical map for *Chlamydomonas* DNA but are somewhat less distant in the DNA from maize and wheat. Moreover, the ribosomal RNA genes are in segments that occur as inverted repeats. The order of the genes in these segments is inverted with respect to the map but is identical within each region. The arrows over these regions, in the physical maps shown in Fig. 6.6, show the direction of transcription of the polycistronic sequence, which is also the same for each.

Not all chloroplast DNA molecules have the same general pattern for ribosomal RNA genes. As shown in Fig. 6.6, the regions coding for ribosomal RNA in *Euglena* chloroplast DNA are positioned differently than on the other three. In *Euglena*, there are three sets of ribosomal RNA genes that occur as tandom, rather than as separate inverted, repeats. Furthermore, not all chloroplast DNA molecules contain multiple genes for ribosomal RNA. For example, in the

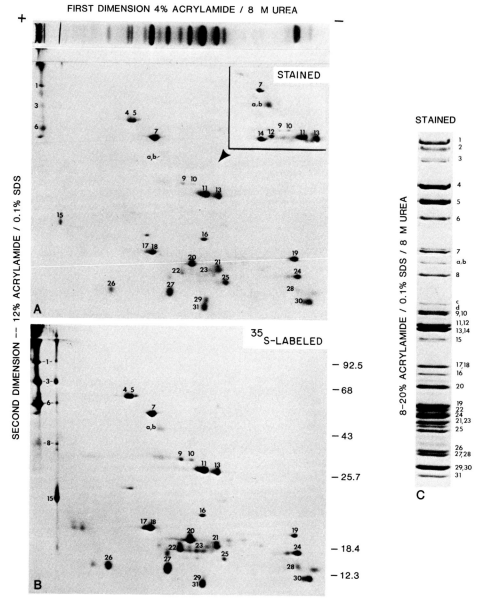

FIG. 6.15. Two-dimensional gel electrophoresis of proteins from chloroplast ribosomal subunits of *Chlamydomonas reinhardtii*. This page shows the pattern for the small (34 S) subunit, whereas page 169 shows the pattern for the large (50 S) subunit. Separation in the first dimension was based on differences in net positive charge as the proteins were subjected to electrophoresis at pH 5.0 in a polyacrylamide gel sufficiently porous to minimize molecular sieving. Electrophoresis in polyacrylamide gels in the presence of sodium dodecyl sulfate (SDS) separated proteins in the second dimension on the basis of size. In each panel, the upper pattern (A) represents gels stained with the dye Coomassie blue to reveal the proteins. A stained first-dimensional gel is shown across the top of the pattern for reference. The positions of the proteins were also revealed by fluorography, in which radioactivity in the proteins was detected with X-ray film. To make the proteins radio-

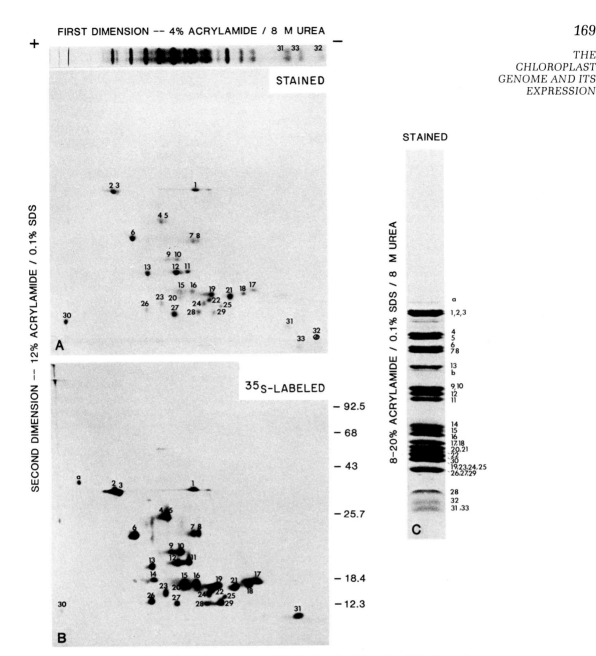

active, cells were grown for 48 hr in the presence of $^{35}SO_4^{2-}$. On the right side of B is the scale for apparent molecular weight $\times 10^{-3}$. A one-dimensional separation in sodium dodecyl sulfate (SDS) is shown in C, which provides the numbering system used to identify each of the ribosomal proteins. A total of 31 and 33 proteins were detected in the small and large subunits, respectively. Several proteins recovered in each subunit appeared to be contaminants and are indicated by letters a, b, c, and d. Proteins 9, 10, and 15–18 in the small subunit are removed by washing with high salt and thus may not be integral ribosomal proteins. (From Schmidt *et al.*, 1983. Reprinted with permission of the Rockefeller University Press.)

broad bean (*Vicia faba*) and pea plants, the chloroplast DNA has only a single gene for each of these RNA molecules.

Transcription of the ribosomal RNA-coding regions occurs in the usual 5′ → 3′ direction, with respect to the RNA product, and produces a single, polycistronic RNA molecule with a molecular weight of about 2.7×10^6. This large precursor RNA species is the major product of RNA synthesis in isolated chloroplasts (Fig. 6.18). The sequences of the 16 S, 23 S, and 5 S RNA molecules are all included within this single molecule, in the order shown in Fig. 6.19, with the 16 S RNA near the 5′ end. Nucleotide sequences of several transfer RNA species also occur within this large precursor. Specific ribonucleases apparently are involved in processing this molecule to the mature RNA species that function in protein synthesis.

An additional interesting feature of the ribosomal genes in *Chlamydomonas* chloroplast DNA is the presence of an **intervening segment** that interrupts the coding sequence of the 23 S RNA gene. Such intervening segments (**introns**) have been found in the primary RNA transcipts of many eukaryotic nuclear genes. These segments are included in the initial transcription product, but during processing of the RNA these segments are cleaved out. The 23 S RNA gene in *Chlamydomonas* contains an intervening segment 870 base-pairs long, which is located near the 3′ end (Fig. 6.19). Specific base sequences in the RNA flank the sites where excision of the intervening sequence occurs, and these sequences may act as signals for the processing enzymes. The remaining pieces are then joined to form the mature 23 S RNA.

The sequences of the 16 S ribosomal RNA genes in maize, tobacco, and *Chlamydomonas* have been determined (see Fig. 6.20). The higher plant sequences are highly homologous with each other (95%) and with the sequence of the 16 S ribosomal RNA from the bacterium *Escherichia coli* (75%). The algal 16 S RNA, 1475 nucleotides in length, is 85% homologous with the higher-plant 16 S RNA and 76% homologous with the bacterial RNA.

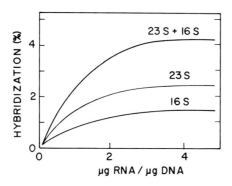

FIG. 6.16. Hybridization of chloroplast ribosomal RNA with chloroplast DNA. Chloroplast DNA was purified from leaves, denatured, and bound to nitrocellulose filters. ^{32}P-labeled ribosomes were isolated from leaves incubated with [^{32}P]orthophosphate, and the large and small ribosomal RNA species were purified by sucrose gradient centrifugation. The level of RNA hybridized to DNA was determined by the amount of radioactivity remaining on the filter after a brief treatment with ribonuclease. From the specific radioactivity (counts/min/μg RNA) it was possible to calculate how much of the DNA was involved in the hybridization reaction. At saturation, 1.4% and 2.5% of the DNA was hybridized to the 16 S and 23 S RNA, respectively. Since the molecular mass of the chloroplast DNA is about 90×10^6 daltons, the hybridization involves 1.26×10^6 and 2.25×10^6 daltons, respectively, for the two ribosomal RNA species. The masses of the 16 S and 23 S RNA are 0.56×10^6 and 1.05×10^6 daltons, and therefore the hybridization data indicate there are two genes for each type of RNA in the DNA molecule. (Adapted from Thomas and Tewari, 1974.)

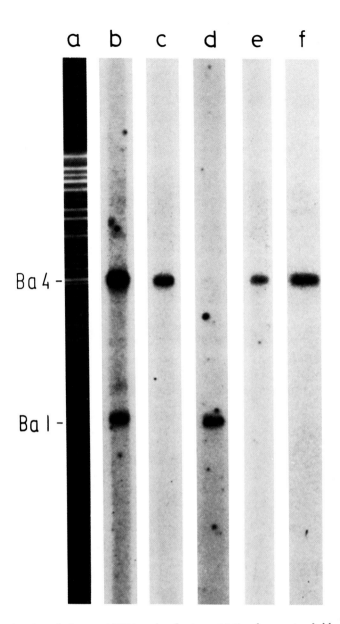

FIG. 6.17. Hybridization of ribosomal RNA molecules to restriction fragments of chloroplast DNA from *Chlamydomonas reinhardtii*. A BamHI endonuclease digest of chloroplast DNA was separated into individual fragments by electrophoresis through a 0.8% agarose gel. After electrophoresis, the DNA fragments were transferred to a sheet of nitrocellulose and incubated with ³²P-labeled ribosomal RNA. Lane a shows the pattern of DNA fragments after electrophoresis. An autoradiogram, which detected positions of bound 23 S (b), 16 S (c), 5 S (d), 3 S (e), and 7 S (f) ribosomal RNA, is shown at the right. A detailed analysis of the hybridization to DNA fragments allowed the positioning of these genes as shown in Fig. 6.19. (From Rochaix and Malnoe, 1978. Copyright, M.I.T. Reprinted with permission.)

B. Transfer RNA

Chloroplast transfer RNA (tRNA) molecules can be separated from the corresponding isoaccepting tRNA species present in the cytoplasmic matrix by chromotographic procedures, which indicates structural differences between tRNA species from the two compartments. Transfer RNA molecules from the chloroplast are active in the synthesis of amino acyl–tRNA complexes only when incubated with the appropriate enzyme from the chloroplast. Furthermore, the amino acyl–tRNA synthetases from the cytoplasm generally are not active with chloroplast tRNA but are, of course, active with the isoaccepting species from the cytoplasmic matrix. Thus, there are two highly specific but separate systems for the synthesis of amino acyl–tRNA complexes for protein synthesis in plant cells.

Transfer RNA molecules are generally 70 to 80 nucleotides in length. The function of these small RNA species is to base-pair, through three nucleotides approximately halfway through their sequence, with a specific code word or **codon** of three nucleotides on messenger RNA (mRNA) and, at the same time, carry at the 3' end an amino acid corresponding to the codon. The ribosome holds the mRNA and two tRNA molecules, one bearing the growing polypeptide chain and the other the entering amino acid (see Section VIII), in proper orientation for the next peptide bond to form. For most amino acids, more than one form of tRNA is present in the chloroplast. For example, three species of tRNA for leucine, serine, and methionine were found in spinach chloroplasts. Since extensive posttranscriptional modification of tRNA molecules occurs, it

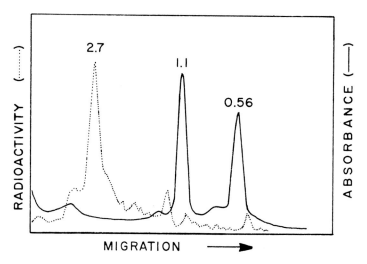

FIG. 6.18. Analysis of the initial products of RNA synthesis in isolated chloroplasts. Purified chloroplasts were incubated for 20 min with [³H]uridine at 20°C. Illumination of the chloroplasts was required for synthesis of RNA. Nucleic acids were extracted and resolved by electrophoresis through a polyacrylamide gel. The solid line (absorbance at 260 nm) shows the position of the mature 23 S (1.1×10^6 daltons) and the 16 S (0.56×10^6 daltons) ribosomal RNA species. The major product synthesized in this system, indicated by the dotted line (radioactivity), was 2.7×10^6 daltons in size. This RNA contained the sequences of both ribosomal RNA molecules, as shown by hybridization studies. (Adapted from Hartley and Ellis, 1973, and Hartley and Head, 1979.)

is not clear as yet whether all of these tRNA species are coded by different genes.

According to the "universal" genetic code, 61 different sets of three nucleotides (triplets) code for amino acids, whereas three code for termination signals during protein synthesis. A comparison of the nucleotide sequence of the gene for the large subunit of ribulose 1,5-bisphosphate carboxylase with the amino acid sequence of this polypeptide (see Fig. 6.24) demonstrated that the "universal" genetic code is used within the chloroplast. (The mitochondrial genetic system employs a simplified code and is therefore a notable exception to the "universal" code.) However, even in this large polypeptide not all possible codons

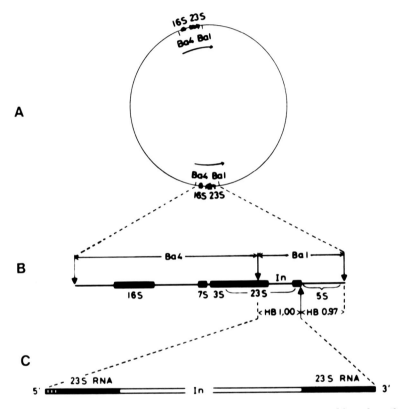

FIG. 6.19. Location and fine-structure map of the intervening sequence in a chloroplast ribosomal RNA gene in *Chlamydomonas reinhardtii*. A: The circle represents the chloroplast genome of *C. reinhardtii* (see Fig. 6.6). The arrow indicates the direction of transcription in the two sets of ribosomal RNA genes. Since these inverted-repeat regions are identical, the sense strand (template strand) in the one region is on the opposite strand from that in the other region. The two sets of 16 S and 23 S RNA genes are contained within the Ba1 and Ba4 fragments generated by restriction enzyme BamHI (see Fig. 6.17). B: An enlargement of the Ba1 and Ba4 fragments shows the location of the ribosomal RNA genes and the intervening sequence or intron (In). Cleavage of the BamHI fragment Ba1 with restriction enzyme HindIII generated two subfragments, HB1.00 and HB0.97. Fragment HB1.00, shown in C, contains the intervening sequence (white) within the coding region of the 23 S RNA gene (black). (From Allet and Rochaix, 1979. Copyright, M.I.T. Reprinted with permission.) Much of the ribosomal RNA genes and surrounding regions have been sequenced (Dron *et al.*, 1982a; Rochaix, and Darlix, 1982.)

are used. Only 40 are used to synthesize this polypeptide in the alga *Chlamydomonas reinhardtii*, whereas 55 and 58 codons are used in spinach and maize, respectively. The relatively low G + C content of chloroplast DNA (Table 6.1) biases the codons in favor of those with A or U in the third, less specific or "wobble" position.

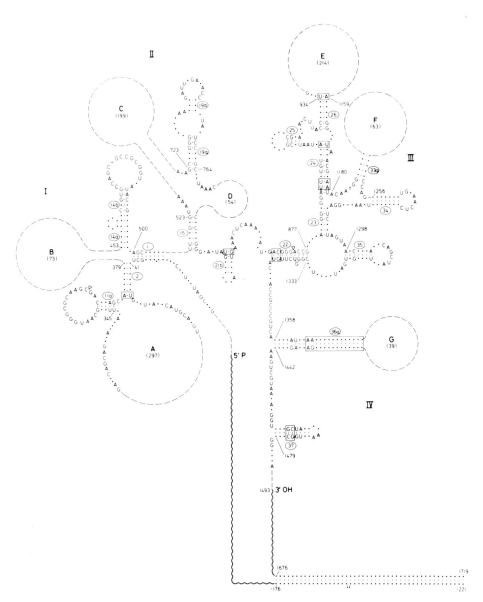

FIG. 6.20. The proposed secondary structure of the 16 S ribosomal RNA of *Chlamydomonas reinhardtii*. The RNA consists of 1475 nucleotides, and the sequence is 85% and 76% homologous with 16 S RNA from tobacco (1486 nucleotides) and *Escherichia coli* (1542 nucleotides), respectively. (From Dron *et al.*, 1982a. This article describes the complete sequence of the RNA. Reprinted with permission.)

Hybridization of tRNA molecules with fragments of DNA generated by restriction enzymes showed that tRNA genes are dispersed around the chloroplast genome. A number of genes for specific tRNA molecules have been located on the physical map of chloroplast DNA, and the sequences of several have been determined. Two tRNA genes are located in the spacer region between the 16 S and 23 S ribosomal RNA genes in chloroplast DNA of maize (Zea mays). Kössel and his colleagues (Koch et al., 1981) determined the base sequence of the entire region and have shown that, as in the prokaryotic cell Escherichia coli, the order of the genes is 5′. . . .16 S—tRNA$_2^{Ile}$—tRNA^{Ala}—23 S . . .3′. Chloroplast DNA from the alga Euglena has the same gene order. Also, the base sequences of the coding segment of chloroplast tRNA genes are very similar to the analogous prokaryotic genes. But in contrast to the prokaryotic genes, these two tRNA genes in Zea mays chloroplast DNA contain large intervening sequences (introns) that interrupt the coding sequences precisely in the middle. The introns are 949 and 806 base-pairs long in the genes for tRNA$_2^{Ile}$ and tRNA^{Ala}, respectively, which themselves are only 72 and 73 nucleotides in length. Since the tRNA sequences are part of the large, polycistronic RNA molecule transcribed from the ribosomal operons (see Fig. 6.19), presumably they are cleaved out and processed by chloroplast ribonucleases along with the ribosomal RNA molecules.

The base sequence for a tRNA^{Val} lies 231 base pairs upstream from the 5′ terminus of the coding sequence for the 16 S ribosomal RNA on the Zea mays chloroplast genome. In contrast to the two tRNA genes mentioned in the last paragraph, no introns exist in this tRNA^{Val} gene. However, a large intron (571 base pairs) exists in the gene for a tRNA^{Val} in tobacco chloroplast DNA (Fig. 6.21). The base sequences of both of the mature chloroplast tRNA^{Val} species are more homologous to their analogous prokaryotic than the eukaryotic counterparts.

A tRNA gene for methionine (tRNA^{Met}) is located near the gene for the ε subunit of coupling factor 1 (see Fig. 6.6) and has been sequenced (Fig. 6.22). This gene, which codes for the form of tRNA^{Met} used during elongation of protein synthesis (see p. 190), does not contain an intron. Fig. 6.22 also shows the sequence of mature forms of tRNA^{Met}, which illustrates the extensive modification of bases that occurs after the initial RNA transcript has been produced. In each case examined so far, the 3′ terminal sequence of -C-C-A, which exists on all mature tRNA molecules, is not present in the gene sequences and consequently is added after transcription and processing of the RNA have occurred.

A particularly surprising reaction was found by Kashdan and Dudock (1982) to occur during processing of a tRNA for isoleucine. Identical copies of the gene for tRNA$_1^{Ile}$ are located near each of the 5′ ends of the inverted repeats in the spinach chloroplast genome. The sequences for the anticodon region of this gene, which does not contain an intron, is

$$5′ . . . C - T - C - A - T - A - A . . . 3′ \quad \text{(noncoding strand)}$$
$$3′ . . . G - A - G - T - A - T - T . . . 5′ \quad \text{(coding strand)}$$

The anticodon in tRNA should, therefore, occur as

5′ . . . C-U-C-A-U-A-A . . . 3′ (anticodon underlined)

and should base-pair, in opposite polarity, with the codon A-U-G (5′ → 3′) in mRNA, a codon for methionine. However, in the mature tRNA$_1^{Ile}$, the cytidine in the third, "wobble" position at the 5′ end of the anticodon is modified such that the tRNA base-pairs instead with the codon for isoleucine, A-U-A (5′ → 3′). Only isoleucine becomes attached to this tRNA molecule, which shows that the amino acyl–tRNA synthetases recognize this tRNA as one for isoleucine rather than methionine.

The number of genes for tRNA molecules for which the sequences are known is rapidly growing. These few examples illustrate that some contain introns and others do not. Whether any general features exist in this family of genes will become apparent as additional sequences become known. However, several characteristics appear to be emerging in the mature tRNA molecules.

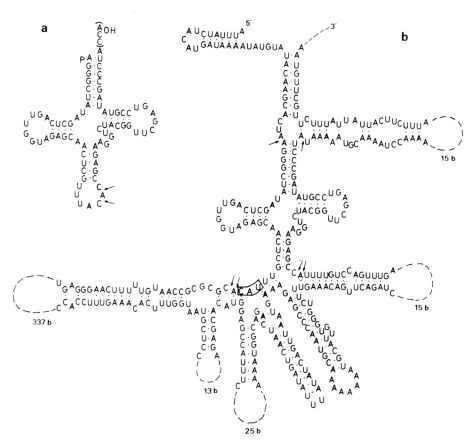

FIG. 6.21. The unmodified sequence of (a) tRNA^{Val} as deduced from (b) the putative RNA precursor, which is transcribed from the tRNA^{Val} gene to tobacco chloroplast DNA. The sequence of the precursor was determined from the DNA sequence. The arrows indicate where the large intron is cleaved out and the 3′ and 5′ extensions of the precursor are removed. The anticodon sequence (UAC) is boxed. (From Deno *et al.*, 1982. Reprinted with permission.)

Common structures found thus far are the methylated ribose in the sequence

177

*THE
CHLOROPLAST
GENOME AND ITS
EXPRESSION*

Common structures found thus far are the methylated ribose in the sequence Gm-G within the dihydrouridine loop (loop I) and the sequence T-ψ-C-A in the pseudouridine loop (loop IV). In general, chloroplast tRNA molecules are significantly more homologous with prokaryotic than eukaryotic isoaccepting species.

C. Messenger RNA

A number of mutant strains, in which the marker trait is inherited uniparentally from the maternal (mt^+) parent, have been isolated that are deficient in photosynthesis and cannot grow on CO_2 as a carbon source. The selection marker for such nonphotosynthetic strains of *Chlamydomonas reinhardtii* is the requirement for acetate as a carbon source for growth or an increase in fluorescence of chlorophyll in photosynthetically incompetent cells. Other mutant strains have alterations in chloroplast ribosomes, which are expressed as resistance to or dependency on antibiotics that normally inhibit chloroplast protein synthesis in wild-type cells. The expression of antibiotic resistance in some cases is accompanied by an impairment in assembly of one of the ribosomal subunits. Since chloroplast DNA codes for ribosomal RNA, it must be considered that some of these phenotypic traits may result from mutations in ribosomal RNA genes. Indeed, genetic mapping of assembly-deficient mutations has shown that they are located in the region of the chloroplast genome that contains the ribosomal RNA genes. But other uniparentally inherited mutations were found that caused alterations in several ribosomal proteins. These structural changes in ribosomal proteins provided the first evidence that chloroplast DNA may contain genes for chloroplast proteins.

Most investigators have concluded that the gene for any protein synthesized on chloroplast ribosomes is contained in chloroplast DNA. Direct evidence that chloroplast DNA contains the information for a sizable number of polypeptides was obtained by the following experiment. Bottomley and Whitfeld (1979) prepared extracts of *Escherichia coli* cells that contained the complete enzymatic activities for both transcription and translation of RNA. The extracts were depleted of endogenous protein synthetic activity by incubation with micrococcal nuclease. Since this nuclease requires Ca^{2+} for activity, the nuclease activity was then abolished by addition of a chelator of Ca^{2+}. When chloroplast DNA subsequently was added to these extracts, the endogenous bacterial RNA polymerase transcribed the DNA into mRNA products, which then were translated into polypeptides on the bacterial ribosomes. Fig. 6.23 shows that a spectrum of discrete polypeptides was synthesized by this system. It is interesting to note that chloroplast DNA from different plants produced somewhat different sets of polypeptides.

Among the known genes that are contained within the chloroplast genome are several that code for thylakoid membrane polypeptides. The major membrane polypeptide synthesized in the chloroplast has a molecular weight of 32,000. This polypeptide is not made by etioplasts, and its mRNA is present only in chloroplasts obtained from illuminated plant cells. *In vitro* translation of the RNA extracted from these chloroplasts provided a product about 35,000 in

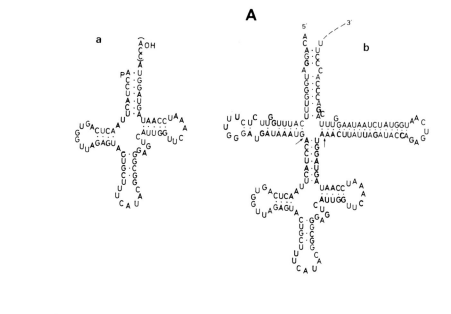

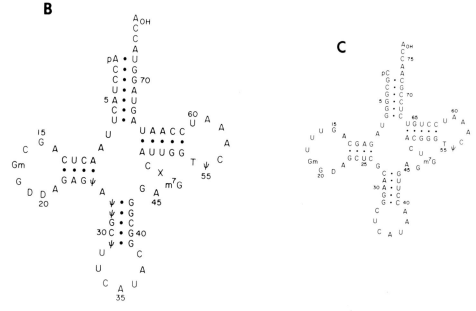

FIG. 6.22. Sequences of chloroplast tRNA^{Met}. A: The unmodified sequence of tRNA$_m^{Met}$ determined from the DNA sequence in the tRNA^{Met} gene in tobacco chloroplast DNA is shown in (a). On the right (b) is the sequence of the putative initial transcript of the gene, the tRNA^{Met} precursor, which does not contain an intron. The arrows indicate where the 3' and 5' extensions of the precursor are removed. The sequence of the tRNA corresponds to that used in elongation steps in protein synthesis. (From Deno *et al.*, 1982. Reprinted with permission.) B: The nucleotide sequence of the mature form of the elongator tRNA^{Met} (tRNA$_m^{Met}$) from spinach chloroplasts. The gene for the tobacco chloroplast tRNA$_m^{Met}$ differs at only two positions (16 and 27) from the coding sequence in spinach. The mature tRNA$_m^{Met}$ illustrates the extensive posttranscriptional modifications that occur

molecular weight. This product was shown to be a precursor of the 32,000-molecular-weight protein made in intact chloroplasts, which indicated that posttranslational processing of the chloroplast protein had occurred. Hybridization of the mRNA to restriction fragments of chloroplast DNA has indicated the approximate location of this gene. The position of this gene, termed **photogene** because light is required for its transcription, is shown on the physical map of maize chloroplast DNA in Fig. 6.6. This 32,000-molecular-weight protein appears to be the same as that found on the reducing side of photosystem 2 (B protein) at the site where CO_2 and many herbicides bind (see p. 89).

The most prominent mRNA extracted from chloroplasts is that for the large subunit of ribulose 1,5-bisphosphate carboxylase. Since this polypeptide is the major product of protein synthesis in chloroplasts, as determined *in vivo* with antibiotics that selectively inhibit chloroplast ribosomes and *in vitro* with isolated chloroplasts, the source of its mRNA was of considerable interest. Edelman and his colleagues found that the mRNA for this polypeptide did not contain a polyadenylate sequence at its 3′ end, in contrast to most mRNA molecules in the cytoplasm (Sagher *et al.*, 1976). The partially purified RNA hybridized to chloroplast DNA, which suggested that the gene for this polypeptide might

D

during processing. The 3′ terminal sequence -C-C-A is added after the tRNA is cleaved from the precursor molecule. Gm = 2′-0-methyl guanosine (the methyl group is on the 2′-hydroxyl of ribose); see part D for structures of other modified bases and their nomenclature. (From Pirtle *et al.*, 1981. Reprinted with premission.) C: The nucleotide sequence of spinach chloroplast methionine initiator tRNA^{Met} (tRNA$_f^{Met}$). Sequences of tRNA$_f^{Met}$ from chloroplasts of several different plants are very similar. Spinach tRNA$_f^{Met}$ is more homologous with prokaryotic tRNA (81–84%) than with the eukaryotic initiator tRNA$_i^{Met}$ (64–69%). (From Calagan *et al.*, 1980. Reprinted with permission.) D: Structures of several modified bases present in chloroplast tRNA molecules.

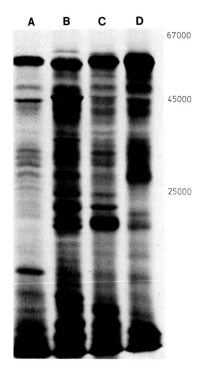

FIG. 6.23. Patterns of polypeptides synthesized in a transcription/translation system prepared from the bacterium *Escherichia coli* and programmed with chloroplast DNA from (A) tobacco, (B) *Oenothera hookeri,* and (C) spinach. In D, spinach chloroplast RNA was added in place of DNA. The polypeptides were labeled by the incorporation of [^{35}S]methionine during synthesis and detected after gel electrophoresis by radioautography with X-ray film. The numbers on the right indicate the molecular weight of polypeptides at those positions. The major band, molecular weight 55,000, is the large subunit of ribulose 1,5-bisphosphate carboxylase. (From Bottomley and Whitfeld, 1979. Reprinted with permission.)

reside within the chloroplast. The location of the gene for the carboxylase large subunit was confirmed by hybridization of the mRNA with fragments of chloroplast DNA generated by restriction endonucleases. Since the arrangement of the restriction fragments is known, as shown by the physical map in Fig. 6.5, this technique permitted determination of the position of the gene.

Another approach to identifying the gene for the carboxylase large subunit was used by Bogorad and his colleagues (Bedbrook *et al.,* 1979), who examined products of a system in which transcription of a fragment of DNA was coupled to translation into a polypeptide. They inserted restriction fragments of chloroplast DNA into small, circular bacterial plasmid DNA molecules *in vitro,* and then introduced the recombinant plasmids back into bacterial cells. The bacterial cells were spread on nutrient agar, and each transformed cell was allowed to grow into a colony. As the bacterial cells grew and divided, the plasmid containing the restriction fragment of chloroplast DNA also was replicated and thus amplified. By extraction and purification of the plasmid from a large batch of cells, a workable quantity of the chloroplast DNA fragment was obtained. Transcription of the DNA to produce mRNA was accomplished *in vitro* with RNA polymerase, as in Fig. 6.23, and the RNA subsequently was translated into protein. An antibody, made by rabbits that were injected with purified large subunits of the carboxylase, was then added to the assay system. A specific antigen–antibody reaction occurred only when the assay system contained a DNA fragment that included the gene for the carboxylase subunit. Again, since the position of the fragment on the physical map of chloroplast DNA was known, the approximate position of the gene for this polypeptide was established.

The location of the single gene for the large subunit of the carboxylase in chloroplast DNA molecules from several plant cells is shown in Fig. 6.6. The position of this gene, with respect to the inverted repeat regions, is similar in many higher plant species. The base sequences of the gene from several plants are shown in Fig. 6.24. No intervening sequences occur in this gene in maize, spinach, or *Chlamydomonas,* but evidence exists for an interruption in the coding sequence in the alga *Euglena.* As a result of this work, the precise position of the gene has been defined unequivocally, without doing a genetic experiment. The amino acid sequence of the polypeptide can also be deduced simply by reading the code in the DNA, thereby obviating the laborious procedures required to obtain this type of information only a few years ago.

A direct correlation between the physical map and a genetic locus has been obtained by Mets and Rochaix and their colleagues (Dron *et al.,* 1983). A uniparentally inherited mutation was discovered in *Chlamydomonas reinhardtii* that affected the structure and activity of the large subunit of the carboxylase. Assuming that a mutation that destroys activity is probably located near the active site of the enzyme, DNA sequences coding for polypeptide segments that make up the active site of the carboxylase were determined in wild-type (wt) and mutant (10-6c) cells. The following result was obtained:

	1 7 0						1 7 5
	l e u	g	l y	c y s	t h r	i l e	l y s
w t	T T A	G	G T	T G T	A C A	A T C	A A A
10-6 c	T T A	G	A T	T G T	A C A	A C T	A A A
		a	s p				

The change involved at G-C to A-T base-pair transition, so that an aspartic acid residue was inserted in the mutant instead of a glycine. This introduction of a negative charge near the essential lysine at position 175 (see Fig. 6.24) drastically alters the activity of the enzyme. Therefore, this work directly connects the uniparental genetic linkage group with chloroplast DNA.

The nucleotide sequences of the large subunit genes are highly homologous between different plant families. The sequences in maize and spinach show a 16% divergence, whereas that in *Chlamydomonas* differs from maize and spinach sequences by 25 and 23%, respectively. However, the amino acid sequences diverge by only 13 to 14%. In particular, the amino acid sequences within the three segments that comprise the active site (boxed areas A1, A2, and A3 in Fig. 6.24) and the CO_2-activation region around lysine 201 (boxed area C; see p. 119 for the activation reaction) are highly conserved.

The continuation of this approach will transform the physical map into a detailed genetic map. The advantage of this approach is that even genetic loci whose mutant forms are difficult to study genetically can be placed with precision on a physical map. The value is particularly relevant to chloroplast DNA, since the multiple copies of DNA within each chloroplast tend to mask the occurrence of mutations.

As the location and nucleotide sequences of other genes are determined, details of the amino acid sequences of proteins coded by chloroplast DNA will become known. Comparisons of amino acid sequences deduced from the nucleotide sequence with chemically determined sequences will provide, in addition, important information regarding posttranslational processing of proteins synthesized in chloroplasts. For example, the amino acid at the N-terminus of the large subunit of the carboxylase purified from barley is alanine. As illustrated in Fig. 6.24, cleavage of the lysine–alanine bond between positions 14 and 15 would remove an N-terminal fragment, exposing alanine as the terminal amino acid in the mature polypeptide. Whether this modification occurs during purification is not certain. In other plants the N-terminal amino group seems to be blocked.

Portions of two other genes that flank the large subunit gene have been sequenced, as shown in Fig. 6.24, but the identities of the protein products are not known. Interestingly, these other genes are oriented with opposite polarity from the large subunit genes, which means that they are coded by the other strand of the DNA.

As described in Chapter 5 (p. 125), C_4 plants contain two cell types that have different chloroplast properties. Although bundle sheath cells contain ribulose 1,5-bisphosphate carboxylase and the normal CO_2-fixation pathway, the mesophyll cells lack this enzyme and fix CO_2 by phosphoenol pyruvate carboxylase. Link *et al.* (1978) found that the absence of ribulose 1,5-bisphosphate carboxylase in mesophyll cells is not caused by a change in chloroplast DNA. Rather, expression of the gene for the carboxylase large subunit seems to be repressed in these cells. The mechanism of this repression presents an intriguing question for future research.

VII. Why Chloroplast DNA?

The data described in the last section provide ample evidence that mRNA molecules for both membrane and soluble (stroma) proteins are transcribed from chloroplast DNA. The question can be asked why a highly refined system for genetic expression has been localized to the chloroplast, even through the

---

FIG. 6.24. The nucleotide sequence of a segment of chloroplast DNA from *Chlamydomonas reinhardtii* that includes the gene for the large subunit of ribulose 1,5-bisphosphate carboxylase (LS gene) and portions of two neighboring genes. The 5′ portion of gene X ranges from nucleotide 261 to 1. The upper and lower lines correspond to the coding and noncoding strands of X, respectively. Gene X possibly codes for the β subunit of CF_1. The 5′ and 3′ ends of the transcripts are marked by two vertical double filled arrows, and a horizontal open arrow near the 5′ ends indicates the direction of transcription, from right to left. The coding region of the LS gene starts at 1191. The gene sequence (noncoding strand) of *C. reinhardtii* is compared with those of spinach and maize (top to bottom). Dots indicate identical nucleotides. The sequence of the mRNA is obtained by substituting U for T. The corresponding amino acid sequences are shown below in the same order. The active sites A1, A2, and A3 and the CO_2 activator region C are boxed. The lysine residue at position 175 involved in catalysis is indicated in A1. Inverted repeats in the 3′ untranslated region of the LS and Y gene transcripts are indicated by filled arrows. The 3′ end of gene Y is at 3306. The upper and lower lines correspond to the coding and noncoding strands of Y, respectively. The codon usage in these genes is typical. Translation of the LS mRNA from each organism is initiated at an A-U-G codon with N-formyl methionine and terminates with U-A-A or U-A-G, resulting in a polypeptide 475 amino acids long. No introns occur in these genes. Recognition sites for restriction endonucleases, which were used to fragment the DNA for determination of the sequence, are shown above the nucleotide sequence. (From Dron *et al.*, 1982b. Reprinted with permission.)

```
                              HPH I INV                                      .60                              .90
              .30
CAA ACC GTC AAC TAC ACG ACC TAA GTA TGC TTC ACC TAC AGG AAT TTC AGC GAT TTT ACC AGT ACA ACG AAC ACG GCT ACC TTC AGT AAT
GTT TGG CAG TTG ATG TGC TGG ATT CAT ACG AAG TGG ATG TCC TTA AAG TCG CTA AAA TGG TCA TGT TGC CGA TGG AAG TCA TTA
LEU GLY ASP VAL ARG GLY LEU TYR ALA GLU GLY VAL PRO ILE GLU ALA ILE LYS GLY THR CYS ARG VAL ARG SER GLY GLU THR ILE
              HPH I INV                              .120                                          TAQ I  ECOR I     .180
TTT TAA ACC ATC ACC TAA TAA TAC CGC ACC TAC GTT GTT TGC TTC TAA GTT AAG TGC ATA AAT ACC TAA AGT ACC ATC TTC GAA TTC AAG TAA
AAA ATT TGG TAG TGG ATT ATT ATG GCG TGG ATG CAA CAA ACG AAG ATT CAA TTC ACG TTA TGG ATT CA TGG TAG AAG CTT AAG TTC ATT
LYS LEU GLY ASP GLY LEU LEU VAL ALA GLY VAL ASN ASN ALA GLU LEU ALA ILE GLY LEU THR GLY ASP GLU PHE GLU LEU LEU
HPH I INV                                                .210            HPH I INV         .240
TTC ACC TGA CAT TGC TTT TTC TAA ACC ATA AAT ACG AGC AAT ACC GTC TAC TTG GAA AAC GAT ACC GAA ATC TAC CAT TTTCACTTCTG
AAG TGG ACT GTA ACG AAA AAG ATT TGG TAT TTA TGC TCG TTA TGG CAG TGG ATG AAC CTT TTG CTA TGG CTT TAG ATG GTA
GLU GLY SER MET ALA LYS GLU LEU GLY TYR ILE ARG ALA ILE GLY ASP GLY VAL GLN PHE VAL ILE GLY PHE ASP VAL MET
                                                                                                    TAQ I
              .300                                    .350
GAGTGTATTGTTCAATTAAATCTTTAATAAGATTACTAAGTTCTTCTGGAGTACGCATTGCCATAAAAAGAAAAAATAAATAAAAGATTAAAAAAAGTTTATTTTAAAATCTTTCTCG
              .400                                    .450                                    .500
AGAATTTTAAATAAGTTTAAAATTCAACAAAAATAGTGAGTGGTAAGACTCACTTGTTAACAAAAGTAATGGTTCACCCTTGTCATATTTAAATACTAAAATTCATTTGCCCGAAGAGGA
              .550                                          .600
CAAATTTATTTATTGCATTAAAATCCCTAAGTTTACTTGCCCGTAAGGGGAGGGGGGGGCGTCCACAGGCGTCGTAAGCAACTAAAGTTTATGACGCCGATTGCTTTGTTAGGAAAAT
              SAUQ6 I  AVA II                      <===== 
              .650                                          .700
ATAAAATATCCCATAGAAAAGGTCCTTTAAAGGTTTTATGGACTAAAATAAAAAAGATAGCATAAGCATTAAAATCATGCAAATTAAAAAAAAAGGTAAATGTATTTATAAAAAGGTAAAT
                                                    CGTA  TAATTT          CCATTT
                                                          -10            -35
              .750                                    .800                                    .850
GTATTTTATATAGTATTTATATATTATAGCATAATAATAAATATATTTATAAAATTGATTGTTCTTAGAGCTAAAAGAGAAGAACAATGGGTTTATAGGTATTTTGAGACCAGTTATAAAAAT
              .900                                    .950
GACTTTTGACGTTTAGGTATATAAACACTGCCTCTAATAAAGTCTACTATATTGGAGAGGAGTGAACAGTGGCCTCGCTTATCCCGACAGGAAATACATGGTTTTAGTAAGTAAACT
              SAUQ6 I  TAQ I          .1050                                    ==> 1100
              .1000                                    
GCGTAAGACGACCGACATATACCTAAAGGCCCTTTCTATGCTGACTGATAAGACAAGTACATAAAATTTGCTAGTTTACATTATTTTTTTATTTCTAAATATAATAATAATATTTTAAAATGTA
              (↓↓)            HPA II        .1150                                    .1200
TTTAAAATTTTTCAACAATTTTTAAATTATATTTCCGGACAGATTTATTTTAGGATCGTCAAAAGAAGTTTACATTTTATTTATATAA      ATG GTT CCA CAA ACA GAA ACT AAA
                                                                                        ... TCA ..  ......   .G  ..  ...
                                                                                        ... TCA ..  ......   ......   ...
                                                                                        MET VAL PRO GLN THR GLU THR LYS
                                                                                            SER
                                                                                            SER
              HINF I    HPA II
              .1230                                    .1260                                    .1290
GCA GGT GCT GGA TTC AAA GCC GGT GTA AAA GAC TAC CGT TTA ACA TAC TAC ACA CCT GAT TAC GTA GTA AGA GAT ACT GAT ATT TTA GCT
...  A.  ..T  .A.  ...  ...  ...  ...  ...  ...  ..T  ...  AAA  .G.  .T  ..T  .T  .G  .T  .A.  ACC  CT.  ...  ...  ...  .C  .G  .A
A..  .T.  ...  ...  .T.  ...  ...  .T.  .G.  .T  AAA  .G  .T  ...  ...  A..  AAC  .AG  ...  ...  .C  .G  .G
ALA GLY ALA GLY PHE LYS ALA GLY VAL LYS ASP TYR ARG LEU THR TYR TYR THR PRO ASP TYR VAL VAL ARG ASP THR ASP ILE LEU ALA
                                                    SER VAL GLU                                          LYS       GLU       GLU THR LEU
                                                    SER VAL GLY                                          LYS       GLU       GLU THR LYS
              HINF I
              .1320                                    .1350                    HINF I         .1380
GCA TTC CGT ATG ACT CCA CTA CAA GGT GTT CCA CCT GAA GAA TGT GGT GCT GCT GTA GCA TCT TCA ACA       GGT ACA TGG ACT
...  ...  ...  A.G  .A.  .G.  ...  T.  ...  .CT  .A  ...  ...  .C  ...  ...  GCA  .G.  ...  .A  ...  ...  ...  .T  .T  ...  ...  ...  .A
...  ...  ...  A.G  .A.  .G.  ...  ...  ...  .C  ...  ...  .G  ...  GCA  .G.  ...  .A  .G.  ...  .G  ...  T.G.T  GCT  ...  ...  .A
ALA PHE ARG MET THR PRO GLN LEU GLN GLY VAL PRO PRO GLU GLU CYS GLY ALA ALA VAL ALA ALA GLU SER SER THR X  GLY THR TRP THR
                                                    VAL SER    PRO            ALA                                      ALA ALA
                                                    VAL THR    LEU            ALA
                                                                            TAQ I          HPA II HPH I
              .1410                                    .1440                                    .1470
ACA GTA TGG ACT GAC GGT TTA ACA AGT CTT GAC CGT TAC AAA GGT CGT TGT TAC GAT ATC GAA CCA GTT CCG GGT GAA GAC AAC CAA TAC
.T.  ...  ...  .C  ...  ...  ...  A.C.T  .C  .AC  ...  .T  ...  ...  .A  .A  ..C  .T  C.C  ...  .G  ..C  ...  .T  .G  .C  CCA  G.T  ...  .T
.T  .T  ...  .C  ...  ...  T.  A.C.T  .C  ...  ...  .T  ...  ...  .A  .A  ..C  .T  C.C  ...  .G  ..C  ...  .T  .G  .C  CCA  G.T  ...  .T
THR VAL TRP THR ASP GLY LEU THR SER LEU ASP ARG TYR LYS GLY ARG CYS TYR ASP ILE GLU PRO VAL PRO GLY GLU ASP ASN GLN TYR
                                                    ASN                                          HIS            ALA       GLU
                                                    SER                                          HIS            PRO ASP PRO ASP
              TAQ I          TAQ I
              .1500                                    .1530                                    .1560
ATT GCT TAC GTA GCT TAC CCA ATC GAC TTC TTC GAA GAA GGT TCA GTA ACT AAC ATG TTC ACT TCT ATT GTA GGT AAC GTA TTC GGT TTC
...  TG.  ..T  ...  ...  .G..T  .T  T.A  ...  C.T  ..T  ...  ...  ...  .T  ...  ...  .T  ...  ...  .C  ...  .G  ...  ...  .T.G  ...
..C  TG.  ...  ...  ...  .G..T  .T  T.A  ...  ...  .G.  ...  .T  ...  ...  ...  ...  .T  ...  ...  .C  ...  .G  ...  ...  .T.G  ...
ILE ALA TYR VAL ALA TYR PRO ILE ASP PHE GLU GLU GLY SER VAL THR ASN MET PHE THR SER ILE VAL GLY ASN VAL PHE GLY PHE
                                                    CYS                      LEU                                                  
                                                    CYS                      LEU
              HIND III
              .1590                                    .1620                                    .1650
AAA GCT TTA CGT GCT CTA CGT CTT GAA GAC CTT CGT ATT CCA CCT GCT TAC GTT AAA ACA TTC GTA GGT CCT CCA CAC GGT ATT CAG GTA
...  .C  .G  ...  ...  ...  ...  ...  ...  T.G  ...  .T  T.G  .A  ..C  .T  GT.  ...  .T  .A  ...  .T  ...  CA.  .G  .G  .T  ...  ...  .C  .A  .T
...  .C  .G  ...  ...  ...  ...  ...  ...  T.G  ...  .T  ...  .A  ..C  .T  CA.  ...  .T  .A  ...  .T  ...  CA.  ...  .G  .G  .T  ...  ...  .C  .A  .T
LYS ALA LEU ARG ALA LEU ARG LEU GLY ASP LEU ARG ILE PRO PRO ALA TYR VAL LYS THR PHE VAL GLY PRO PRO HIS GLY ILE GLN VAL
                                                    VAL                      PRO                      GLN                      ARG       MET
                                                                            SER                      GLN
              .1680                          |75          .1740
GAA CGT GAC AAA TTA AAC AAA TAT CGT CGT GGT CTT TTA GGT TGT ACA ATC AAA CCT AAA TTA GGT CTT TCA GCT AAA AAC TAC GGT CGT
.G  A.A  ..T  ...  .G  ...  ...  ...  CCC  ..A  .G.  .A  ..C  .T  ...  ...  ...  ...  ...  ...  .T  .A  ..C  ...  ...  .T  ...  A.A
..  A.G  .T  .G  ..G  ...  ..C  .C  ...  ...  CC.  T.A  .G.  .A  ...  .T  ..T  .A  ...  .G  .A  T.A  ..C  .A  ...  ...  .T  ...  A.A
GLU ARG ASP LYS LEU ASN LYS TYR [GLY ARG GLY LEU LEU GLY CYS THR ILE] [LYS] [PRO LYS] LEU GLY LEU SER ALA LYS ASN TYR GLY ARG
                                  ASN          PRO              A1
                                              PRO
              .1770                                    .1800                                    .1830
GCA GTT TAT GAA TGT TTA CGT GGT CTT GAC TTT ACT AAA GAC GAC GAA AAC GTA AAC TCA CAA CCA TTC ATG CGT TGG CGT GAC CGT
...  ...  ...  ...  ...  ...  ...  .C.T  ..A  ...  .A  ...  .T  ...  ..C  ...  ...  .T  ...  ...  .G  ...  .C  .G  ..G  .T  ...  ...  A.A  ...
.G  TG.  ...  ..G  .C  ...  .A  ...  ...  ...  .T  ...  ...  ...  .T  ...  ...  ...  ...  ...  .T  ...  ...  ...  .C  ...  A.A  ...
ALA VAL TYR GLU CYS LEU ARG [GLY GLY LEU ASP PHE THR] [LYS] [ASP ASP GLU ASN VAL ASN SER GLN PRO] PHE MET ARG TRP ARG ASP ARG
          CYS                                                          C
              HPH I
              .1860                                    .1890                                    .1920
TTC CTT TTC GTT GCT GAA GCT ATT TAC AAA GCT CAA GCA GAA ACA GGT GAA GTT GGT CAC TAC TTA AAC GCT ACT GCT GGT ACT TGT
...  .G.  .T  TG.  ..C  ...  ...  .A  ...  ...  ...  .T.A  ...  ...  ...  A.C  ...  .G  .T  ...  .G  .T  ...  .C  .G  ...  .A  .C
...  G.C  .T  TG.  ..C  ...  ...  .A  ...  ...  ...  .T.A  ...  ...  ...  A.C  ...  .G  .T  ...  ...  .G  .A  ...  .A  .C
PHE LEU PHE VAL ALA GLU ALA ILE TYR LYS ALA GLN ALA GLU THR GLY GLU VAL GLY HIS TYR LEU ASN ALA THR ALA GLY THR CYS
          CYS                                    LEU                      ILE
          CYS                                    SER                      ILE
              .1950                                    .1980                                    .2010
GAA GAA ATG ATG AAA CGT GCA GTA TGT GCT AAA GAA TTA GGT GTA CCT ATT ATT ATG CAC GAC TAC TTA ACA GGT GGT TTC ACA GCT AAC
...  .T  ...  ...  ...  A.G  .A  .T  ...  .T.  .C.G  ...  ...  .G.C  .T  ...  ...  G.A  ...  .T  ...  ...  .G  .A  ...  .T  .A  .T
.T  ...  .T  ...  T.G  .G.A  .T  ...  .T.  .A  .GG  C...  ...  .G.C  .T  ...  ...  G.A  ...  .T  ...  ...  .G  .A  ...  .T  ...  ...  .T
GLU GLU MET MET LYS ARG ALA VAL CYS ALA LYS GLU LEU GLY VAL PRO ILE ILE MET HIS ASP TYR LEU THR GLY GLY PHE THR ALA ASN
          ASP                      PHE          ARG                      VAL
ASP GLU    ILE      GLY            PHE      ARG GLN                      VAL      ASP
```

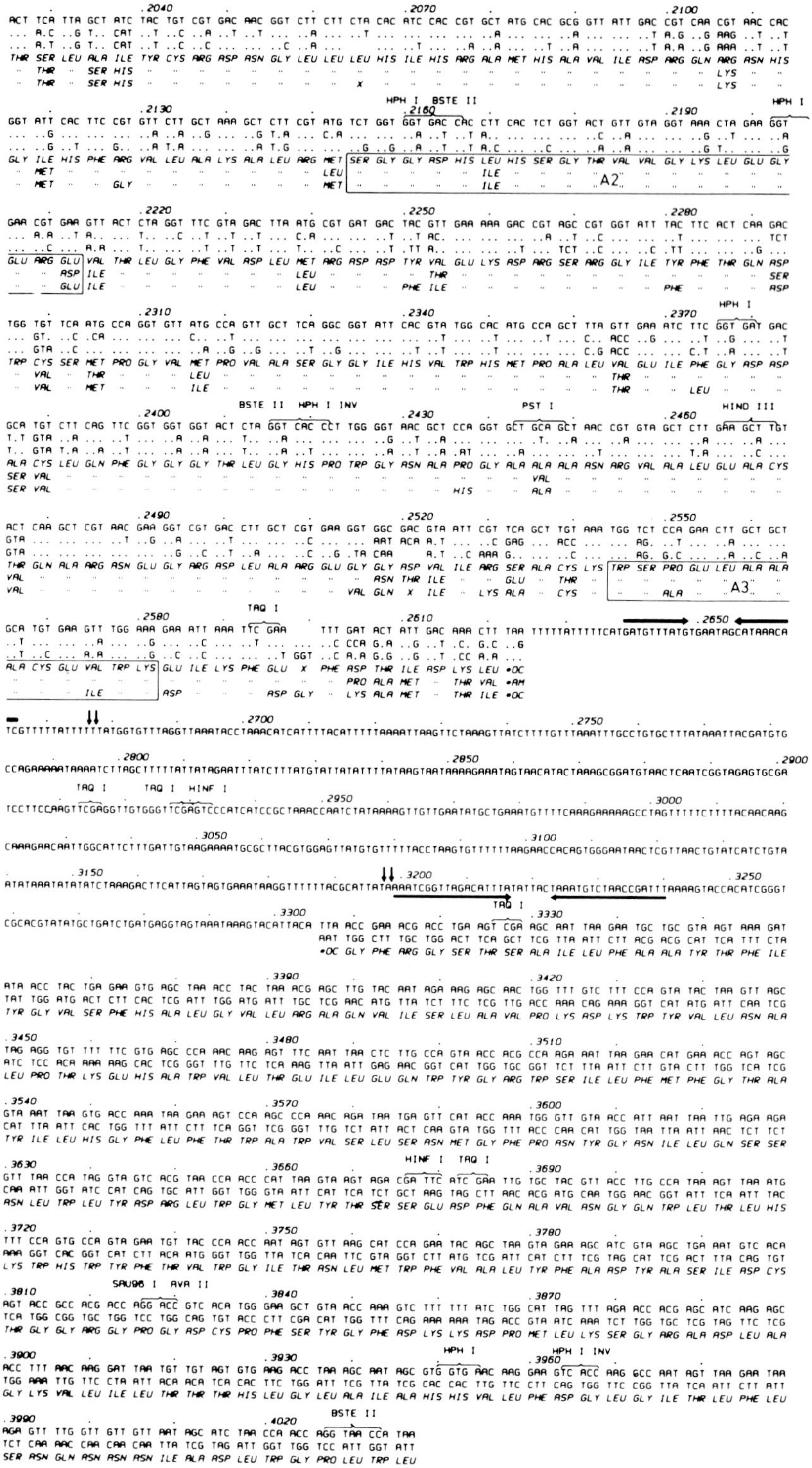

FIG. 6.24 (Continued)

chloroplast makes relatively few of its own proteins. The answers probably lie both in the evolutionary heritage of the chloroplast (see Chapter 9) and in the nature of the proteins. It seems probable that proteins made within the organelle must reside, for their function, in a position or orientation that cannot be achieved by proteins made in the cytoplasm. For example, as diagramed in Fig. 4.11, the chloroplast thylakoid membrane is asymmetric, and its assembly may require that proteins approach the membrane from a particular side. Furthermore, since most of these proteins are relatively hydrophobic, to achieve their site of function on one side of the membrane may necessitate synthesis on the correct side. Therefore, the asymmetry of the membrane, which is essential for function, may depend on cooperation of the chloroplast and the nuclear–cytoplasmic systems for its formation.

An interesting aspect of cooperation between chloroplastic and cytoplasmic protein synthesizing systems is the production of ribulose 1,5-bisphosphate carboxylase, the most abundant protein in green plant tissue. This enzyme is found within the stroma of the chloroplast. As described in the last section, the large subunit of the enzyme, 55,000 daltons in mass, is synthesized on chloroplast ribosomes with mRNA that is transcribed from chloroplast DNA. This RNA lacks a polyadenylate addition at the 3′ end. On the other hand, the small subunit of this enzyme, 12,000–16,000 daltons in mass (depending on the source), is coded by genes in nuclear DNA. The mRNA for the small subunit is polyadenylated and is translated on cytoplasmic ribosomes. Moreover, the small subunit is synthesized as a precursor that has a molecular weight about 5000 larger than that of the mature form found in the enzyme. The additional segment of the polypeptide, at the N-terminal end, is required for transfer of the subunit into the chloroplast and is cleaved off as the polypeptide is transported into the organelle. The complete protein apparently is assembled within the chloroplast stroma.

Regulation of the synthesis of the carboxylase occurs at several levels. Transcription of the mRNA for the small subunit in the nucleus is stimulated by light. Although multiple copies of the large subunit gene exist, one in each chloroplast DNA molecule, a family of only about five small subunit genes exists within the haploid plant nucleus. Which subunit is made at a limiting rate is not known, but accumulation of the two subunit polypeptides is coordinated, so that an excess of either one is not observed. Specific proteolytic systems seem to degrade any of the polypeptides in excess of that which can be assembled into the holoenzyme. (See Chapter 8 for more details.)

VIII. Protein Synthesis

In the green algae and the higher plants, most of the proteins of the chloroplast are synthesized outside the organelle by the cytoplasmic ribosomes. Most soluble enzymes in the chloroplast matrix, most membrane polypeptides, including the major membrane components, and most of the ribosomal proteins are synthesized outside of the organelle. However, many studies have demonstrated that active synthesis of proteins within the chloroplast occurs and is necessary for development of a functional photosynthetic system. The process of chloroplast protein synthesis, in terms of its mechanism as well as the products of this system, has attracted high interest from investigators for many years. An important aspect of understanding how the chloroplast functions is a knowledge of which proteins are made within the organelle.

A characteristic feature of chloroplast protein synthesis is its sensitivity to inhibitors (antibiotics) that also inhibit bacterial ribosomes. Some antibiotics

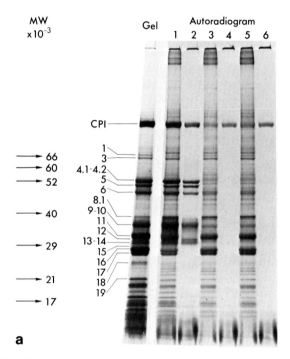

FIG. 6.25. Sites of synthesis of thylakoid membrane polypeptides in the wild-type strain of *Chlamydomonas reinhardtii* as deduced by selective inhibition of protein synthesis. Chloramphenicol and spectinomycin were used to inhibit chloroplast ribosomes, and anisomycin was used as the inhibitor of cytoplasmic ribosomes. Proteins were labeled *in vivo* by providing [^{14}C]acetate, which was metabolized by the cells to amino acids. Thylakoid membranes were isolated from labeled cells and the polypeptides were separated by electrophoresis in the presence of sodium dodecyl sulfate. During preparation of the samples for electrophoresis, those shown in panel (a) were kept at room temperature, whereas those for panel (b) were heated in boiling water for 1 min. The chlorophyll *a*-protein complex (CP1), present in (a), was destroyed by heating. Radioactivity in each polypeptide was detected by autoradiography of the polyacrylamide gel. In each panel, the lane marked "Gel" shows the Coomassie blue-stained pattern of membrane polypeptides. Lanes 1–6 show autoradiograms for control cells (both chloroplast and cytoplasmic products, lane 1) and for cells labeled in the presence of anisomycin (chloroplast products, lane 2), chloramphenicol (cyto-

that have been used to inhibit protein synthesis in chloroplasts are chloramphenicol, lincomycin, spectinomycin, streptomycin, erythromycin, and tetracycline. Cytoplasmic protein synthesis is not sensitive to these compounds but is sensitive to other inhibitors such as cycloheximide and anisomycin. The differential action of these antibiotics provided the basis for extensive studies on identifying which of the chloroplast proteins are made within the organelle.

An elegant example of the use of this approach to determine where each thylakoid membrane polypeptide is synthesized is shown in Fig. 6.25. Chua and

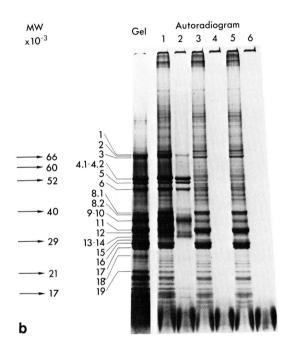

plasmic products, lane 3), anisomycin plus chloramphenicol (lane 4), spectinomycin (cytoplasmic products, lane 5) and anisomycin plus spectinomycin (lane 6). Polypeptides are numbered consecutively, starting with the highest-molecular-weight component. The positions of molecular weight markers (known standard proteins) are indicated on the left. Lane 1 shows that the pattern of radioactivity for control cells was similar to the polypeptide pattern visualized by the stain. Polypeptides labeled by chloroplast protein synthesis (lane 2) include the apoprotein of CP1 (polypeptide 2) and the α and β subunits of the coupling factor (polypeptides 4.1 and 4.2). Synthesis of these polypeptides was inhibited by chloramphenicol and spectinomycin (lanes 3 and 5). Lanes 3 and 5 show the polypeptides synthesized on cytoplasmic ribosomes, which include the major chlorophyll a/b-binding protein (polypeptides 11, 16, and 17). When inhibitors of both types of ribosomes were present, no labeling of polypeptides occurred, which indicated that protein synthesis was necessary for incorporation of ^{14}C (lanes 4 and 6). Labeling of CP1 in lanes 4 and 6 in panel (a) was due to synthesis of chlorophyll from [^{14}C]acetate. (From Chua and Gillham, 1977. Reprinted with permission of The Rockefeller University Press.)

Gillham (1977) labeled *Chlamydomonas reinhardtii* cells with [^{14}C]acetate, which was metabolically converted to amino acids *in vivo*. Incorporation of amino acids into proteins in the presence of chloramphenicol or of anisomycin was studied. Thylakoid membranes were isolated from the labeled cells, and the polypeptides of the membrane were separated by electrophoresis through a polyacrylamide gel. In this analysis, the proteins were treated with the detergent sodium dodecyl sulfate to form detergent–protein complexes that each had a mass proportional to the size of the polypeptide. The detergent, by conferring a large number of negative charges to the complexes, caused all polypeptides to migrate toward the anode. With this procedure, the polypeptides were separated on the basis of their differences in size, since small polypeptide–detergent complexes move through the porous gel more rapidly than those of large polypeptides. Labeling of each separated polypeptide then was monitored by radioautography of the gel with X-ray film. As Fig. 6.25 shows, those polypeptides labeled in the presence of chloramphenicol, and thus made on cytoplasmic ribosomes, are clearly distinguished from those labeled in the presence of anisomycin. This latter group, the products of chloroplast protein synthesis, is relatively small in number but contains several distinct components.

Unequivocal proof that a protein is made in the chloroplast, however, should be possible by demonstrating its synthesis in the isolated organelle. The ability to obtain chloroplasts that are active in protein synthesis required a prior, slow development of technical procedures for isolating intact chloroplasts. Success in obtaining chloroplasts that were quite active in protein synthesis was achieved by Blair and Ellis (1973). They found that isolated chloroplasts, if carefully prepared so that the envelope was intact, needed only to be suspended in a suitable buffer containing 0.2 M KCl, which served both as an agent to balance the osmotic pressure across the envelope and as a source of the proper ionic environment for protein synthesis. Replacement of NaCl for KCl completely inhibited the system. Usually 5–7 mM MgCl$_2$ is required for stability of the ribosomes and for activity of several of the enzymes involved. The energy sources required for protein synthesis, as ATP and GTP, can be generated *in situ* by photophosphorylation. In the dark, an exogenous supply of these nucleotides is necessary.

The course of protein synthesis in these chloroplasts, measured by the incorporation of radioactive amino acids, is illustrated in Fig. 6.26. Incorporation of amino acids stopped after about 60 min of incubation, for reasons that are not understood, yet sufficient amounts of labeled polypeptides were synthesized to allow an analysis of the products made by these systems. The fact that synthesis of protein in this system required light, in the absence of an added energy source, confirmed that the activity resided in chloroplasts.

Only relatively few of the polypeptides synthesized within the chloroplast have been identified. Products recovered in the soluble, stroma fraction include the major product, the large subunit of ribulose 1,5-bisphosphate carboxylase. Other identified soluble products are the elongation factors that function in protein synthesis. Analysis of the sedimentable, membrane fraction identified several thylakoid polypeptides, including the α, β, and ϵ subunits of coupling factor CF$_1$, cytochromes f and b-559, the apoprotein of the photosystem 1 antenna chlorophyll a-protein complex, and the 32,000-molecular-weight product of the "photogene." This latter polypeptide is the major integral membrane polypeptide made by isolated chloroplasts. The polypeptides of the phycobilisomes of red algae also are made within the chloroplast.

An analysis of products of protein synthesis by two-dimensional gel electrophoresis resolved 80 to 90 different polypeptides. It is not known whether all these polypeptides are complete products of different genes. However, since a polypeptide of average size contains 300 to 400 amino acids, a simple calculation indicates, assuming that each of the products is coded by a different gene, that perhaps two-thirds of the total coding capacity of the DNA codes for proteins. The total number of polypeptides may be even greater than experimental results indicate, if any made in small amounts escaped detection. Nevertheless, the potential information contained in a molecule of DNA the size of the chloroplast genome is obviously limited.

There are great differences in the amounts of the various polypeptides made in the chloroplast. The mechanisms that regulate the amount of each polypeptide produced are not yet known. Transcription of RNA molecules by RNA polymerase is initiated at specific DNA sequences called **promoters.** Variations in these sequences, or other regulatory sites in DNA, may affect the ability of the polymerase to initiate transcription. The abundance of mRNA species, consequently, could be determed by these effects.

As stated in Section VI.A.1, chloroplast ribosomes are the same size as those found in bacterial cells and are sensitive to the same inhibitors. The mechanism of protein synthesis in chloroplasts, at the level of detail that is currently known, also resembles the process as it occurs on the 70 S ribosomes in

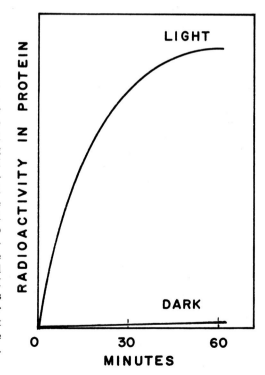

FIG. 6.26. The time course of light-driven protein synthesis in isolated chloroplasts. Pea seeds (Pisum sativum) were grown under a 12-hr-light, 12-hr-dark regime for 7–10 days. The apical leaves were homogenized for 4 sec in a sterile ice slurry containing 0.35 M sucrose, 25 mM HEPES-NaOH (sodium N-2-hydroxyethylpiperazine-N′-2-ethanesulfonate, a buffer), 2 mM EDTA, and 2 mM sodium isoascorbate (pH 7.6). The homogenate was immediately strained through eight layers of muslin cloth to remove large particulate material and centrifuged at 2500 $\times$ g for 1 min at 0°C. The pellet was resuspended in sterile 0.2 M KCl containing 66 mM tricine-KOH buffer (pH 8.3) and 6.6 mM MgCl$_2$. The chloroplasts were incubated with either [^{14}C]leucine or [^{35}S]methionine at 20°C under red light at 4000 lux. Protein was extracted and the incorporated radioactivity was determined. (Adapted from Blair and Ellis, 1973.)

prokaryotic cells. Although many of the details have not been studied in chloroplasts, it has been anticipated, and even assumed, that the general mechanism is similar to the process in bacterial cells.

A. Initiation

At least two tRNA molecules exist that are specific for methionine (see Fig. 6.22). One of these allows the addition of a formyl group onto the amino group of methionine attached to it, whereas the other does not. The formylation reaction produces **N-formyl methionyl–tRNA** (Fig. 6.27) and is catalyzed by the enzme **transformylase.** This enzyme is not present in the cytosol of plant cells. Therefore, protein synthesis begins in the chloroplast with N-formyl methionine as the amino acid incorporated into the first position at the N-terminal end, but in the cytosol protein synthesis begins with an unmodified methionyl-tRNA. The formyl group in this reaction is donated by N^{10}-formyl tetrahydrofolic acid, a derivative of the vitamin folic acid.

The initiating codon on chloroplast mRNA is A-U-G (5′ → 3′) (See Table 6.4). By analogy to the bacterial system, protein synthesis in chloroplasts presumably begins with formation of a complex between mRNA, N-formyl methionyl–tRNA, and the small ribosomal subunit. These initiation steps are catalyzed by specific **initiation factors.** The functions of these proteins have been defined in the bacterial system, but the chloroplast initiation factors have not been extensively studied. The final step in the process of initiaion is addition of the large ribosomal subunit to the complex, to form the complete functional translation unit (refer to Fig. 6.28).

B. Elongation

After initiation is completed, the second amino acyl–tRNA is added to the ribosome, and formation of the first peptide bond occurs. The ribosome then pulls the mRNA through it like a computer tape, adding amino acids to the growing polypeptide chain one at a time according to the codon sequence. Two proteins **(elongation factors)** have been found in chloroplasts that are involved

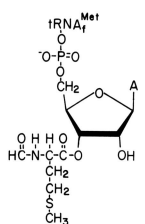

FIG. 6.27. The structure of N-formyl-L-methionyl-tRNA$_f^{Met}$, showing the attachment of the amino acid, through an ester linkage, to the 3′-hydroxyl of the terminal adenylate residue in the tRNA. Hydrogen atoms on carbons 1′–4′ of the ribose moiety are not included.

in the elongation reactions. The factor EF-T$_{Chl}$ (elongation factor T of the chloroplast) is required for binding of each successive amino acyl–tRNA to the acceptor site on the ribosome, as directed by the codon on the mRNA. The second factor, EF-G$_{Chl}$, is required for translocation of the peptidyl–tRNA—mRNA complex back to the donor site after synthesis of each peptide bond (Fig. 6.28). These elongation factors are specific for 70 S ribosomes and are not active with plant cytoplasmic or other 80 S-type ribosomes.

The chloroplast elongation factors have been purified (Fig. 6.29). EF-Tu$_{Chl}$ has a molecular weight of 45,000 and that of EF-G$_{Chl}$ is about 77,000. Both values are nearly identical to the analogous proteins from bacterial cells, and the chloroplast and bacterial proteins are functionally interchangeable.

C. Amino Acyl–tRNA Synthetases

Since the tRNA, and not the amino acid, "reads" the codons on mRNA, it is imperative for accurate synthesis of a protein that the correct amino acids are attached to the tRNA molecules. The enzymes that perform this task are extraordinarily accurate. Each of these enzymes is specific for ATP, the source of energy required for formation of the "high-energy" amino acyl–tRNA linkage, as well as an amino acid and all tRNA species for that amino acid. These enzymes catalyze the general reaction shown in reactions 6.1 and 6.2.

TABLE 6.4. The Universal Genetic Code
Used in Chloroplasts[a]

First position (5′ end)	Second position				Third position (3′ end)
	U	C	A	G	
U	Phe	Ser	Tyr	Cys	U
	Phe	Ser	Tyr	Cys	C
	Leu	Ser	Stop	Stop	A
	Leu	Ser	Stop	Trp	G
C	Leu	Pro	His	Arg	U
	Leu	Pro	His	Arg	C
	Leu	Pro	Gln	Arg	A
	Leu	Pro	Gln	Arg	G
A	lle	Thr	Asn	Ser	U
	lle	Thr	Asn	Ser	C
	lle	Thr	Lys	Arg	A
	Met	Thr	Lys	Arg	G
G	Val	Ala	Asp	Gly	U
	Val	Ala	Asp	Gly	C
	Val	Ala	Glu	Gly	A
	Val	Ala	Glu	Gly	G

[a]The codons refer to triplet sequences in mRNA. Although not all codons are used in every chloroplast gene, the usage is the same as that in prokaryotic and the eukaryotic nuclear-cytoplasmic systems. The codon 5′-AUG-3′ is the initiation point and also serves as the codon for methionine at internal positions in the polypeptide. UAA, UAG, and UGA are termination signals. If the coding strand in DNA can be determined, the amino acid sequence of any protein can be deduced from the nucleotide sequence in DNA.

A. INITIATION

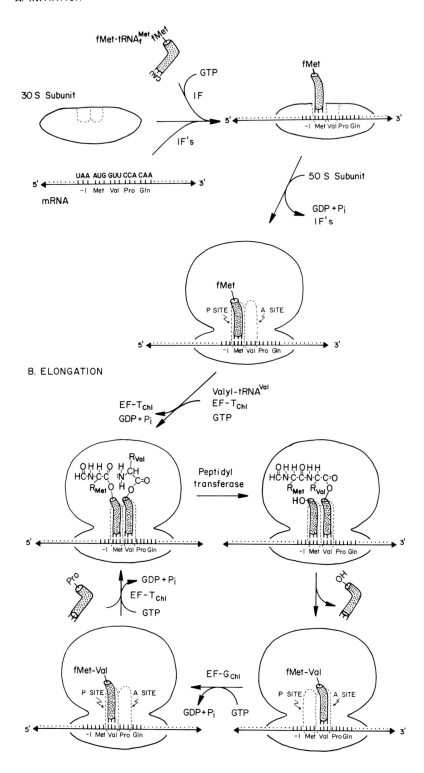

B. ELONGATION

$$\underset{NH_3^+}{\overset{H}{R_1-\overset{|}{\underset{|}{C}}-CO_2^-}} + ATP + Enz_1 \rightleftharpoons$$

$$\left[\underset{NH_3^+}{\overset{H}{R_1-\underset{|}{\overset{|}{C}}}} - \overset{O}{\overset{\|}{C}}-O-\underset{O_-}{\overset{O}{\overset{\|}{P}}}-O-\text{ribose}-\text{Ad} \right] Enz_1 + PP_i \qquad [6.1]$$

$$\left[\underset{NH_3^+}{\overset{H}{R_1-\underset{|}{\overset{|}{C}}}} - \overset{O}{\overset{\|}{C}}-O-\underset{O_-}{\overset{O}{\overset{\|}{P}}}-O-\text{ribose}-\text{Ad} \right] Enz_1 + tRNA^1 \rightleftharpoons$$

$$\underset{NH_3^+}{\overset{H}{R_1-\underset{|}{\overset{|}{C}}}} - \overset{O}{\overset{\|}{C}}-O-tRNA^1 + Enz_1 + AMP \qquad [6.2]$$

FIG. 6.28. The mechanism of protein synthesis in chloroplasts. A: Initiation, the first stage in protein synthesis, is the formation of the functional translation complex. Both mRNA and the initiating amino acyl-tRNA, in this case N-formyl methionyl-tRNA$_f^{Met}$, bind to the small ribosomal subunit. The order of addition in chloroplasts is not known, but either can bind first in other systems. Formation of this initial complex requires the activities of several proteins, called **initiation factors** (IF's). The large subunit then joins the complex to form the complete ribosome. Energy in the form of guanosine 5′-triphosphate (GTP) is required for initiation; hydrolysis of GTP facilitates release of the initiation factors. At the completion of initiation, two sites exist on the ribosome for binding amino acyl-tRNA. One is occupied by N-formyl methionyl-tRNA, which is based-paired to an AUG codon in the mRNA. This site is referred to as the "peptidyl" or P site. The other site, the "amino acyl" or A site, contains the codon in the mRNA for the next amino acyl-tRNA. B: Elongation, the second stage in synthesis of a protein, adds amino acids one at a time to the growing polypeptide, each one determined in turn by the next codon in the mRNA as it appears in the A site. The insertion of each amino acid can be represented as a cycle. First, the amino acyl-tRNA binds the A site in a reaction facilitated by **elongation factor** EF-T$_{Chl}$. Which one of the 20 amino acyl-tRNA molecules that bind is determined by the codon in the mRNA. Second, the amino acid or peptide attached to tRNA in the P site is transferred to the amino acyl-tRNA in the A site, forming a peptide bond and extending the length of the peptide by one amino acid. In this step, the amino group, with its unshared electrons, on the amino acyl-tRNA in the A site attacks the reactive ester bond between the amino acid (or peptide) and the tRNA in the P site in a S_N2-type reaction. The enzyme that catalyzes this step, **peptidyl transferase,** is an integral component of the large ribosomal subunit. Third, the elongated peptidyl-tRNA:mRNA complex must be translocated so that the peptidyl-tRNA resides in the P site and the A site contains the codon for the next amino acid. The translocation step is facilitated by **elongation factor** EF-G$_{Chl}$. The process repeats itself for the insertion of each amino acid. EF-T$_{Chl}$ and EF-G$_{Chl}$ are soluble proteins. Hydrolysis of GTP is required for the activity of both these proteins. Elongation continues until a termination codon, UAA, UAG, or UGA, appears in the A site. Since these codons represent no amino acid, no amino acyl-tRNA will enter the A site. Consequently, the now completed polypeptide is freed by hydrolysis from the tRNA that brought in the final, carboxyl-terminal amino acid, and the polypeptide is released from the ribosome. The termination reaction requires several proteins called **termination factors.**

Work in Weil's laboratory has shown that the chloroplast amino acyl–tRNA synthetases are different proteins than the analogous enzymes located in the cytosol (Dietrich *et al.*, 1983). For example, the molecular weight of the chloroplast leucyl-tRNA synthetase is 120,000 whereas that of the cytoplasmic enzyme is 130,000. Although this is not a large difference in size, the amino acid compositions of the two enzymes are also different and they do not share any antigenic determinants, since antibodies raised against the one do not react with the other. The chloroplast enzyme is active with tRNA^{Leu} extracted from chloroplasts but not with the isoaccepting species from the cytosol. The cytoplasmic enzyme has the opposite specificity. The remarkable substrate specificity of these enzymes is particularly interesting in view of the fact that at least most, and probably all, of the chloroplast synthetases are synthesized on cytoplasmic ribosomes and afterward transferred into the chloroplast. The differences in the chloroplast and cytoplasmic synthetases suggest that they are coded by different genes and do not simply arise from posttranslational modification of the same gene product.

IX. Association of Ribosomes with Membranes

Studies in several laboratories have shown that a large portion of the chloroplast ribosomes are bound to thylakoid membranes during synthesis of proteins. The factor that usually determines whether ribosomes will bind to mem-

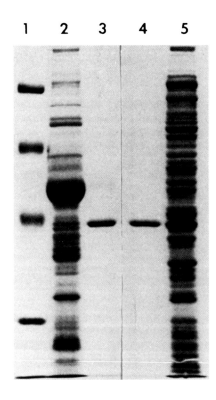

FIG. 6.29. Gel electrophoresis of purified EF–Tu from spinach chloroplasts and comparison with EF–Tu from *Escherichia coli*. Lane 1, molecular weight standards (top to bottom: phosphorylase B, 94,000; bovine serum albumin, 68,000; ovalbumin, 45,000; carbonic anhydrase, 30,000). Lane 2, crude chloroplast extract. Lane 3, purified EF–Tu$_{Chl}$. Lane 4, purified EF–Tu from *E. coli*. Lane 5, crude extract of *E. coli*. (From Tiboni and Ciferri, 1982. Reprinted with permission.)

branes is the polypeptide product they are engaged in making. The ribosome-membrane association is favored by a sequence within the polypeptide product that contains a relatively high concentration of hydrophobic amino acids. This hydrophobic portion of the polypeptide has an affinity for and dissolves into the hydrophobic membrane, thereby pulling the ribosome onto the membrane. In chloroplasts, the polypeptide products of these membrane-bound ribosomes have not been identified, but apparently they are membrane components. Since most membrane proteins are rich in hydrophobic amino acids, or at least have segments containing hydrophobic amino acid sequences, it seems likely that this suggestion is correct.

X. Summary

Chloroplasts contain their own DNA, which exists as multiple copies of a relatively small, circular DNA molecule about 140,000 base-pairs in length. All DNA molecules within a single chloroplast are identical. The DNA contains generally, but not universally, two sets of genes for chloroplast ribosomal RNA molecules, which are located in separate, inverted repeat regions. Chloroplast DNA also contains a complete set of genes for transfer RNA molecules and contains the genes for a limited number of proteins. Several thylakoid membrane polypeptides are coded by chloroplast DNA and synthesized on chloroplast ribosomes. One major soluble polypeptide made within chloroplasts has been identified as the large subunit of ribulose 1,5-bisphosphate carboxylase. Several minor soluble polypeptide products have been detected but, with the exception of the elongation factors in protein synthesis, have not been identified. Most messenger RNA molecules made in the chloroplast, and in particular that for the carboxylase large subunit, do not have a polyadenylate sequence added at the 3' end.

Chloroplast ribosomes have a sedimentation coefficient of about 70 S and contain a 23 S and a 5 S RNA in the large (50 S) subunit and a 16 S RNA in the small (30 S) subunit. These ribosomes function in protein synthesis by mechanisms apparently similar to those of bacterial ribosomes. Chloroplast ribosomes are inhibited by the same antibiotics that inhibit bacterial ribosomes. Although most chloroplast proteins are synthesized outside of the organelle on cytoplasmic ribosomes, the protein synthetic system with the chloroplast is necessary for development of a complete, functional photosynthetic apparatus.

Literature Cited

Allet, B., and Rochaix, J.-D. (1979) Structure analysis at the ends of the intervening DNA sequences in the chloroplast 23S ribosomal genes of *C. reinhardii*, *Cell* **18**:55–60.

Bedbrook, J. R., Coen, D. M., Beaton, A. R., Bogorad, L., and Rich, A. (1979) Location of the single gene for the large subunit of ribulosebisphosphate carboxylase in the maize chloroplast chromosome, *J. Biol. Chem.* **254**:905–910.

Blair, G. E., and Ellis, R. J. (1973) Protein synthesis in chloroplasts, I. Light-driven synthesis of the large subunit of fraction I protein by isolated pea chloroplasts. *Biochim. Biophys. Acta* **319**:223–234.

Boffey, S. A., and Leech, R. M. (1982) Chloroplast DNA levels and the control of chloroplast division in light-grown wheat leaves, *Plant Physiol.* **69**:1387–1391.

Bogorad, L. (1981) Chloroplasts, *J. Cell Biol.* **91**:256s–270s.

Bolen, P. L., Grant, D. M., Swinton, D., Boynton, J. E., and Gillham, N. W. (1982) Extensive methylation of chloroplast DNA by a nuclear gene mutation does not affect chloroplast gene transmission in *Chlamydomonas, Cell* **28**:335–343.

Bottomley, W., and Whitfeld, P. R. (1979) Cell-free transcription and translation of total spinach chloroplast DNA, *Eur. J. Biochem.* **93**:31–39.

Brügger, M., and Boschetti, A. (1975) Two-dimensional gel electrophoresis of ribosomal proteins from streptomycin-sensitive and streptomycin-resistant mutants of *Chlamydomonas reinhardi, Eur. J. Biochem.* **58**:603–610.

Burton, W. G., Grabowy, C. T., and Sager, R. (1979) Role of methylation in the modification and restriction of chloroplast DNA in *Chlamydomonas, Proc. Natl. Acad. Sci. USA* **76**:1390–1394.

Calagan, J. L., Pirtle, R. M., Pirtle, I. L., Kashdan, M. A., Vreman, H. J., and Dudock, B. S. (1980) Homology between chloroplast and prokaryotic initiator tRNA. Nucleotide sequence of spinach chloroplast methionine initiator tRNA, *J. Biol. Chem.* **255**:9981–9984.

Chiang, K.-S. (1971) Replication, transmission and recombination of cytoplasmic DNAs in *Chlamydomonas reinhardi,* in *Autonomy and Biogenesis of Mitochondria and Chloroplasts* (N. K. Boardman, A. W. Linnane, and R. M. Smillie, eds.), Elsevier/North Holland, Amsterdam, pp. 235–249.

Chiang, K.-S., and Sueoka, N. (1967) Replication of chloroplast DNA in *Chlamydomonas reinhardi* during vegetative cell cycle: its mode and regulation, *Proc. Natl. Acad. Sci. USA* **57**:1506–1513.

Chua, N.-H., and Gillham, N. W. (1977) The sites of synthesis of the principal thylakoid membrane polypeptides in *Chlamydomonas reinhardtii, J. Cell Biol.* **74**:441–452.

Crouse, E. J., Schmitt, J. M., Bohnert, H.-J., Gordon, K., Driesel, A. J., and Herrmann, R. G. (1978) Intramolecular compositional heterogeneity of *Spinacia* and *Euglena* chloroplast DNAs, in *Chloroplast Development* (G. Akoyunoglou and J. H. Argyroudi-Akoyunoglou, eds.), Elsevier/North Holland, Amsterdam, pp. 565–572.

Deno, H., Kato, A., Shinozaki, K., and Sugiura, M. (1982) Nucleotide sequences of tobacco chloroplast genes for elongator tRNA^{Met} and tRNA^{Val} (UAC): The tRNA^{Val}(UAC) gene contains a long intron, *Nucleic Acids Res.* **10**:7511–7520.

Dietrich, A., Souciet, G., Colas, B., and Weil, J.-H. (1983) *Phaseolus vulgaris* cytoplasmic leucyl-tRNA synthetase. Purification and comparison of its catalytic, structural, and immunological properties with those of the chloroplastic enzyme. *J. Biol. Chem.* **258**:12386–12393.

Dron, M., Rahire, M., and Rochaix, J.-D. (1982a) Sequence of the chloroplast 16S rRNA gene and its surrounding regions of *Chlamydomonas reinhardii, Nucleic Acids Res.* **10**:7609–7620.

Dron, M., Rahire, M., and Rochaix, J.-D. (1982b) Sequence of the chloroplast DNA region of *Chlamydomonas reinhardii* containing the gene of the large subunit of ribulose bisphosphate carboxylase and parts of its flanking genes, *J. Mol. Biol.* **162**:775–793.

Dron, M., Rahire, M., Rochaix, J.-D., and Mets, L. (1983) First DNA sequence of a chloroplast mutation: A missense alteration in the ribulosebisphosphate carboxylase large subunit gene. *Plasmid* **9**:321–324.

Goodenough, U. W., and Weiss, R. L. (1975) Gametic differentiation in *Chlamydomonas reinhardtii,* III. Cell wall lysis and microfilament-associated mating structure activation in wild-type and mutant strains, *J. Cell Biol.* **67**:623–637.

Grant, D. M., Gillham, N. W., and Boynton, J. E. (1980) Inheritance of chloroplast DNA in *Chlamydomonas reinhardtii, Proc. Natl. Acad. Sci. USA* **77**:6067–6071.

Hartley, M. R., and Ellis, R. J. (1973) Ribonucleic acid synthesis in chloroplasts. *Biochem. J.* **134**:249–262.

Hartley, M. R., and Head, C. (1979) The synthesis of chloroplast high-molecular-weight ribosomal ribonucleic acid in spinach, *Eur. J. Biochem.* **96**:301–309.

Hoober, J. K., and Blobel, G. (1969) Characterization of the chloroplastic and cytoplasmic ribosomes of *Chlamydomonas reinhardi, J. Mol. Biol.* **41**:121–138.

Howe, C. J., Bowman, C. M., Dyer, T. A., and Gray, J. C. (1982) Localization of wheat chloroplast genes for the beta and epsilon subunits of ATP synthesis, *Mol. Gen. Genet.* **186**:525–530.

Howe, C. J., Bowman, C. M., Dyer, T. A., and Gray, J. C. (1983) The genes for the alpha and proton-translocating subunits of wheat chloroplast ATP synthase are close together on the same strand of chloroplast DNA, Mol. Gen. Genet. **190**:51–55.

Howell, S. H., and Walker, L. L. (1976) Informational complexity of the nuclear and chloroplast genomes of Chlamydomonas reinhardi, Biochim. Biophys. Acta **418**:249–256.

Kashdan, M. A., and Dudock, B. S. (1982) The gene for a spinach chloroplast isoleucine tRNA has a methionine anticodon, J. Biol. Chem. **257**:11191–11194.

Koch, W., Edwards, K., and Kössel, H. (1981) Sequencing of the 16S–23S spacer in a ribosomal RNA operon of Zea mays chloroplast DNA reveals two split tRNA genes, Cell **25**:203–213.

Kolodner, R., and Tewari, K. K. (1972) Molecular size and conformation of chloroplast deoxyribonucleic acid from pea leaves, J. Biol. Chem. **247**:6355–6364.

Kolodner, R., and Tewari, K. K. (1975) The molecular size and conformation of the chloroplast DNA from higher plants, Biochim. Biophys. Acta **402**:372–390.

Link, G., Coen, D. M., and Bogorad, L. (1978) Differential expression of the gene for the large subunit of ribulose bisphosphate carboxylase in maize leaf cell types, Cell **15**:725–731.

Martin, N. C., and Goodenough, U. W. (1975) Gametic differentiation in Chlamydomonas reinhardtii, I. Production of gametes and their fine structure, J. Cell Biol. **67**:587–605.

Passavant, C. W., Stiegler, G. L., and Hallick, R. B. (1983) Location of the single gene for elongation factor Tu on the Euglena gracilis chloroplast chromosome, J. Biol. Chem. **258**:693–695.

Pirtle, R., Calagan, J., Pirtle, I., Kashdan, M., Vreman, H., and Dudock, B. S. (1981) The nucleotide sequence of spinach chloroplast methionine elongator tRNA, Nucleic Acids Res. **9**:183–188.

Rochaix, J.-D. (1978) Restriction endonuclease map of the chloroplast DNA of Chlamydomonas reinhardii, J. Mol. Biol. **126**:597–617.

Rochaix, J.-D., and Darlix, J.-L. (1982) Composite structure of the chloroplast 23S ribosomal RNA genes of Chlamydomonas reinhardii. Evolutionary and functional implications, J. Mol. Biol. **159**:383–395.

Rochaix, J.-D., and Malnoe, P. (1978) Anatomy of the chloroplast ribosomal DNA of Chlamydomonas reinhardii, Cell **15**:661–670.

Sager, R. (1972) Cytoplasmic Genes and Organelles, Academic Press, New York.

Sager, R. (1977) Genetic analysis of chloroplast DNA, Adv. Genet. **19**:287–338.

Sager, R., and Grabowy, C. (1983) Differential methylation of chloroplast DNA regulates maternal inheritance in a methylated mutant of Chlamydomonas, Proc. Natl. Acad. Sci. USA **80**:3025–3029.

Sagher, D., Grosfeld, H., and Edelman. M. (1976) Large subunit ribulosephosphate carboxylase messenger RNA from Euglena chloroplasts. Proc. Natl. Acad. Sci. USA **73**:722–726.

Sano, H., Grabowy, C., and Sager, R. (1981) Differential activity of DNA methyltransferase in the life cycle of Chlamydomonas reinhardi, Proc. Natl. Acad. Sci. USA **78**:3118–3122.

Schmidt, R. J., Richardson, C. B., Gillham, N. W., and Boynton, J. E. (1983) Sites of synthesis of chloroplast ribosomal proteins in Chlamydomonas, J. Cell Biol. **96**:1451–1463.

Swinton, D. C., and Hanawalt, P. C. (1972) In vivo specific labeling of Chlamydomonas chloroplast DNA, J. Cell Biol. **54**:592–597.

Tewari, K. K., Kolodner, R. D., and Dobkin, W. (1976) Replication of circular chloroplast DNA, in Genetics and Biogenesis of Chloroplasts and Mitochondria (T. Bücher, W. Neupert, W. Sebald, and S. Werner, eds.), Elsevier/North-Holland, Amsterdam, pp. 379–386.

Thomas, J. R., and Tewari, K. K. (1974) Ribosomal RNA genes in the chloroplast DNA of pea leaves, Biochim. Biophys. Acta **361**:73–83.

Tiboni, O., and Ciferri, O. (1982) A rapid procedure for the purification of elongation factor Tu (EF-Tu/chl) from spinach chloroplasts, FEBS Lett. **146**:197–200.

Additional Reading

Bogorad, L., Gubbins, E. J., Krebbers, E., Larrinua, I. M., Mulligan, B. J., Muskavitch, K. M. T., Orr, E. A., Rodermel, S. R., Schantz, R., Steinmetz, A. A., Vos, G. D., and Ye, K. K. (1983) Cloning and physical mapping of maize plastid genes, Meth. Enzymol. **97**:524–554.

Buetow, D. (1982) Molecular biology of chloroplasts, in Photosynthesis: Development, Carbon

Metabolism, and Plant Productivity, Vol. 2 (Govindjee, ed.), Academic Press, New York, pp. 43–88.

Capel, M. S., and Bourque, D. P. (1982) Characterization of *Nicotiana tabacum* chloroplast and cytoplasmic ribosomal proteins, *J. Biol. Chem.* **257:**7746–7755.

Edelman, M., Hallick, R. B., and Chua, N.-H. (eds.) *Methods in Chloroplast Molecular Biology*, Elsevier, Amsterdam.

Ellis, R. J. (1977) Protein synthesis by isolated chloroplasts, *Biochim. Biophys. Acta* **463:**185–215.

Ellis, R. J. (1981) Chloroplast proteins: Synthesis, transport and assembly, *Annu. Rev. Plant Physiol.* **32:**111–137.

Gillham, N. W. (1978) *Organelle Heredity*, Raven Press, New York.

Hoober, J. K. (1976) Protein synthesis in chloroplasts, in *Protein Synthesis* (E. H. McConkey, ed.), Marcel Dekker, New York, pp. 169–248.

Jolly, S. O., McIntosh, L., Link, G., and Bogorad, L. (1981) Differential transcription *in vivo* and *in vitro* of two adjacent maize chloroplast genes: The large subunit of ribulosebisphosphate carboxylase and the 2.2-kilobase gene, *Proc. Natl. Acad. Sci. USA* **78:**6821–6825.

Keller, S. J., and Ho, C. (1981) Chloroplast DNA replication in *Chlamydomonas reinhardtii*, *Internatl. Rev. Cytol.* **69:**157–190.

Lake, J. A. (1981) The ribosome, *Sci. Am.* **245:**(2)84–97.

Palmer, J. D., Singh, G. P., and Pillay, D. T. N. (1983) Structure and sequence evolution of three legume chloroplast DNAs, *Mol. Gen. Genet.* **190:**13–19.

Potter, J. W., and Black, C. C., Jr. (1982) Differential protein composition and gene expression of leaf mesophyll cells and bundle sheath cells of the C4 plant *Digitaria sanguinalis* (L.) Scop., *Plant Physiol.* **70:**590–597.

Rochaix, J.-D. (1981) Organization, function and expression of the chloroplast DNA of *Chlamydomonas reinhardii*, *Experientia* **37:**323–332.

Schwarz, Zs. and Kössel, H. (1980) The primary structure of 16S rDNA from *Zea mays* chloroplast is homologous to *E. coli* 16S rRNA, *Nature (London)* **283:**739–742.

Whitfeld, P. R., and Bottomley, W. (1983) Organization and structure of chloroplast genes, *Annu. Rev. Plant Physiol.* **34:**279–309.

7

Development of Chloroplasts

Structure and Function

I. Introduction

As a developmental system, the chloroplast has several exceptional features. First, its developmental path spans a wide structural range, from the extremely primitive to the complex. During the process, a large amount of the eventual major membrane system in the plant cell is produced. Thus, the morphological changes are striking and are readily described by electron microscopy. Second, several biochemical changes are equally as impressive as the structural changes. Extensive synthesis of lipids, including chlorophyll, is required for formation of the membrane. Proteins that become integrated into the growing membrane are synthesized both inside and outside the organelle. And photosynthetic activities develop in parallel with the structural changes. As described in Chapter 3, chloroplasts and thylakoid membranes are relatively easy to purify, which makes possible careful biochemical analysis. Third, a feature that has attracted numerous investigators is that both the nuclear–cytoplasmic and chloroplastic systems of genetic expression in plant cells are involved in this developmental process. Key questions that have been examined refer to the role of each genetic system and where each component of the chloroplast is synthesized. Chloroplast and cytoplasmic ribosomes are structurally different and can be specifically inhibited to examine the source of a protein. Confirmation that a protein is made inside the chloroplast can be obtained by examining products synthesized by the isolated organelle. And it is now feasible to prove that the gene for a protein exists in chloroplast DNA by hybridization of mRNA with fragments of the DNA. The nucleotide sequence of the DNA is easily determined, and, when compared with only a partial amino acid sequence of the protein, the exact position of the gene on the chloroplast genome can be located. Consequently, the development of chloroplasts is interesting to molecular biologists. And finally, the system is amenable to investigation. Chloroplast development can be manipulated easily by several means. In particular, since the process requires light, the most effective means of probing the process is simply to alter the quality and intensity of light.

II. Etiolation of Chloroplasts

Chloroplast development requires the synthesis of chlorophyll. In lower eukaryotic plants species (i.e., the algae), chlorophyll synthesis and membrane formation generally occur in the light or dark. The chloroplasts enlarge and divide as the cells grow and divide. In these cells, the chloroplasts consequently are always maintained in a relatively highly developed state, whether or not the cells are grown in the light. In contrast, a few species of algae and most higher plants (angiosperms) require absorption of light energy for the synthesis of chlorophyll. If the initial growth of these latter types of cells occurs in continuous darkness, development of the chloroplast is arrested at an early stage. These plant cells lack chlorophyll and are white or pale green to yellow in color. When subsequently exposed to light, they are capable of making chlorophyll and of developing a mature chloroplast.

A simple experiment can demonstrate this phenomenon. When seeds are germinated in complete darkness, the young plant grows rapidly for several days utilizing food reserves in the seed. However, the young seedling usually is nearly white, with only traces of a green pigment. When transferred to light, the young plant will gradually become fully green over a period of two to three days (Fig. 7.1). The major change that occurs in the plant as it "greens" is development of the chloroplast.

In those plant cells that require light for development of the chloroplast, one of the light-mediated reactions resides within the pathway of chlorophyll synthesis. Promotion of the synthesis of chlorophyll by light in higher plants is a powerful controlling factor in the overall process of chloroplast differentiation and is, in fact, the cardinal event. The presence of chlorophyll is a prerequisite for the assembly of thylakoid membranes as well as for their function.

Plant cells that contain little or no chlorophyll are referred to as **etiolated** (from the French word meaning bleached or whitened). Among the algae, only a few species, such as *Euglena* and *Ochromonas*, normally become etiolated during growth in the dark. However, some strains of other algal species have been rendered etiolatable by mutation. The latter include mutant strains of several green algae, notably *Chlamydomonas*, *Chlorella*, and *Scenedesmus*. These strains, therefore, are analogous to the higher plants, in that light is required for synthesis of chlorophyll. Because of the ease of handling cultures of algal cells, they have provided excellent systems for studying processes involved in the development of the chloroplast. Since the green chlorophylls are found only within thylakoid membranes, the process of chloroplast development is commonly referred to as **greening.**

III. Morphology of Chloroplast Development

A. Higher Plants

The youngest and most primitive form of plastids, which is found in etiolated proliferating cells at the base of the plant leaf, is the **proplastid.** This structure, about 1 μm in diameter, is simply a double-membrane-bounded vesicle with little if any internal membranous material. The course of development of a proplastid into a chloroplast can take two alternate paths, depending on when seedlings are exposed to light (Fig. 7.2). The most commonly studied path involves germination and initial growth of the seedling in the dark for about 1 wk prior to exposure to light. During this period in the dark, chlorophyll is not synthesized, but an extensive and organized network of intersecting tubules accumulates within the plastids (Fig. 7.3, also Fig. 2.14). This structure, called a **prolamellar body,** resides within a plastid form designated the **etioplast.** Development of etioplasts can be observed by examining a leaf from base to tip in a dark-grown seedling. The position of a cell along the length of the leaf is a function of its age. At the base, the site of cell proliferation, the youngest cells contain small and relatively undifferentiated proplastids. The prolamellar body within etioplasts becomes progressively more highly developed in older cells toward the tip of the leaf. Each leaf, therefore, contains the developmental sequence.

Prolamellar bodies initially were thought to contain material that was used to form the first **lamellae** or thylakoid membranes. This notion has been modified by additional work. Isolated prolamellar bodies are rich in lipid (about

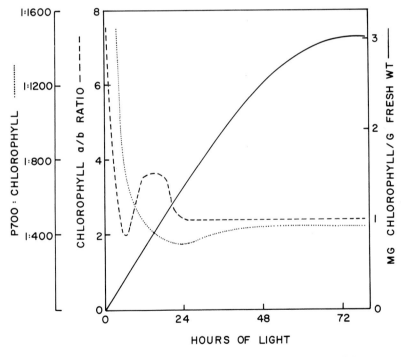

FIG. 7.1. The linear time course of greening of seven-day-old seedlings of jack bean. Seeds were germinated and grown in the dark for seven days and then exposed to incandescent light. Leaves were collected and extracted with 80% acetone for the analysis of chlorophyll *a* and *b*. P700 was determined by light-induced absorbance changes in purified thylakoid membranes that were solubilized in the detergent sodium dodecyl sulfate. (Redrawn from Alberte *et al.*, 1972.)

75% by weight) and contain relatively little protein (about 25% by weight)(Lütz and Nordmann, 1983). One major protein, or only a few, has been detected in the purified structure. The major lipids are mono- and digalactosyl diglycerides, which occur in approximately a 2:1 ratio. Although these lipids are the major thylakoid lipids as well, it is clear that not all thylakoid components, particularly proteins, are stored in the etioplast.

The formation of prolamellar bodies apparently is caused by a persistant synthesis of these glycolipids in dark-grown plants. The highly ordered structures seem to form as an outgrowth of preexisting membranes, in the sense of

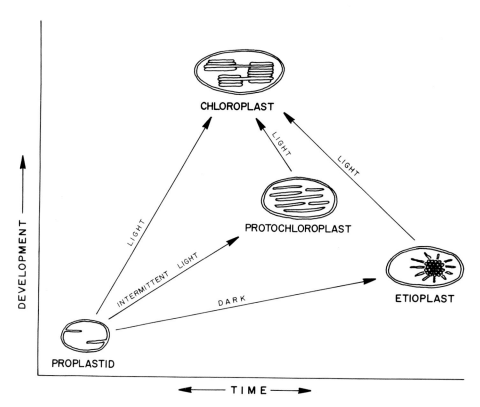

FIG. 7.2. The pathways of chloroplast development in higher plants. The organelle in the youngest cells of a leaf exists as a simple, double-membrane vesicle. Often as proplastids begin to develop, invaginations of the inner membrane are found. In continuous light, the amount of thylakoid membranes steadily increases. No prolamellar bodies form. Early in development thylakoids begin differentiation into grana. In contrast, in seedlings initially grown in the dark, the proplastid develops into an etioplast, which usually contains one or two prolamellar bodies. Upon transfer of the dark-grown plant to continuous light, the prolamellar body in the etioplast disperses, and the plastid gradually fills with thylakoid membranes. This stage of development is similar to chloroplast development in continuous light as just above. In a third pathway, proplastids in plants grown in light–dark cycles (e.g., 2-min-light–98 min-dark) develop slowly into protochloroplasts, which contain a few thylakoids, no chlorophyll b, and no grana. Exposure of protochloroplasts to continuous light results in a rapid development into the mature chloroplast.

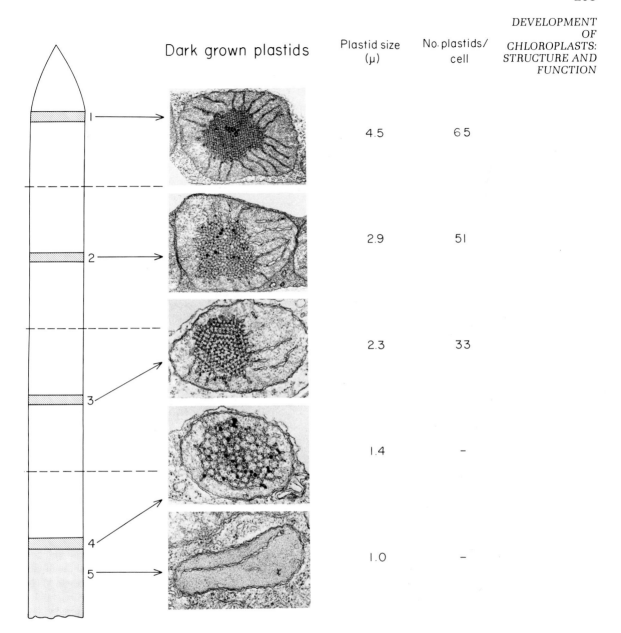

	Plastid size (μ)	No. plastids/ cell
1	4.5	65
2	2.9	51
3	2.3	33
4	1.4	–
5	1.0	–

Dark grown plastids

FIG. 7.3. Development of etioplasts in dark-grown barley seedings. The age of the cells is related to their positions in the leaf, with the oldest cells at the tip and the youngest at the base. The increase in complexity of the prolamellar bodies with age in clearly illustrated. The size and number of plastids increase with age of the cells. (From Robertson and Laetsch, 1974. Reprinted with permission.)

an overflow of membrane lipids (Fig. 7.3). Extending from the prolamellar bodies are rudimentary membranes referred to as **prothylakoids.** The etioplast contains no chlorophyll but does contain a small amount of protochlorophyll(ide),* a precursor of chlorophyll (see p. 227). The pale green color of some etiolated plants is caused by the presence of small amounts of protochlorophyll(ide).

Lütz *et al.* (1981) have obtained evidence that the bulk of the protochlorophyll(ide) (about 75%) and most of the enzyme protochlorophyllide reductase are located in the prothylakoid membranes in etioplasts from oat seedlings. This distribution also occurs in other plant species, but it may not be universal. Murakami and Ikeuchi (1982) have found the opposite distribution in squash seedlings.

The function of the prolamellar body and its role in development of chloroplasts are still uncertain and remain an important area of research. Upon exposure of etiolated seedlings to continuous light, protochlorophyll(ide) is converted to chlorophyll *a*, and the material in the prolamellar body disperses. The major polypeptides are degraded, and the lipid material is probably recruited for thylakoid membrane assembly as the leaf begins to green. Synthesis *de novo* of chlorophyll is then initiated. Synthesis of chlorophyll *b* begins shortly thereafter and often coincides with fusion of thylakoids to form grana. During the early stage of greening, the chlorophyll *a/b* ratio drops rapidly, as chloroplyll *b* is synthesized, until the final ratio of 2 to 4 is achieved (Fig. 7.1). The end result of this pathway, after about two to three days of continuous illumination, is the mature, functional chloroplast.

A perhaps more natural path of development occurs in plants grown from germination in the light. Again, the developmental sequence is contained within each leaf of a seedling. Fig. 7.4 shows parameters for wheat leaves in plants grown under a 16-hr-light-8-hr-dark regime. As the cells move outward from the base, they cease to divide. However, chloroplast division continues for two more days at a decreasing rate, as also occurs in the dark (see Fig. 7.3). As a result, there is a severalfold increase in the number of plastids per cell as the cells mature. Synthesis of chloroplast DNA continues as long as organelle division proceeds. Each chloroplast contains multiple copies of the organellar genome–in some plants more than a hundred—and thus the mature cell may have several thousand total chloroplast DNA molecules.

The amount of chlorophyll per plastid increases progressively with age of the cell and reaches a maximum of nearly 1×10^{-12} g (about 5×10^8 molecules) (Fig. 7.4). In the young chloroplasts, near the base of the leaf, the chlorophyll *a/b* ratio is 10 to 15. As chlorophyll accumulates, the ratio drops to the value characteristic of mature chloroplasts of 2 to 4. The amount of thylakoid membranes within a developing plastid increases steadily with time and in parallel with the increase of chlorophyll (Fig. 7.5). Single thylakoids appear first. The appression of thylakoids into grana is correlated with the drop in the chloro-

---

*Protochlorophyllide *a*, with the suffix *-ide*, refers to the structure lacking the long-chain, isoprenoid alcohol **phytol.** In chlorophyll, without the *-ide* ending, phytol is esterified to the propionic acid group on ring D. When the specific form is not known, the *-ide* is placed in parentheses.

phyll *a/b* ratio and the appearance of the light-harvesting chlorophyll *a/b*–protein complex in the membranes. Addition of the light-harvesting complex to the photosystem 2 reaction centers also generates the large, 18-nm-diameter particles revealed by freeze–fracture on the E face of thylakoid membranes (see p. 23). As the functional complexes form, the membranes expand until the chloroplast is fully developed with respect to both structure and function.

Variations of these paths occur when the environmental conditions are manipulated. For example, exposure of etiolated seedlings to intermittent light, such as 2-min pulses of white light separated by 98-min periods of darkness, allows a slow rate of chlorophyll synthesis. Etioplasts are converted under these conditions to **protochloroplasts,** an early intermediate stage in development (Fig. 7.2). During this regimen, the prolamellar body disperses and primary thylakoids form that are photosynthetically functional. However, the membranes that form are devoid of both chlorophyll *b* and the polypeptides of the light-harvesting chlorophyll *a/b*-protein complex (Fig. 7.6). In this respect, plants that are greened under intermittent light resemble chlorophyll *b*-deficient mutant strains. In these mutant strains, chloroplasts also usually contain fewer grana, and the thylakoids are deficient in the chlorophyll *a/b*-protein

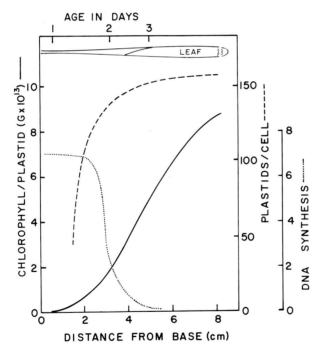

FIG. 7.4. The relationship between chloroplast parameters and cell age in wheat seedlings. A drawing of a primary wheat leaf is shown at the top of the graph, with a scale for the age in days of cells in the leaf. Very little chlorophyll is present in young cells at the base of the leaf, but chlorophyll accumulates rapidly as the cells move outward. The number of plastids per cell also increases rapidly between day 1 and day 2. Plastid division during this period is supported by a high rate of chloroplast DNA synthesis, which ceases after day 2. (Adapted from Boffey *et al.*, 1980.)

complex. In normal plants the chlorophyll a/b–protein complex is concentrated in granal regions of the membrane, and the evidence indicates that this complex plays a role in formation of grana.

B. Algae

Several species of algae can be transformed from green to etiolated cells by continued growth and cell division in culture in the dark. Since light is not available for photosynthesis, dark-grown algae must be provided with a source of organic carbon as acetate or glucose. In those algae that become etiolated normally, such as *Euglena* and *Ochromonas*, the chloroplast reverts to a pro-

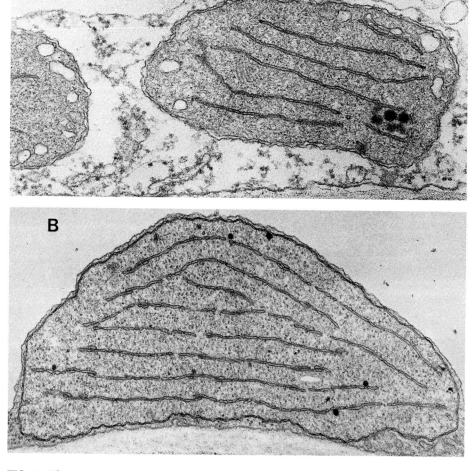

FIG. 7.5. Electron micrographs of chloroplasts at different stages of development in barley leaves. A: A plastid in the earliest stage of development. Thylakoid formation has begun, but all exist as single vesicles. B: A plastid in an intermediate stage of development, with very short regions of

plastidlike structure during growth in the dark. However, in strains of *Chlamydomonas* and *Chlorella*, rendered etiolatable by mutation, and therefore unable to make chlorophyll in the dark, the plastid decreases only slightly in volume but loses nearly all thylakoid membranes (Fig. 7.7A). Extensive prolamellar bodies generally are not found in the algae.

Development of plastids upon return to light of etiolated algal cells, such as those of *Chlamydomonas reinhardtii*, morphologically entails simply an increase of thylakoid membranes with time in parallel with the accumulation of chlorophyll (Fig. 7.7B,C). Membrane fusion to form grana begins at an early stage, and there is less differentiation between granal and stromal thylakoids than is found in the higher plants. The rate of development varies widely among the algal organisms. Etiolated *Euglena* and *Ochromonas* require two to three days to complete development of the chloroplasts, whereas in a mutant strain of *Chlamydomonas reinhardtii* only several hours are needed to perform the complete greening process (Fig. 7.8).

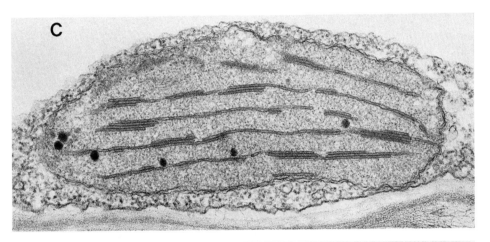

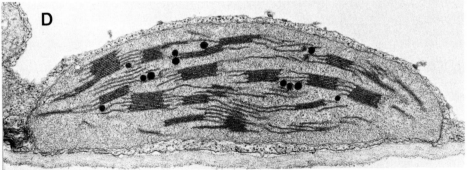

thylakoid fusion to form the first, rudimentary grana. C: A plastid at a later, nearly fully developed, stage. Grana formation by thylakoid appression is now extensive. D: A mature plastid with abundant thylakoid membranes, which are distinctly differentiated into granal and stromal thylakoids. This developmental process requires between 24 and 48 hr of continuous light. (From Robertson and Laetsch, 1974. Reprinted with permission. Micrographs courtesy of W. M. Laetsch.)

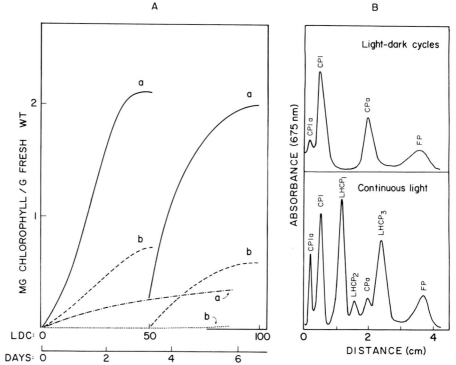

FIG. 7.6. Development of thylakoid membranes in intermittent light. Etiolated 5-day-old seedlings were exposed to either continuous or intermittent white light. A: In those seedlings exposed to intermittent light (cycles of 2-min light–98-min-dark), the amount of chlorophyll *a* increased slowly (–·–··), but chlorophyll *b* was not detected until 4–5 days of treatment (······). In contrast, seedlings exposed to continuous light, either from time 0 or after 50 light-dark cycles (LDC), rapidly accumulated chlorophyll *a* (——) and *b* (----) at an *a*/*b* ratio of 3. B: Thylakoid membranes were purified from the greening seedlings and subjected to electrophoresis in polyacrylamide gels in the presence of sodium dodecyl sulfate under conditions that preserved chlorophyll-protein complexes. Small complexes migrate more rapidly through the gel than larger complexes. The gels were scanned with 625-nm light to determine the distribution of chlorophyll. Under light–dark cycles, as shown in the upper panel, CP1 and CPa, the core complexes of photosystems 1 and 2, respectively, were formed. (CP1a is an oligomeric form of CP1.) Membranes containing these core complexes were photosynthetically active. Free pigment, i.e., chlorophyll free of protein, migrated as the peak labeled FP.

The lower panel shows that when chlorophyll *a* and *b* were made in continuous light, whether from time 0 or after light-dark cycles, photosystem 2 light-harvesting chlorophyll-protein complexes (LHCP) also accumulated and became the predominant chlorophyll-protein species. (LHCP$_1$ and LHCP$_2$ are oligomeric forms of LHCP$_3$.) The light-harvesting complexes surround the core complexes and enhance photosynthetic efficiency. The completed structures give rise to the large 18-nm particle on the E face of the membrane. LHCP$_3$ is also called CP2. (Refer to Fig. 3.19 for additional information on chlorophyll-protein complexes.) (Adapted from Akoyunoglou and Argyroudi-Akoyunoglou, 1978, 1979.)

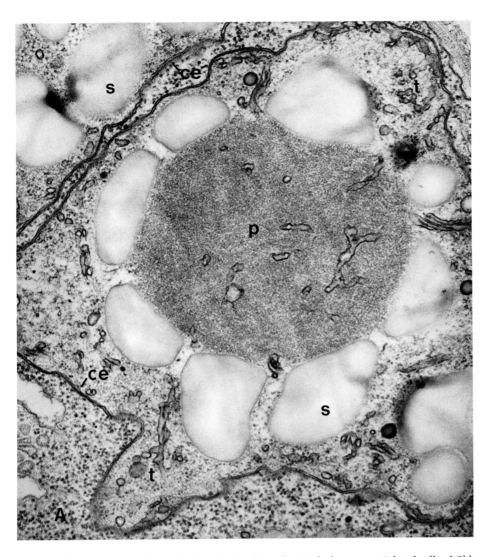

FIG. 7.7. Electron micrographs of the developing chloroplast in dark-grown, etiolated cells of *Chlamydomonas reinhardtii* y-1. Etiolated cells were produced by growth of green cells in the dark for 4–5 days. Neither chlorophyll nor thylakoid membranes are made in the dark; they are diluted among the progeny as the cells divide. A: The chloroplast in etiolated cells, at the time of exposure to light, contains only small, vestigial thylakoids (t). The central pyrenoid body (p) is surrounded by starch particles (s), which are also dispersed throughout the stroma. The double-membrane chloroplast envelope (ce) encloses the organelle. B: Thylakoid membrane formation occurs in parallel with chlorophyll synthesis. At 5 hr of continuous white light, as shown in this micrograph, membrane formation is already extensive, and the developing thylakoids have fused to form grana. The amount of starch markedly decreases during the first several hours in light. C: The progression of membrane and grana formation continues until the fully developed chloroplast, characteristic of light-grown, green cells, is achieved by 8–10 hr in light. In such cells, the organelle is nearly filled with thylakoid membranes. These micrographs correspond to the greening profile at 25°C shown in Fig. 7.8. (From Ohad *et al.*, 1967. Reprinted with perimission of The Rockefeller University Press.)

IV. Site of Assembly of Thylakoid Membranes

Since many of the thylakoid polypeptides are synthesized on cytoplasmic ribosomes, whereas chlorophyll and the membrane lipids are synthesized in the chloroplast, an important question is where these components come together. Analysis by electron microscopy of the early stages of chloroplast development in a number of higher plants and some algae suggested that invaginations of the inner membrane of the chloroplast envelope formed during periods of rapid thylakoid assembly. These observations could be interpreted

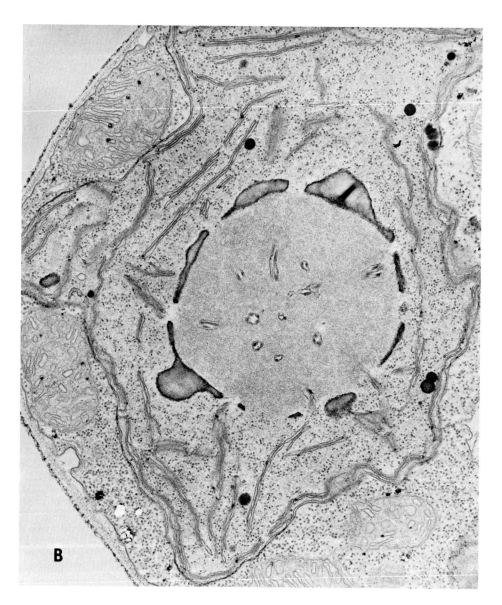

FIG. 7.7 (Continued)

as an indication that the inner membrane of the plastid envelope is a site of membrane assembly and that growth of the thylakoid membranes is accomplished by a flux of material as small vesicles from the inner envelope membrane. However, whether proteins made in the cytoplasm are incorporated into the inner membrane or pass through, to be integrated into thylakoid membranes from the stroma, has not been established.

As the first step in assessing biochemically the role of the envelope in thylakoid assembly, Douce and his colleagues obtained the two membranes of the envelope in highly purified form (Block *et al.*, 1983a, 1983b). Table 7.1 shows

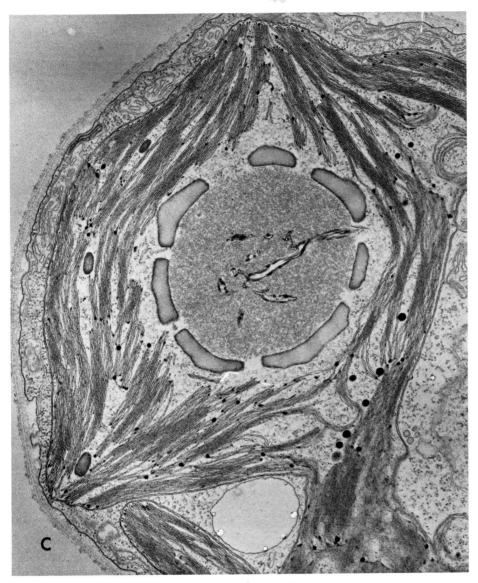

FIG. 7.7 (*Continued*)

the polar lipid composition of these membranes. Both outer and inner membranes are practically devoid of chlorophyll and both have a yellow color as the result of their carotenoid components. About half of the polar lipid in the inner membrane is monogalactosyl diglyceride (see p. 61), whereas the outer membrane contains less of this lipid but much more phosphatidyl choline (see p. 61). In general, the lipid composition of the inner envelope membrane is remarkably similar to that of thylakoid membranes. The enzymes involved in synthesis of thylakoid lipids, such as **phosphatidic acid phosphatase,** which provides the diacylglycerol moiety of the lipids, and **UDP-galactose:diacylglyerol galactosyltransferase,** which synthesizes monogalactosyl diglyceride, are both located on the inner membrane. Thus, the envelope possibly is a major source of thylakoid lipids.

The polypeptide compositions of the chloroplast envelope, thylakoid membranes, and the soluble stroma are quite distinct (Fig. 7.9). If the envelope were a major site of assembly of the thylakoid membranes, the presence of some

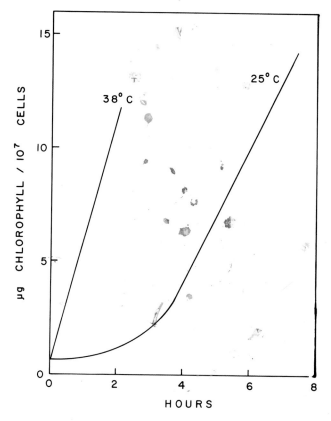

FIG. 7.8. The time course of greening of etiolated *Chlamydomonas reinhardtii* y-1 cells. At 25°C, chlorophyll synthesis exhibits a lag period of 1–2 hr after etiolated cells are transferred to light, but thereafter chlorophyll accumulates and reaches the level in fully green, light-grown cells (25–30 μg chlorophyll per 10⁷ cells) by 8–10 hr. Photosynthetic activity increases in parallel with chlorophyll. At higher temperatures, however, chlorophyll is synthesized more rapidly. When etiolated cells are preincubated 1–2 hr in the dark at 38°C, upon subsequent exposure to light the kinetics of chlorophyll synthesis are linear, without an initial lag. Under these conditions the fully green state would be achieved in 3–4 hr.

TABLE 7.1. Polar Lipid Composition of Membranes from Spinach Chloroplasts

Membrane	MGDG	DGDG	SL	PC	PG	PI	Lipid/protein
	(Weight percent fatty acids)						(mg/mg)
Whole envelope	36	29	6	18	9	2	1.4
Outer membrane	17	29	6	32	10	5	2.5
Inner membrane	49	30	5	6	8	1	0.8
Thylakoids	52	26	6.5	4.5	9.5	1.5	0.4

MGDG = monogalactosyl diglyceride; DGDG = digalactosyl diglyceride; SL = sulfolipid; PC = phosphatidyl choline; PG = phosphatidyl glycerol; and PI = phosphatidyl inositol. (Adapted from Block *et al.*, 1983b.)

thylakoid polypeptides would be expected. Since thylakoid polypeptides cannot be detected in envelope membranes, either there is a remarkably efficient transfer of membrane material from the envelope to thylakoids or the proteins pass through the envelope and become incorporated into membranes inside the organelle. In fact, since many of the proteins in the stroma of the chloroplast also are synthesized in the cytoplasm, a path through the envelope may be common to both soluble and membrane proteins. The destination inside the chloroplast perhaps is then determined by the properties of the individual proteins (discussed in more detail in Chapter 8).

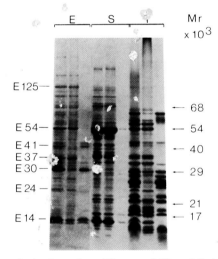

FIG. 7.9. Electrophoretic analysis of envelope (E), stromal (S), and thylakoid (T) polypeptides from spinach chloroplasts. The chloroplasts were isolated by sedimentation through a density gradient of silica sol and then lysed in hypotonic medium. Envelope and thylakoid membranes were separated by sedimentation through a sucrose gradient. A portion of each sample was extracted with chloroform/methanol (2:1,v/v), which extracts the light-harvesting chlorophyll *a/b*-binding protein (LHCP) and several other membrane polypeptides. The polypeptides were separated by electrophoresis through a 7.5–15% polyacrylamide gel at 4°C in the presence of lithium dodecyl sulfate. For each sample, the lanes left to right contained the total fraction, the chloroform/methanol-insoluble polypeptides and the chloroform/methanol-soluble polypeptides. The LHCP, the major polypeptide of 26,000 molecular weight in thylakoid membranes, cannot be detected in the envelope or stromal fractions. The major polypeptides of 54,000 and 14,000 molecular weight in the stromal fraction are the large and small subunits, respectively, of ribulose 1,5-bisphosphate carboxylase. The major polypeptide (E30) in the envelope fraction is the phosphate translocator, which has a molecular weight of 30,000 and is a component of the inner membrane. (From Joyard *et al.*, 1982. Reprinted with permission.)

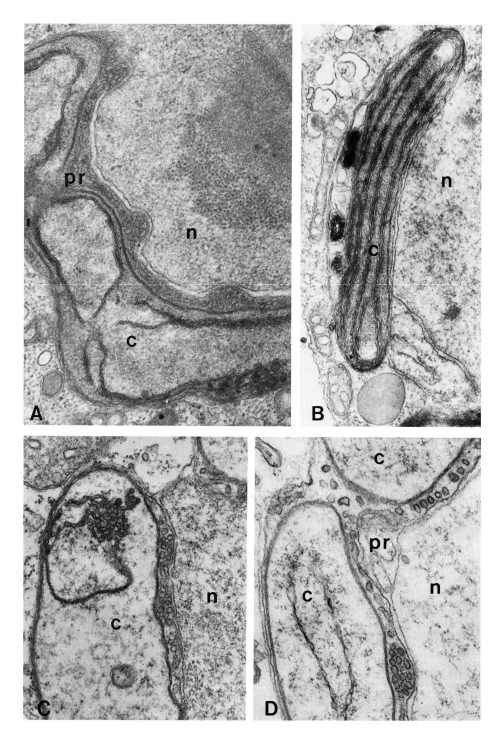

FIG. 7.10. The effects of inhibitors of chloroplast or cytoplasmic ribosomes on chloroplast development and the periplastidal reticulum in the chrysophycean alga *Ochromonas danica*. A: A section through the nucleus and adjacent chloroplast (c) in an etiolated cell exposed to light for 24 hr in the presence of chloramphenicol (300 μg/ml), an inhibitor of chloroplast ribosomes. Thylakoid membrane production was severely inhibited, but the periplastidal reticulum (pr) became filled

V. Chloroplast Development and the Chloroplast Endoplasmic Reticulum

A special problem seems to exist for the transport of cytoplasmically synthesized plastid proteins across the chloroplast envelope in those types of algae that have additional membranes surrounding the organelle. In *Euglena* and the dinoflagellates, a third membrane encloses the chloroplast. This membrane may have been derived from the plasma membrane during enclosure of chloroplasts into endocytic vesicles at the time chloroplasts entered the cell (See Fig. 2.17). Nevertheless, at least in *Euglena*, the cooperation of chloroplast and cytoplasm is required to form chloroplast structures. Ribulose 1,5-bisphosphate carboxylase is synthesized in *Euglena* by cooperation between the cellular compartments as in green algae and higher plants. Also, the major chlorophyll *a*/*b*-binding protein is made on cytoplasmic ribosomes. And membrane assembly follows the same general pattern as that in the green algae and higher plants.

Chloroplast development in *Euglena* requires about two days of continuous light for completion and in this respect is similar to the process in higher plants but slower than the process in green algae. Whether the additional membrane slows the rate of flux of cytoplasmically made polypeptides into the chloroplast, which would result in an extended greening time, it not known. The additional membrane around the chloroplast, however, does not seem to be a significant barrier in this process.

In most classes of algae, such as those in the Cryptophyta and Chrysophyta classes, the chloroplast is surrounded by two additional membranes. These additional membranes, which form the **chloroplast endoplasmic reticulum** (see p. 29), would seem to present a particular barrier to the flux of proteins into the chloroplast. Chloroplast development has not been studied extensively in these organisms, and the extent of influx of proteins from the cytoplasm is not known. Gibbs (1979) has studied in some detail the possible function of the chloroplast endoplasmic reticulum in *Ochromonas danica*, a Chrysophyte alga. She has observed cytoplasmic ribosomes attached to the outer, cytoplasmic surface and has suggested that the completed proteins synthesized by these ribosomes are vectorially released into the lumen of the chloroplast endoplasmic reticulum. During chloroplast development, vesicles appear between this structure and the chloroplast envelope, which suggests that they shuttle material between these two membrane systems (Fig. 7.10). In fact, vesicles appear to fuse

with vesicles. B: A section through the chloroplast and adjacent nucleus of an etiolated cell exposed to light for 48 hr in the presence of cycloheximide (8 µg/ml), an inhibitor of cytoplasmic ribosomes. In contrast to A, in this cell the chloroplast appears nearly normal for this stage of development. The periplastidal reticulum is devoid of vesicles. C: In cells treated with chloramphenicol for only 6 hr, the hypertrophy of the periplastidal reticulum is already quite pronounced. D: A section through a cell treated 24 hr with spectinomycin (100 µg/ml), another inhibitor of chloroplast ribosomes. As with chloramphenicol, chloroplast development was severely retarded, and the periplastidal reticulum contained an abundance of vesicles. Possibly, vesicles in this compartment are involved in transport of proteins from the cytoplasm to the chloroplast. Inhibition of chloroplast activities may block delivery into the organelle. Interestingly, the most extensive accumulation of vesicles occurred between the chloroplast and the nucleus. (From Gibbs, 1979. Reprinted with permission. Micrographs courtesy of S. P. Gibbs.)

with the chloroplast envelope. In cycloheximide-treated cells, the vesicles disappear, as would be expected if production of their contents is inhibited. Moreover, in chloramphenicol-treated cells, the utilization of this material seems to be blocked and the vesicles accumulate (Fig. 7.10). These interesting morphological observations should prompt an extensive biochemical analysis of the development of chloroplasts in these organisms.

VI. Chromatophore Development in Photosynthetic Bacteria

Invagination of the envelope inner membrane may not account for the major flux of cytoplasmically made polypeptides to thylakoid membranes, but this is the process by which chromatophores form in photosynthetic bacteria. These organisms have photosynthetic activity only when grown anaerobically. In the purple photosynthetic bacteria (e.g., *Rhodospirillum* and *Rhodopseudomonas*, see Fig. 2.6A), biosynthesis of bacteriochlorophyll occurs only at very low partial pressures of oxygen. However, light is not required for the formation of chromatophores. The number of chromatophores per cell varies depending on the growth conditions. Phototrophically grown cells, which require photosynthesis for growth, adapt to the intensity of light by adjusting the number of chromatophores and the size of the photosynthetic units, in a manner similar to "sun" and "shade" plants. The greatest development of chromatophores occurs in cells grown anaerobically at low light intensity.

Formation of chromatophores is initiated at localized sites on the cell membrane. Bacteriochlorophyll and specific polypeptides enter the membrane simultaneously at these sites, until the number of photosynthetic units achieves a "threshold" level. Above this number, an invagination of the localized area of the membrane occurs. As soon as these differentiated structures form, subsequent additions of pigments and polypeptides for new photosynthetic units occur exclusively on these invaginations. Thus growth of the chromatophores proceeds by the coordinated assembly of functional units within the intraacytoplasmic membrane. These membrane segments, however, remain continuous with the cell membrane.

VII. Development of Function

Thylakoid membranes are structurally asymmetric. Most protein components, and perhaps some of the lipids, are arranged so that they reside functionally only on one side of the membrane (see p. 95). Function, therefore, develops as the components achieve their proper orientation. The photochemical reaction center activities can be detected shortly after the accumulation of chlorophyll becomes evident. Photosystem 1 activity is usually detectable within 15 min after the start of illumination of etiolated leaves, and photosystem 2 activity, measured by evolution of oxygen, is evident by 30 min of light. The efficiency of photosynthetic electron transport is initially low but rapidly achieves the characteristics of mature membranes. The photosynthetic activities then continue to increase in parallel with chlorophyll and membrane assembly. Although membrane assembly is a process of temporally random insertion of individual components into the expanding membrane, the function increases in parallel with the developing structure. This process of membrane assembly includes mechanisms for checks and balances in the production of each component and thus is a highly coordinated process.

The activity of photosystem 2 commonly is assayed with artificial electron acceptors that intercept electron flow on the reducing side of photosystem 2 reaction centers, a reaction first described by Hill (1951). Water is the electron donor in this assay. A simple spectrophotometric assay employs the light-dependent reduction of dichlorophenolindophenol (Fig. 7.11), a dye that when reduced becomes colorless, by broken-cell samples or membrane preparations. Since oxygen is also liberated in this reaction, a polarographic assay for the evolved oxygen provides another means to measure activity of photosystem 2.

Photosystem 1 activity is measured spectrophotometrically by the light-dependent reduction of methyl viologen (Fig. 7.11), a dye that has a violet color when reduced. This dye has a low oxidation–reduction potential and cannot be reduced by photosystem 2. Alternatively, the ability of preparations to photo-reduce $NADP^+$ can be used to measure photosystem 1 activity. In the latter case, ferredoxin must also be added, to ensure that electron transfer between the reaction center and $NADP^+$ is not a rate-limiting factor. If the activity of only photosystem 1 is to be assayed, an artificial electron donor must be provided, which often is reduced dichlorophenolindophenol in the presence of an excess of ascorbic acid as a reductant. This mixture feeds electrons into the electron transport chain near the oxidizing side of photosystem 1 (refer to Fig. 4.7).

The "reaction centers," in which the photochemical oxidation–reduction reactions occur, are composed of specific proteins complexed with chlorophyll a. Mutant strains, which have defects inherited in a uniparental pattern, have been isolated that are deficient in either photosystem 1 or 2 reaction center activities. These strains lack proteins whose synthesis in normal cells is inhibited by chloramphenicol. This evidence suggests that these reaction center proteins are made in the chloroplast.

The photosystems develop initially as "core complexes," which contain 40–60 chlorophyll molecules/reaction center. In photosystem 1, the major core complex polypeptide has a molecular weight of 66,000–68,000 and binds only

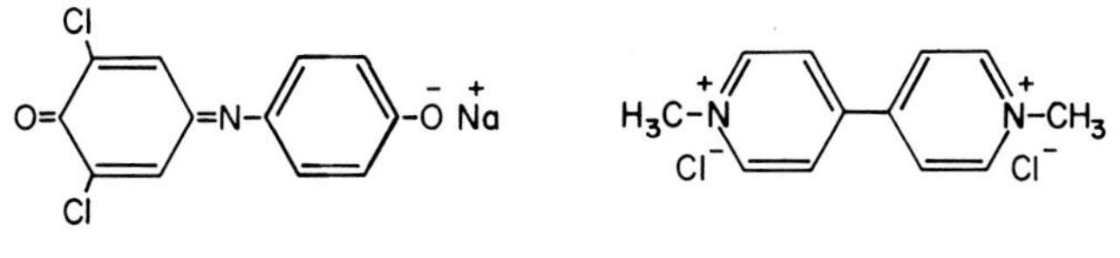

2,6-Dichlorophenolindophenol
(DCPIP, $E_o^1 = +0.22V$)

Methyl viologen
(MV, $E_o^1 = -0.44V$)

FIG. 7.11. Structures of dyes used in assays of the photosystems. Photosystem 2 can be assayed by the photoreduction of 2,6-dichlorophenolindophenol. Methyl viologen is reduced by photosystem 1, but if H_2O is the electron donor, activity of both photosystems is required for dye reduction. To assay specifically for photosystem 1, electrons must be fed directly into photosystem 1 by exogenous electron donors.

chlorophyll *a*. This polypeptide is synthesized on chloroplast ribosomes. The core complexes are progressively surrounded by "light-harvesting complexes." Several polypeptides found in the light-harvesting complex of photosystem 1 have molecular weights in the range of 21,000–24,000 and bind chlorophyll *a* and *b* with an *a/b* ratio of about 4.

The major light-harvesting complex of thylakoid membranes surrounds the core complex of photosystem 2. These light-harvesting complexes contain a polypeptide of 26,000–30,000 in molecular weight, which is the predominant polypeptide in the membrane. Probably between three and six molecules of each of chlorophyll *a* and *b* are bound to each polypeptide, with an *a/b* ratio of 1. This chlorophyll *a/b*–protein complex accounts for one-half to two-thirds of the total chlorophyll and nearly one-half of the integral protein of the membrane. Nearly all chlorophyll *b* is associated with this complex. Although the chlorophylls are made in the chloroplast, the polypeptide is synthesized on cytoplasmic ribosomes. The light-harvesting complexes are important for increasing the efficiency of photosynthesis, but they are not required for photosynthetic activity.

VIII. Development of Photosynthetic Units

Emerson and Arnold carried out experiments in 1932 that set the stage for much of the subsequent work on the functional arrangement of photosynthetic complexes. They asked the question that, if an intense but brief (10 μsec) flash of light were provided such that the photosynthetic apparatus was saturated with light but could turn over the photochemical reaction only once, how much O_2 would be evolved per chlorophyll molecule? By summing the response of cells of the alga *Chlorella* to many flashes, they determined that one O_2 molecule was evolved per about 2400 molecules of chlorophyll. From this result emerged the concept that groups of chlorophyll molecules act cooperatively to evolve O_2 and reduce CO_2. These groups were referred to as *photosynthetic units*.

As knowledge of the mechanisms of photosynthetic reactions developed, it became clear that for each O_2 molecule evolved or CO_2 molecule reduced, four electrons were transferred one at a time through two photosystems and a connecting electron transport chain. Thus, in Emerson and Arnold's experiments, rather than 2400 chlorophyll molecules acting as a unit, only about 600 were involved in the functional transfer of each electron through the complete system, or 300 per reaction center of photosystem 1 or 2. Indeed, to produce one molecule of O_2, a photosystem 2 reaction center must turn over four times, with four successive flashes (Joliot and Kok, 1975). Therefore, the functional photosynthetic unit is currently defined as a reaction center with its surrounding light-harvesting complexes. The size of the photosynthetic unit is determined by the total number of chlorophyll molecules within the functional complex. This number is variable and depends upon the type of plant and its environmental conditions.

The fluorescence properties of chlorophyll has provided a powerful tool for analysis of the assembly and size of photosynthetic units. In tightly inte-

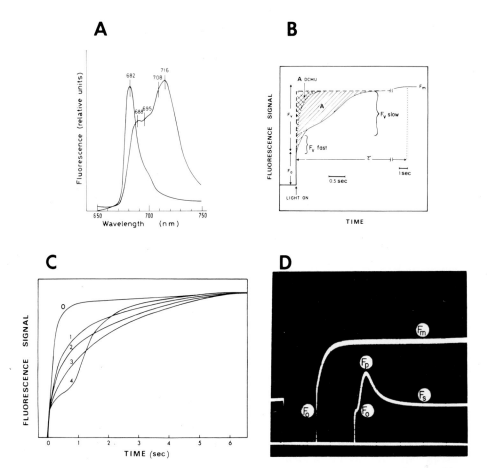

FIG. 7.12. Changes in fluorescence properties during development of the chloroplast in *Chlamydomonas reinhardtii* y-1. A: Unorganized, residual chlorophyll in etiolated cells, when excited with blue light, emits fluorescent light with maximal intensity at 682 nm at 77°K. Under the same conditions, chlorophyll in greened cells, which is organized into photosystems, emits light from several spectral components (see text), with maximal intensity at 716 nm. (From Gershoni and Ohad, 1980. Reprinted with permission of The Rockefeller University Press.) B: A schematic representation of a typical fluorescence induction curve in the absence (——) or presence (-----) of DCMU. The initial level of fluorescence upon illumination is F_0. The increase in fluorescence intensity with time to the maximal level (F_m) is characterized by variable components (F_v), including fast and slow elements. The area above the fluorescence curve (A) is an expression of the induction time. In the presence of DCMU, only the fast component (F_v fast) exists, which reflects the rapid reduction of Q, the primary electron acceptor of photosystem 2. C: Changes in the fluorescence induction curves for purified membranes obtained from greening *Chlamydomonas* cells. The curves were normalized to equal F_v/chlorophyll levels. Curves 0, 1, 2, 3, and 4 represent successive stages in greening (0.8, 1.8, 2.3, 3.5, and 6.4 μg chlorophyll per 10^7 cells, respectively). As membrane development progresses, photosynthetic activity increases and the F_v component becomes more pronounced. (From Cahen *et al.*, 1976. Reprinted with permission.) D: Fluorescence induction curves for intact, greened cells. In the trace on the right, chlorophyll fluorescence rises rapidly from F_0 to reach a peak level (F_p) but then declines to a steady-state level (F_s). In the presence of DCMU (trace on left), photosynthetic electron transport is inhibited after photosystem 2, and only a fast rise to F_m is observed. The steady-state level (F_s) reflects the cells' ability to use absorbed energy to drive electrons into nonmembrane components (such as NADPH and glucose). (From Gershoni *et al.*, 1982. Reprinted with permission.)

grated, functionally efficient units, most of the absorbed light energy is trapped in electron transfer reactions and only a minor portion escapes as heat or fluorescence (see page 83). However, if electron transfer is impeded because the chlorophyll molecules are not integrated into functional units, or if the primary electron acceptors become fully reduced because electron flow is blocked by an inhibitor, then much more of the absorbed energy is reemitted as light (fluorescence). An analysis of the kinetics of fluorescence induction, therefore, provides a sensitive assessment of the development of photosynthetic function.

During initial stages of greening of etiolated plants and algae, before photosynthetic activity has developed, most of the light absorbed by chlorophyll is reemitted as light. This high fluorescence yield indicates a nonfunctional arrangement. Moreover, the maximal level of fluorescence (F_m) is achieved almost immediately upon illumination. The emission spectrum of the reemitted light shows a maximum near 682 nm (Fig. 7.12A). As greening proceeds, and as functional units develop, the emission spectrum shifts to that characteristic of the green cell. As Fig. 7.12A shows, this latter spectrum has maxima at 688, 695, and 716 nm, which represent the light-harvesting chlorophyll *a* of photosystem 2, the chlorophyll *a* of photosystem 2 reaction centers, and the chlorophyll *a* of photosystem 1 reaction centers, respectively. Chlorophyll *b* in light-harvesting complexes is not fluorescent, since energy absorbed by this pigment is efficiently transferred to the longer-wavelength-emitting component of the complex, chlorophyll *a*.

As greening proceeds, the rate at which F_m is achieved becomes slower and the kinetics more complex. As diagramed in Fig. 7.12B, a small but immediate rise (F_0) is contributed by chlorophyll molecules not integrated into functional units. The relative amount of such chlorophyll molecules grows smaller as greening proceeds. But if the absorbed energy can be transferred over a pool of organized chlorophyll molecules and subsequently used to reduce the plastoquinone pool, F_m is delayed until all electron acceptors have been fully reduced. Thus, the more highly integrated the chlorophyll molecules are into electron transport systems, the slower the rate of induction of this variable fluorescence (F_v). The shape of the fluorescence induction curve, then, is a functional marker of chloroplast development. This is illustrated by the curves

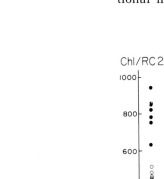

FIG. 7.13. Sizes of photosynthetic units for photosystem 2 in various shade (●) and sun (○) plants. The points represent the numbers of chlorophyll molecules associated with each photosystem 2 reaction center. A typical shade plant, *Trillium ovatum*, contains about 810 chlorophyll molecules per reaction center. Tobacco (*Nicotiana tabacum*), spinach (*Spinacia oleracea*), and pea (*Pisum sativum*), which are typical sun plants, contain 510, 450, and 420 chlorophyll molecules per reaction center, respectively. (Adapted from Malkin and Fork, 1981.)

shown in Fig. 7.12C, which were obtained with membranes purified at various times during greening of *Chlamydomonas* cells.

In intact chloroplasts or whole cells, after an initial fast rise to an initial level (F_0), as shown in Fig. 7.12D the fluorescence increases more slowly to a peak value (F_p) and then gradually declines to a steady-state level (F_s). This decrease is not observed in preparations of thylakoid membranes, which have been separated from their associated electron-consuming reactions (see Fig. 7.12C). But in whole cells, the steady-state level represents only the fraction of the absorbed energy that is not captured during photosynthesis. The kinetic pattern for whole cells also changes characteristically during the early stages of greening of etiolated plants, in a manner similar to that described in the last paragraph for membrane preparations. Consequently, a sensitive means is available to investigate the process.

The addition of DCMU (see p. 89), an inhibitor of electron transport at the primary acceptor, Q, of photosystem 2, causes the fluorescence induction to revert back to primarily the fast-rise component. The shape of the curve in the presence of DCMU is illustrated in Fig. 7.12B, and an actual trace is shown in Fig. 7.12D. In the presence of the inhibitor, the number of chlorophyll molecules per primary acceptor (i.e., per reaction center) can be determined from the kinetics of fluorescence induction. This ratio, the size of the photosynthetic unit, for photosystem 2 is shown in Fig. 7.13. The size of photosystem 2 units varies from 200 to 500 in sun plants and 600 to 1000 in shade plants.

Additional Reading

A list of sources for additional reading in chloroplast development is provided at the end of Chapter 8.

8

Development of Chloroplasts

Biosynthetic Pathways and Regulation

I. Introduction

Development of the chloroplast involves an intricate interplay between both the chloroplast and the nuclear–cytoplasmic synthetic systems. The lipids of the membranes, including chlorophyll, are made within the organelle. However, the genetic information for the synthesis of most, if not all, of the enzymes involved in these processes and also of most of the membrane proteins is contained in nuclear DNA. Moreover, these chloroplast-confined proteins are synthesized on cytoplasmic ribosomes. Although active chloroplast ribosomes are required for maximal rates of chlorophyll synthesis and for full development of photosynthetic activity, only fragmentary knowledge is available on which of the proteins are synthesized within the organelle. In Section II–IV, synthesis of the major components of thylakoid membranes will be described. Subsequently, the regulation of these synthetic processes will be discussed.

II. Chlorophyll

A. Pathway of Biosynthesis

The general pathway for the synthesis of chlorophyll is known, although most of the enzymes have not been characterized, and many of the mechanisms of the reactions are still unclear. The regulatory controls over the process also are not clearly defined. The first compound committed to the synthesis of chlorophyll, as well as of other porphyrins and related compounds, is **δ-aminolevulinic acid** (Fig. 8.1). The pathway for synthesis of δ-aminolevulinic acid for chlorophyll synthesis in chloroplasts is not known with certainty, but the available evidence has established that the carbon skeleton is derived intact from glutamic acid. The two most likely pathways from glutamic acid are the following. The scheme in Fig. 8.2 shows one pathway, currently the most favored, in which glutamic acid is reduced to glutamic semialdehyde, which then by either an inter- or intramolecular transamination reaction is converted to δ-aminolevulinic acid. A second pathway may also occur, in which oxidative deamination or transamination reactions convert glutamic acid to α-ketoglutaric acid.

223

Reduction of the α-carboxyl group, through reactions similar to those illustrated in Fig. 8.2, then forms the dicarbonyl compound 4,5-dioxovaleric acid, which is converted to the final product by addition of an amino group by another transamination reaction. However, recent evidence by Harel and Ne'eman (1983) suggests that 4,5-dioxovaleric acid is not a substantial precursor and indeed may be a degradative product of δ-aminolevulinic acid metabolism in the cytoplasm.

In animal mitochondria and bacteria, δ-aminolevulinic acid is synthesized by condensation of glycine and succinyl-coenzyme A, catalyzed by the enzyme **δ-aminolevulinic acid synthetase:**

$$H_3N-CH_2-\overset{\overset{O}{\|}}{C}-O^- + CoA-S-\overset{\overset{O}{\|}}{C}-CH_2CH_2-\overset{\overset{O}{\|}}{C}-O^-$$

Glycine   Succinyl-coenzyme A

$$\rightarrow \overset{+}{H_3}N-CH_2-\overset{\overset{O}{\|}}{C}-CH_2CH_2-\overset{\overset{O}{\|}}{C}-O^- + CoA-SH + CO_2 \qquad [8.1]$$

δ-Aminolevulinic acid

Although this reaction also occurs in plant cells, its contribution to the pool of δ-aminolevulinic acid for chlorophyll synthesis appears to be minor. Furthermore, labeled [14C]glycine is incorporated into chlorophyll much less efficiently than is [14C]glutamic acid. Consequently, the glutamic-acid-to-δ-aminolevulinic-acid pathway is the principal source of precursors for synthesis of chlorophyll in most green plants and algae.

The next reaction on the path to chlorophyll is condensation of two molecules of δ-aminolevulinic acid to form **porphobilinogen,** the compound that provides the basic pyrrole units of the porphyrins, in a reaction catalyzed by **δ-aminolevulinic acid dehydrase** (Fig. 8.3). This enzyme will also bind structural analogues of δ-aminolevulinic acid (Fig. 8.1), which thus act as competitive inhibitors of the enzyme. When greening cells are treated with these inhibitors, δ-aminolevulinic acid and, to a much lesser extent, dioxovaleric acid accumulate.

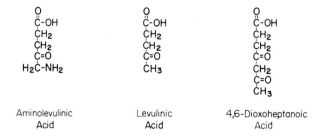

FIG. 8.1. Structures of two analogues of δ-aminolevulinic acid that block its utilization by inhibiting the enzyme δ-aminolevulinic acid dehydrase. In the presence of these compounds, δ-aminolevulinic acid accumulates in plant cells.

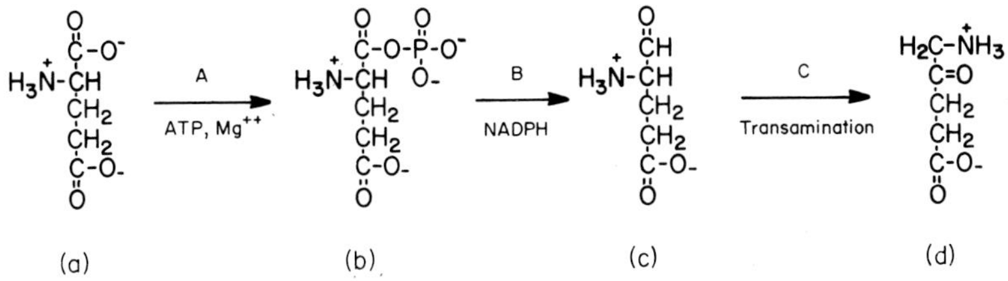

FIG. 8.2. A proposed pathway for the conversion of (a) glutamic acid to (d) δ-aminolevulinic acid. Glutamate is reduced in a two-step reaction, in which a phosphorylation of the α-carboxyl group is followed by a reduction with NADPH, to (c) glutamic acid 1-semialdehyde. Then, either an intra-molecular or two intermolecular transamination reactions transfer the amino group to the 5-position to form δ-aminolevulinate. Note that the 1-carbon of glutamate becomes the 5-carbon of δ-aminolevulinate. (Adapted from Kannangara et al., 1978.)

Four porphobilinogen molecules are linked into a cyclic tetrapyrrole inter-mediate, **uroporphyrinogen III** (Fig. 8.4). All natural porphyrin pigments are derived from this precursor. An interesting but puzzling feature of the synthesis of this porphyrin is that three porphobilinogen units are arranged "head to tail," whereas the fourth, which contributes ring D of the porphyrin, is an iso-por-phobilinogen. Insertion of the latter unit results in the opposite positioning of the side chains on ring D as compared to the others.

Two enzymes are required for conversion of porphobilinogen into uropor-phyrinogen III. The first is **porphobilinogen deaminase,** which condenses four monopyrrole units head to tail, with the release of ammonia, into a linear tetra-pyrrole. Simple chemical closure of this tetrapyrrole into a ring produces uro-porphyrinogen I, an unnatural isomer, in which all four pyrrole units are ori-ented in the same direction. In biological systems, an additional protein, **uroporphyrinogen III cosynthetase,** is required for synthesis of uroporphyrin-ogen III. In the presence of this protein, closure of the tetrapyrrole ring is accompanied by rearrangement of the last porphobilinogen unit. The mecha-nism of this reaction catalyzed by the cosynthetase is not clear. It is interesting to note that the "backward" porphobilinogen unit (ring D) is the one that becomes reduced in the conversion of protochlorophyllide to chlorophyllide and to which the long-chain alcohol phytol is added.

The propionic acid and acetic acid residues on the porphobilinogen units are processed into the various side chains that appear on the final porphyrin

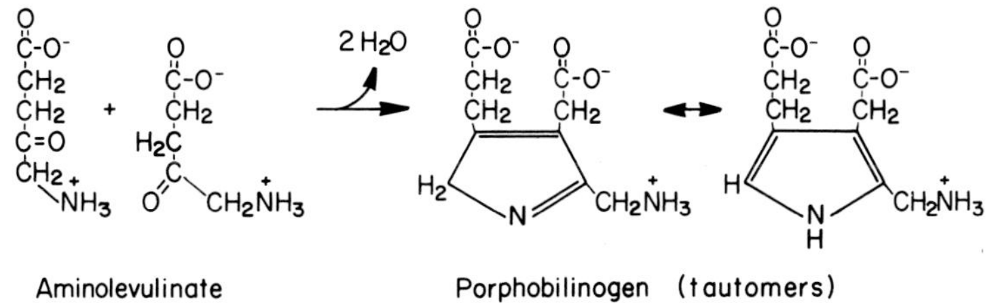

Aminolevulinate   Porphobilinogen (tautomers)

FIG. 8.3. The conversion of two molecules of δ-aminolevulinate to porphobilinogen, the reaction catalyzed by δ-aminolevulinic acid dehydrase.

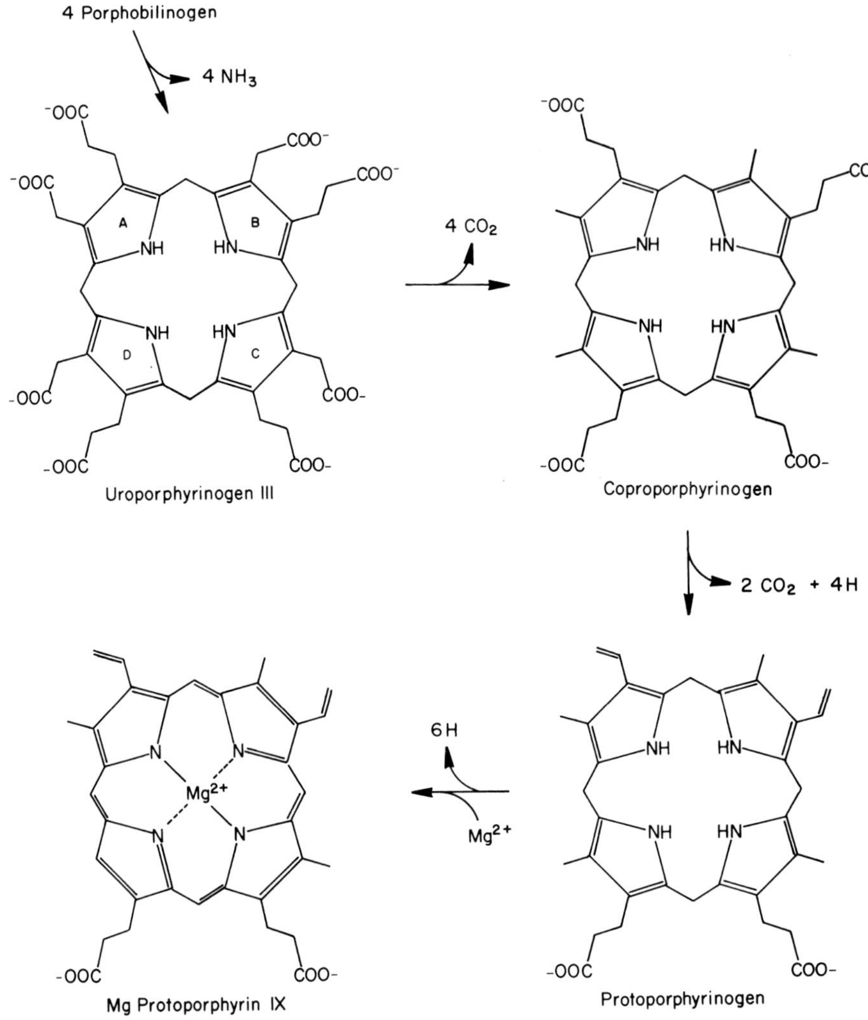

FIG. 8.4. The synthesis of chlorophylls from porphobilinogen. The first macrocyclic tetrapyrrole intermediate formed is uroporphyrinogen III, which through successive decarboxylation reactions is converted to protoporphyrinogen. Oxidation of protoporphyrinogen generates the conjugated, aromatic ring system in protoporphyrin IX. Insertion of Mg^{2+} and formation of the fifth, isocyclic ring produces divinyl protochlorophyllide. Between protoporphyrin IX and protochlorophyllide, several alternate routes are possible, in that the intermediates occur as the divinyl derivatives (as shown) or as the monovinyl forms, which have the vinyl side chain on ring B reduced to an ethyl group. Protochlorophyllide is reduced by NADPH to chlorophyllide a. This reduction requires light in most higher plants and some algae. Esterification of the propionate side chain on ring D occurs by condensation of the carboxyl group with geranylgeranyl pyrophosphate (GGPP). Reduction of the geranylgeranyl moiety to the phytol form completes the synthesis of chlorophyll a. There is some evidence for the synthesis of chlorophyll b from chlorophyll a, but such a conversion has not been established. The reactions that introduce the formyl group on ring B, to generate chlorophyllide b, are obscure. Note that the early intermediates in the pathway are highly charged and therefore soluble in water. As the intermediates pass through the pathway, decarboxylation reactions remove the ionized carboxyl groups, and the subsequent products become less water soluble. Finally, after esterification of the porphyrin to the long-chain alcohol, the chlorophylls are essentially water insoluble.

structures. By a series of reactions, the carboxyl groups of the acetic acid residues are removed, forming methyl groups. Two of the propionic acid residues are converted to vinyl groups. Further desaturation of the developing molecule results in electronic rearrangements that allow a fully conjugated system of double bonds in the product, **protoporphyrin IX.**

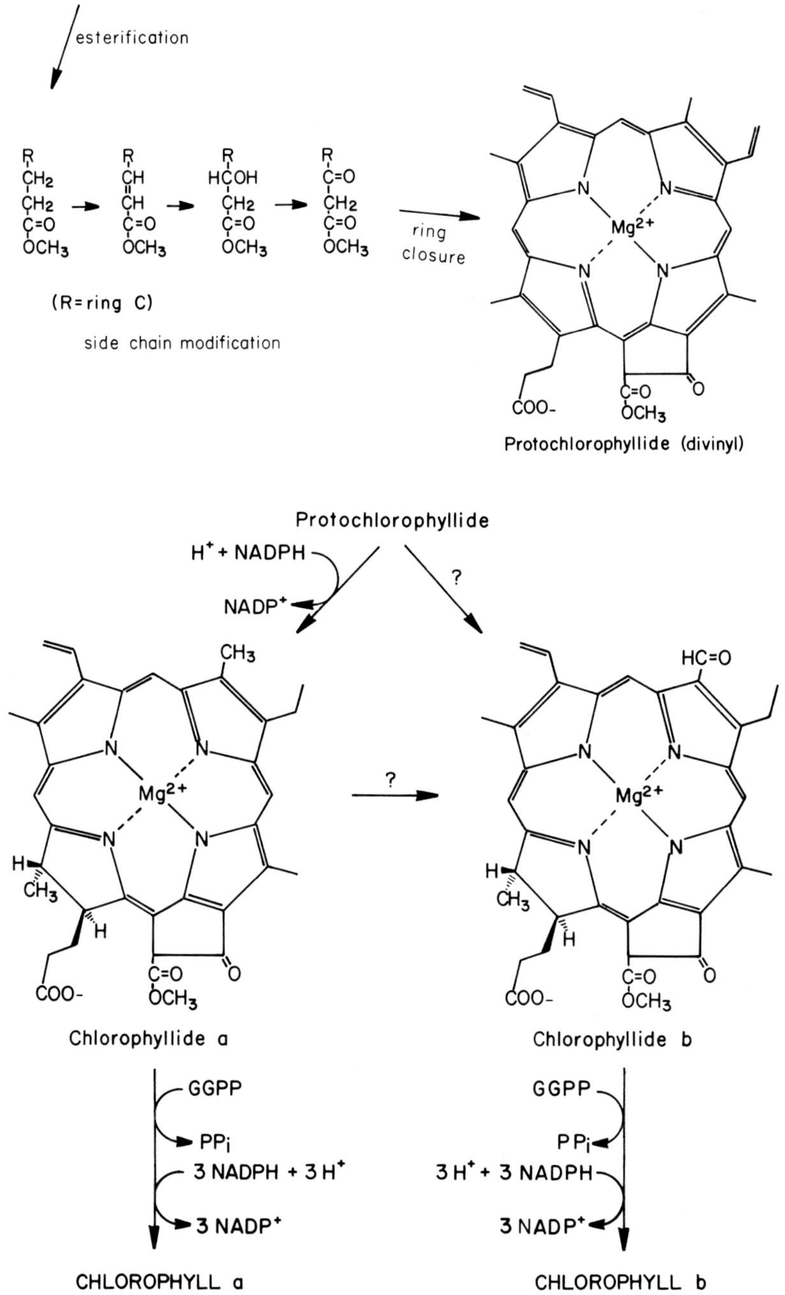

Protochlorophyllide (divinyl)

Chlorophyllide a

Chlorophyllide b

CHLOROPHYLL a

CHLOROPHYLL b

Protoporphyrin IX is a pivotal compound in the pathway of porphyrin synthesis. By chelation with Fe^{2+} it is directed toward heme. By chelation with Mg^{2+} it is directed toward chlorophyll. Or by an opening of the macro ring, it becomes the precursor of the accessory phycobilin pigments (see p. 64). Mg–Protoporphyrin IX is converted to Mg-2,4-divinyl pheoporphyrin a_5 (**protochlorophyllide**) by a series of modifications to the propionic acid side chain on ring C and formation of the additional five-membered ring. Protochlorophyllide also occurs as the monovinyl derivative, in which the vinyl side chain on ring B has been reduced to an ethyl group, but in most plants divinyl protochlorophyllide is the predominant form.

Many, but not all, of the algae and a few types of higher plants (the gymnosperms) can continue the pathway on to chlorophyll in the dark. However, the remaining species form chlorophyll only in the light, and in these latter organisms protochlorophyll(ide) is the end product of the pathway in the dark. Conversion to chlorophyllide a requires reduction of ring D, a reaction that occurs enzymatically in the dark in cells with this capability. But for organisms lacking this activity, the absorption of light energy is required for this step. The

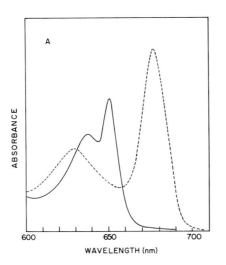
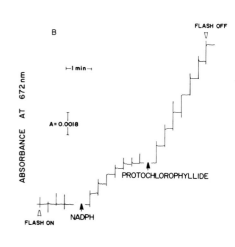

FIG. 8.5. A: The absorption spectrum at 77°K of etiolated bean leaves before ———— and immediately after (- - - -) a 1-msec flash of light. "Phototransformable" protochlorophyllide in unilluminated leaves has a maximum absorption in red light at 650 nm and is associated with a specific protein termed the **holochrome** or protochlorophyllide reductase. The action spectrum for the reduction of protochlorophyllide corresponds to the absorption spectrum of this form. Protochlorophyllide not associated with the reductase absorbs light at shorter wavelengths (630–640 nm). As reduction occurs in the light, these latter forms serve as substrate and bind to the enzyme as the chlorophyllide product leaves. [Adapted from Dujardin, 1982.] B: An assay of the activity of protochlorophyllide reductase. Etioplast membranes were isolated and incubated with protochlorophyllide and NADPH. At 20-sec intervals, saturating light flashes (0.1 msec in duration) were provided to reduce protochlorophyllide associated with the enzyme. During the dark periods, the reduced product dissociated from the active site on the enzyme, and unreduced substrate refilled the site. Activity was measured by the increase in absorbance at 672 nm, the absorption maximum for chlorophyllide a in this system. No chlorophyllide a was formed if NADPH was omitted. A small amount of chlorophyllide a was made, before protochlorophyllide was added, from the endogenous substrate in the membrane preparation. (Adapted from Griffiths, 1978.)

photoreceptor for the reduction is protochlorophyll(ide) itself, as indicated by coincidence of the action spectrum for chlorophyll *a* formation with the *in vivo* absorption spectrum of protochlorophyll(ide).

Griffiths (1978) has developed an *in vitro* system to study this interesting reduction reaction (Fig. 8.5). While in an excited state, protochlorophyllide is reduced in ring D by electrons from NADPH. One hydrogen atom also is transferred from NADPH to the porphyrin, and the other presumably ultimately comes from the solvent. The reduction results in a stereospecific *trans* addition of the hydrogen atoms across the double bond (Fig. 8.4). The enzyme that catalyzes this reaction is **NADPH:protochlorophyllide oxidoreductase** or simply **protochlorophyllide reductase.** The enzyme contains a polypeptide chain about 36,000 daltons in mass, which is synthesized on cytoplasmic ribosomes and is the predominant polypeptide in prolamellar bodies.

The final step in the synthesis of chlorophyll is esterification of the propionic acid side chain on ring D of the porphyrin with a long-chain alcohol. This reaction, catalyzed by the enzyme **chlorophyll synthetase,** proceeds by displacement of the pyrophosphate moiety of geranylgeranyl pyrophosphate (Fig. 8.6) by the porphyrin carboxyl group to form the ester linkage. Subsequently, three of the four double bonds in the geranylgeranyl portion are reduced by NADPH to produce the phytol residue.

The pathway of chlorophyll synthesis from δ-aminolevulinic acid occurs entirely within the chloroplast. Reactions between δ-aminolevulinic acid and protoporphyrin IX occur in the stroma, catalyzed by soluble enzymes. The enzymes involved in the latter part of the pathway are membrane bound.

B. Regulation of Chlorophyll Synthesis

The availability of δ-aminolevulinate is generally thought to be the rate-limiting factor for chlorophyll biosynthesis during greening in illuminated plants. Synthesis of δ-aminolevulinate is regulated by feedback inhibition, but the levels of one or more of the enzymes may also be subject to control. In the dark, when the pathway of chlorophyll biosynthesis is inhibited, very small amounts of intermediates in the pathway accumulate. Accumulation of proto-

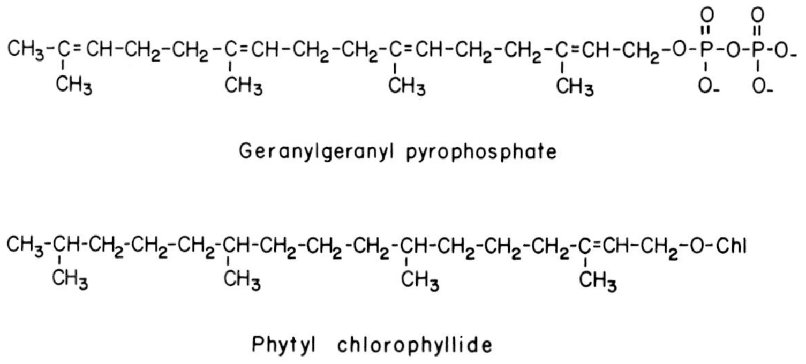

$$CH_3-C=CH-CH_2-CH_2-C=CH-CH_2-CH_2-C=CH-CH_2-CH_2-C=CH-CH_2-O-\overset{\overset{O}{\|}}{P}-O-\overset{\overset{O}{\|}}{P}-O-$$
$$\quad\ \ \underset{CH_3}{|}\qquad\qquad \underset{CH_3}{|}\qquad\qquad \underset{CH_3}{|}\qquad\qquad \underset{CH_3}{|}\qquad\quad \underset{O-}{|}\ \ \underset{O-}{|}$$

Geranylgeranyl pyrophosphate

$$CH_3-CH-CH_2-CH_2-CH_2-CH-CH_2-CH_2-CH_2-CH-CH_2-CH_2-CH_2-C=CH-CH_2-O-Chl$$
$$\quad\ \ \underset{CH_3}{|}\qquad\qquad \underset{CH_3}{|}\qquad\qquad\quad \underset{CH_3}{|}\qquad\qquad \underset{CH_3}{|}$$

Phytyl chlorophyllide

FIG. 8.6. Structures of the isoprenoid geranylgeranyl pyrophosphate and the product of its reduction, phytol, after esterification to chlorophyllide.

chlorophyll(ide) in the dark is very slight, amounting to less than 1% of the final green cell level of chlorophyll. In some higher plants the amount of protochlorophyll(ide) increases somewhat with age of etiolated cells, along with development of prothylakoid membranes associated with the prolamellar bodies within the etioplasts.

There are several possible candidates for the feedback inhibitor. When etiolated cells are exposed to light, protochlorophyllide is converted to chlorophyllide. Consequently, the level of protoporphyrin IX falls as it is converted to protochlorophyllide. The concentration of free heme also decreases under these conditions. Thus, as indicated in Fig. 8.7, perhaps all three porphyrins may be involved as feedback regulators of the synthesis of δ-aminolevulinic acid.

A simple experiment has demonstrated that synthesis of δ-aminolevulinic acid is a major site of regulation of chlorophyll synthesis. When etiolated seedlings or algal cells were fed δ-aminolevulinic acid in the dark, to circumvent the control point, a severalfold increase in the amount of protochlorophyll(ide) resulted. Protochlorophyll(ide) occurred in two forms in these treated cells, one form with an absorption maximum at 650 nm and the second that absorbed maximally near 630 nm. In δ-aminolevulinate-fed cells, the second form is the one that accumulated. Free protochlorophyll(ide) absorbs near 630 nm, and absorption at 650 nm signifies association with protochlorophyllide reductase (also called the **holochrome**). Only the 650-nm form is functionally phototransformed to chlorophyll in the light. However, even in the presence of δ-aminolevulinic acid, the increase in protochlorophyll(ide) is held in check by addi-

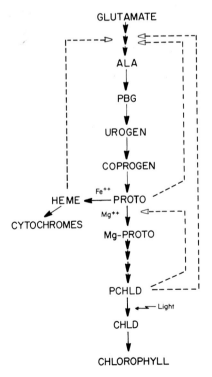

FIG. 8.7. A schematic representation of the biosynthetic pathway of chlorophyll and heme and the proposed regulatory loops. ALA, δ-aminolevulinate; PBG, porphobilinogen; UROGEN, uroporphyrinogen; COPROGEN, coproporphyrinogen; PROTO, protoporphyrin IX; Mg-PROTO, magnesium protoporphyrin IX; PCHLD, protochlorophyllide; CHLD, chlorophyllide. At the branch point, protoporphyrin IX is converted either to heme by chelation with Fe^{2+} or on to chlorophyll by chelation with Mg^{2+}. The pathway is controlled primarily at the synthesis of δ-aminolevulinate by feedback inhibition. Possible feedback inhibitors of this step include protoporphyrin IX, heme, and protochlorophyllide (the last soluble intermediate in the pathway). Exposure of cells to light relieves the inhibition, possibly by a reduction in the concentrations of intermediates as protochlorophyllide is reduced to chlorophyllide. Some evidence also is available for a short feedback loop in which protochlorophyllide regulates its synthesis from protoporphyrin IX. (Adapted from Wang et al., 1977.)

tional control points, probably between protoporphyrin IX and protochlorophyllide (Fig. 8.7).

As shown in Fig. 8.8, exposure of etiolated plant cells to light resulted in a dramatic increase in the rate of synthesis of δ-aminolevulinic acid. In this experiment, accumulation of δ-aminolevulinic acid was achieved by inhibition of its subsequent metabolism through the use of the analogue levulinic acid (see Fig. 8.1). Of course, in the absence of levulinic acid, δ-aminolevulinic acid would be rapidly converted to chlorophyll. An interpretation of these results is that, upon exposure of the cells to light, the level of the regulatory feedback inhibitor decreased, and thereby inhibition of synthesis of δ-aminolevulinic acid was relieved. Rapid synthesis of δ-aminolevulinic acid then ensued, catalyzed by enzymes present constitutively. Alternatively, light may be required for production of ATP and NADPH, which are essential for synthesis of δ-aminolevulinic acid (Fig. 8.2). This second suggestion implies that synthesis of porphyrins is regulated by concentrations of substrates in addition to regulatory factors. The dependence of this process on light is illustrated by the fact that, when greening cells were returned to the dark, the pathway was rapidly and essentially completely shut down (Fig. 8.8).

There is much yet to be learned about regulation of the pathway of chlorophyll biosynthesis. The pathway has a high capacity for porphyrin production, yet in the dark it is blocked. Why light is required for reduction of protochlorophyllide in etioplasts, in view of the occurrence of this reaction in gymnosperms (e.g., evergreens) and many algae in the dark, is not well understood. Both genetic and environmental factors are involved in the regulation. Mutant strains of barley and of the alga *Chlamydomonas reinhardtii* have been

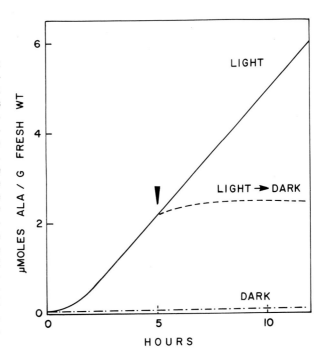

FIG. 8.8. The effect of light on the synthesis of δ-aminolevulinate (ALA) in greening corn leaves. The analysis of synthesis of δ-aminolevulinate was permitted by inhibition of its utilization by treating leaves with 20 mM levulinic acid (see Fig. 8.1). The rate of δ-aminolevulinate synthesis was nearly zero in the dark. Upon exposure of the leaves to light, the activity increased after a short lag. However, upon return to the dark, the rate again dropped to zero. Since chlorophyll synthesis was not completely inhibited by levulinic acid, this type of experiment underestimates the total synthesis of δ-aminolevulinate. (Adapted from Harel and Klein, 1972).

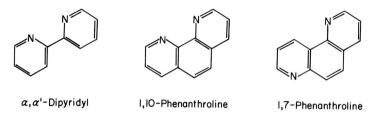

α,α'-Dipyridyl   1,10-Phenanthroline   1,7-Phenanthroline

FIG. 8.9. Structures of α, α'-dipyridyl and 1,10-phenanthroline, which chelate Fe^{2+} ions, and the nonchelating analogue, 1,7-phenanthroline. These compounds stimulate porphyrin synthesis in plant cells in the dark by an unknown mechanism. In higher plants, Mg-protoporphyrin IX was the predominant product, whereas in *Chlamydomonas* chlorophyllide b was the major product (see Fig. 8.10).

isolated that are defective in the regulation of synthesis of δ-aminolevulinic acid, and these strains accumulate precursors of chlorophyll, principally protoporphyrin IX and protochlorophyllide, in the dark. These effects can be mimicked in etiolated higher plants by chelators of iron, such as α, α'-dipyridyl and 1,10-phenanthroline (Fig. 8.9). These chelators possibly act by lowering the level of heme and thereby relieve feedback inhibition of δ-aminolevulinic acid synthesis (see Fig. 8.7). However, the nonchelating isomer 1,7-phenanthroline (Fig. 8.9) induced in etiolated *Chlamydomonas* cells as high a rate of porphyrin synthesis in the dark as its iron-chelating isomer.

Moreover, both these compounds were as effective as light (Fig. 8.10). The porphyrins produced in the algal cells under these conditions in the dark were

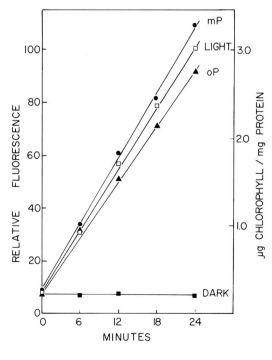

FIG. 8.10. Comparison of the rates of porphyrin synthesis in the dark in response to 10 mM 1,7-phenanthroline (mP) or 1,10-phenanthroline (oP) by cells of *Chlamydomonas reinhardtii* y-1 with that in untreated cells exposed to light. Etiolated cells were incubated 1 hr in the dark at 38°C prior to initiation of porphyrin synthesis. (Data of D. Bednarik and J. K. Hoober.)

reduced chlorophyllides similar to chlorophyll *b*, as determined by their spectra (Fig. 8.11). There are as yet no biochemical explanations for these effects.

III. Lipids

Fatty acids are synthesized by soluble enzymes, **fatty acid synthetases,** in the chloroplast stroma. The products of this system are the acyl–coenzyme A derivatives of the fatty acids. A fatty acid-desaturating system also exists in the stroma to produce the highly unsaturated fatty acids found in thylakoid lipids (see p. 62).

In yeast and animals, the steps in fatty acid synthesis are catalyzed by a large, soluble, polyfunctional protein containing a covalently bound acyl–carrier polypeptide. In prokaryotic cells, however, the acyl–carrier protein, to which the growing fatty acid is attached by a thioester linkage, and the enzymes that catalyze each step are separate, discrete proteins. In C_3 plants, the fatty acid-synthesizing system is exclusively located in the chloroplast stroma. Shi-

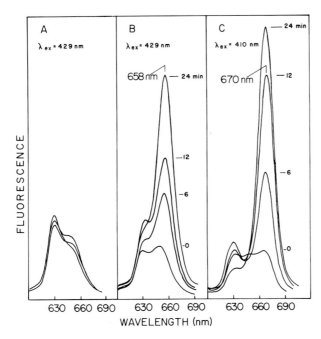

FIG. 8.11. Fluorescence emission spectra of porphyrins in extracts of etiolated *Chlamydomonas reinhardtii* y-1 cells treated as described in Fig. 8.10. A: Spectra for control cells showed that over a 30-min period untreated cells were not making porphyrins at a significant rate in the dark. The maximum at 630 nm represents protochlorophyll(ide). B: Spectra of porphyrins induced by 1,10-phenanthroline exhibit an emission maximum at 658 nm, which is characteristic of chlorophyll *b*. C: Spectra of porphyrins made in cells exposed to light exhibit a maximum at 670 nm, which corresponds to chlorophyll *a* as the major component. The numbers to the right of the spectra indicate the time of treatment. (Adapted from Hoober *et al.*, 1982.)

makata and Stumpf (1982) demonstrated that the chloroplast system consists of separate enzymatic components, similar to the prokaryotic system.

The synthesis of the membrane lipids, however, is catalyzed by membrane-bound enzymes. Block *et al.* (1983b) have shown that a major site of synthesis of the galactosyl diglycerides is the chloroplast envelope. The reactions in the synthesis of mono- and digalactosyl diglycerides (see Fig. 3.11 for structures) are as follows:

Uridine diphospho-galactose (UDP-gal)
$$+ \text{ diacyl glycerol} \rightarrow \text{monogalactosyl diglyceride (MGDG)}$$
$$+ \text{ uridine diphosphate (UDP)} \quad [8.2]$$

$$\text{UDP-gal} + \text{MGDG} \rightarrow \text{digalactosyl diglyceride (DGDG)} + \text{UDP} \quad [8.3]$$

Diacyl glycerol is also produced in the chloroplast envelope by the following reactions:

Fatty acyl-coenzyme A
$$+ \text{ sn-glycerol 3-phosphate} \rightarrow \text{lysophosphatidic acid} + \text{CoA} \quad [8.4]$$

Lysophosphatidic acid
$$+ \text{ fatty acyl-coenzyme A} \rightarrow \text{phosphatidic acid} + \text{CoA} \quad [8.5]$$

$$\text{Phosphatidic acid} \rightarrow \text{diacyl glycerol} + \text{inorganic phosphate} \quad [8.6]$$

The first fatty acid is added to the 1-position of the glycerol 3-phosphate (reaction 8.4). The second fatty acid, added to the 2-position, generally is the most highly unsaturated (reaction 8.5).

Because of their insolubility in water, membrane lipids generally are synthesized *in situ*, and synthesis of new lipid thereby contributes to growth of the membrane. In some systems, transfer of lipids between membranes can occur at low rates, but this exchange requires the presence of a protein carrier. In chloroplasts, the evidence available does not allow conclusions about whether the galactolipids made in the envelope are transferred to thylakoids as membrane vesicles or individually on protein carriers.

Synthesis of polar lipids occurs rapidly during periods of thylakoid membrane expansion. In cultures of *Chlamydomonas reinhardtii*, induced to grow synchronously by 12-hr light–12-hr-dark cycles, maximal rates of lipid synthesis occurred during the middle of the light period. The peak of lipid synthesis slightly preceded the maximal period of chlorophyll synthesis (Fig. 8.12). However, in the dark, the rate of synthesis of these lipids was quite low. Apparently the activities of enzymes in the biosynthetic pathway for lipids are also regulated by light. Alternatively, since lipid synthesis requires formation of "high-energy" precursors, and thus a supply of ATP, perhaps these reactions cannot proceed in the dark in the absence of photophosphorylation.

IV. Synthesis of Chloroplast Proteins

After the discovery and characterization of chloroplast ribosomes during the 1960s, an extensive amount of work was devoted to determination of the sites of synthesis of chloroplast proteins. It was soon established that relatively few of the chloroplast proteins, including both stromal and membrane proteins, are made within the chloroplast. With the recent determination of the size of plastid DNA has come the realization that the genetic capacity of the organelle is quite limited. Consequently, the task then became to elucidate how the majority of the chloroplast proteins, which are synthesized on cytoplasmic ribosomes from mRNA transcribed in the nucleus of the cell, enter the chloroplast. In particular, an important question is how proteins cross the double-membrane plastid envelope. This is not a trivial question in view of the amount of protein involved in this process. Evidence is beginning to accumulate regarding how it is accomplished, which started with work on ribulose 1,5-bisphosphate carboxylase.

Presumably, most, if not all, of the chloroplast proteins synthesized on cytoplasmic ribosomes are made as precursors and undergo proteolytic procession during transport into the chloroplast. Those precursors examined thus far have extensions at the N-terminus of the polypeptide. In recognition of its apparent function, the peptide extension that is removed from the precursors during uptake by the chloroplast is designated the **transit sequence.** Examples of this process are described in Sections IV.A–IV.D.

A. Ribulose 1,5-Bisphosphate Carboxylase

A particularly interesting protein, from the standpoint of cooperation between both chloroplast and cytoplasm for its production, is the CO_2-fixing enzyme ribulose 1,5-bisphosphate carboxylase. This large protein contains 16 subunits, eight that have a mass of 52,000–55,000 daltons (the "large" subunit) and eight that have a mass of 12,000–16,000 daltons (the "small" subunit). The large subunit is made within the chloroplast, and it is now known that the gene for this polypeptide resides in the chloroplast DNA (see Fig. 6.6). The mRNA for this polypeptide does not contain a polyadenylate [poly(A)] extension at its 3′ end.

The nucleotide sequence of the gene for the large subunit and the sequence of the 475 amino acids in the polypeptide were shown in Fig. 6.24. The nuclear gene for the small subunit has also been sequenced. In contrast to the gene for the large subunit, which contains no introns, the small subunit gene includes two relatively short introns, as shown in Fig. 8.13. The introns presumably are transcribed into the initial RNA product of this gene but subsequently are cleaved out, leaving an mRNA that contains the coding sequence for 123 amino acids (for the polypeptide in pea) plus the transit sequence extension at the N-terminal end. During processing of the small subunit mRNA in the nucleus, ATP is polymerized onto the 3′ end to produce a poly(A) tail. This feature is typical of mRNA molecules made in the cell nucleus.

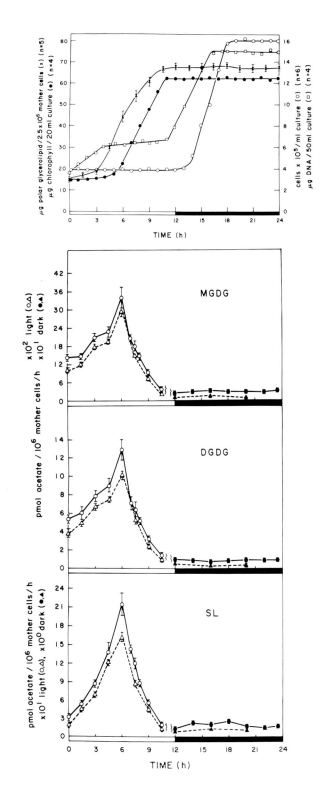

In both higher plants and algae, the small subunit is synthesized as a precursor about 5000–6000 daltons larger than the mature subunit polypeptide. In pea, the N-terminal extension is 57 amino acids long (Fig. 8.13), whereas in *Chlamydomonas reinhardtii* it contains only 44 amino acids (Fig. 8.14). Furthermore, there is no significant homology in the transit sequences from the two organisms. Careful studies involving cell fractionation indicated that the precursor polypeptide is synthesized on cytoplasmic ribosomes that are unattached to membranes. After synthesis, the small subunit precursors are released into the cytosol and subsequently transported into the chloroplast. The N-terminal transit sequence is then immediately removed by a proteolytic activity located in the chloroplast stroma. Within the organelle, the mature small subunits combine with large subunits to form the complete holoenzyme.

With an elegant experiment, Highfield and Ellis (1978) and Dobberstein *et al.* (1977) demonstrated the assembly of the carboxylase *in vitro*. Poly(A)-rich RNA was prepared from total cellular RNA with a column of oligo(dT)-cellulose. In this procedure, as the RNA passes through the column in a medium of relatively high ionic strength (to minimize charge repulsion of nucleic acids), RNA with a poly(A) extension on the 3′ end hybridizes with the short oligo(dT) strands attached to the cellulose. Consequently, only mRNA containing poly(A) is retained by the column. The bulk of the cellular RNA, including the poly(A)-lacking ribosomal and tRNA, passes through. The poly(A)-rich RNA is then eluted by lowering the ionic strength, causing charge repulsion to disrupt hydrogen bonds between the dT:A base-pairs.

Addition of the poly(A)-rich RNA to a cell-free translation system, prepared with extracts of wheat germ that contain ribosomes, enzymes, an energy source, and the other components needed for protein synthesis, resulted in production of polypeptides. Under these conditions, the precursor of the carboxylase small subunit is the most prominent product of *in vitro* protein synthesis

---

FIG. 8.12. Increases in cell number, chlorophyll, DNA, and polar glycerolipids in synchronous cultures of wild type *Chlamydomonas reinhardtii*. Synchrony was induced by cycles of 12-hr-light/12-hr-dark. As shown in the top panel, each cell produced four daughter cells (o) during the cell cycle; division occurred midway during the dark period. The increase in chlorophyll (●) during the light phase represented the course of thylakoid membrane synthesis. Accumulation of the polar glycerolipids (x), components of thylakoid membranes, followed closely the accumulation of chlorophyll. The increase in DNA (□) during the early part of the light period represented chloroplast DNA replication. Nuclear DNA was replicated shortly after cells entered the dark phase. The lower panels show the rates of synthesis of several polar glycerolipids and their incorporation into thylakoid membranes during the cell cycle of *C. reinhardtii*. The rate of incorporation of ^{14}C from acetate into the total amount of each lipid was measured at various times (o, light; ■, dark). At each time point, thylakoid membranes were purified, and the amount of ^{14}C in each lipid in the membranes was also determined (△, light; ▲, dark). The data show that the maximal rates of synthesis of the lipids, and entry into the membranes, corresponded to the time of maximal membrane assembly, which peaked midway in the light period (see top panel). Rates of lipid synthesis were very low in the dark. MGDG, monogalactosyl diglyceride; DGDG, digalactosyl diglyceride; SL, sulfoquinovosyl diglyceride. Synthesis of other membrane lipids, and of the major thylakoid polypeptides, follows a similar pattern. (From Janero and Barnett, 1981. Reprinted with permission of the Rockefeller University Press.)

```
SS3.6      GTACTAGGCAGTAGCTAATTACCACAATATTAAGACCATAATATTGGAAATAGATAAATAAAA
SS8.0         TG      C ATC       T     A C              A     C T


                                                                MetAla
     ACATTATATATAGCAAGTTTTAGCAGAAGCTTTGCAATTCATACA-----------GAAGTGAGAAAAATGGCTT
              A            T C                     AC  CAAGAACTAACA     C


     SerMetIleSerSerSerAlaValThrThrValSerArgAlaSerArgGlyGlnSerAlaAlaValAlaProPhe
     CTATGATATCCTCTTCCGCTGTGACAACAGTCAGCCGTGCCTCTAGGGGGCAATCCGCCGCAGTGGCTCCATTCG
                                          T         T         G
                                                   val


     GlyGlyLeuLysSerMetThrGlyPheProValLysLysValAsnThrAspIleThrSerIleThrSerAsnGly
     GCGGCCTCAAATCCATGACTGGATTCCCAGTGAAGAAGGTCAACACTGACATTACTTCCATTACAAGCAATGGTG
     G


     GlyArgValLysCysMetGln
     GAAGAGTAAAGTGCATGCAGGTGACAGAAACATATACATATATATATATAGTTGAATATCAGTAATGATTCAAGT
                                     --------C T         C


                              ValTrpProProIleGlyLysLysLysPheGluThrLeuSerTyrLeu
     TTGTTAACCGTTTATGTTGAATATTTAGGTGTGGCCTCCAATTGGAAAGAAGAAGTTTGAGACTCTTTCCTATTT
              TC  AT


      ProProLeuThrArgAspGlnLeuLeuLysGluValGluTyrLeuLeuArgLysGlyTrpValProCysLeuGlu
     GCCACCATTGACGAGAGATCAATTGTTGAAAGAAGTTGAATACCTTCTGAGGAAGGGATGGGTTCCATGCTTGGA
                          G


      PheGluLeuGlu
     ATTTGAGTTGGAGGTTTCATATTCATTCCTTTTTTCAATGATTATATAAATACTTTTGTTTGAAACCGTAATGAG


                           LysGlyPheValTyrArgGluHisAsnLysSerProArgTyrTyrAspGly
     TTGATTTTGACTGTTTGGTTGCAGAAAGGATTTGTGTACCGTGAGCACAACAAGTCACCAAGATACTATGATGGA


     ArgTyrTrpThrMet TrpLysLeuProMetPheGlyThrThrAspAlaSerGlnValLeuLysGluLeuAspGlu
     AGATACTGGACAATGTGGAAGCTTCCTATGTTTGGTACCACTGATGCTTCTCAAGTCTTGAAGGAGCTTGATGAA


     ValValAlaAlaTyrProGlnAlaPheValArgIleIleGlyPheAspAsnValArgGlnValGlnCysIleSer
     GTTGTTGCCGCTTACCCTCAAGCTTTCGTTCGTATCATCGGTTTCGACAACGTTCGTCAAGTTCAATGCATCAGT
                           C           C


     PheIleAlaHisThrProGluSerTyr
     TTCATTGCACACACACCAGAATCCTACTAAGTTTGAGTATTATGGCATTGGAAAAGCTGTTTCTCTTGTACCATT
               C                                      A  T     A


     TGTTGTGCTTGTAATTTACTGTGTTTTTTTTTTCGGTTTTTGGTTTCGGACTGTAAAATGGAAATGGATGGAGAA
       C A                           A -  -------------          AC       T


     GAGTTAATGAATGATATGGTCCTTTTGTTCATTCT
          C        A       C  AA  CAAG
```

FIG. 8.13. The nucleotide sequences of genes from pea for the small subunit polypeptide of ribulose 1,5-bisphosphate carboxylase. The complete sequence is for one of the family of small subunit genes (SS3.6). The sequence of another member of the family (SS8.0) is indicated below only where it differs from SS3.6. Deleted nucleotides in a gene are indicated by - - - - -. The predicted amino acid sequence is given above the nucleotide sequence. Gaps in the amino acid sequence indicate

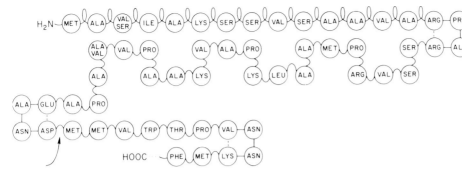

FIG. 8.14. The transit amino acid sequence at the N-terminus of the precursor polypeptide of the small subunit of ribulose 1,5-bisphosphate carboxylase from *Chlamydomonas reinhardtii*. An additional sequence of 44 amino acids exists in the primary product of translation, which is cleaved from the polypeptide, at the site indicated by the arrow, during transport of the polypeptide into the chloroplast. A proposed secondary structure is indicated, which includes regions of α-helix ($\bigcirc\!\!\!\!\!\bigcirc$) and random coil ($\bigcirc\bigcirc$) separated by β-turns ($\bigcirc\!\!-\!\!\bigcirc$). The site of cleavage is within a β-turn. The transit sequence has no unusual hydrophobic character but does contain clusters of basic amino acids. (From Schmidt *et al.*, 1979. Reprinted with permission of the Rockefeller University Press.)

(Fig. 8.15). Then if to the translation system was added a preparation of purified chloroplasts after synthesis of protein was completed, the precursor of the small subunit was taken up by the chloroplasts and processed to the size of the mature polypeptide. Finally, some of the *in vitro* synthesized small subunit was recovered in the holoenzyme subsequently purified from the chloroplasts.

This experiment demonstrated several important features of the flux of polypeptides from cytoplasm to chloroplast. First, the polypeptide precursors themselves are the agents of uptake, and synthesis of the polypeptides on cytoplasmic ribosomes bound to the envelope membrane is not required. Second, only the precursor form of the small subunit of the carboxylase is able to enter the chloroplast. If the mature form is isolated from the enzyme ribulose 1,5-bisphosphate carboxylase and added to intact chloroplasts, no uptake by the chloroplast occurs. Third, a highly specific endopeptidase is involved in the processing of these precursors. The proteolytic activity that performs the conversion of precursor to mature subunit has been detected in extracts of chloroplasts. Purification of this enzyme should make possible a better understanding of the requirements for cleavage of the transit sequence.

It is not known how many polypeptides also are made as precursors *in the chloroplast*. Some evidence exists that the N-terminal amino acid alanine in the carboxylase large subunit in barley corresponds to code position 15 in the genetic sequence. Whether a 14-amino acid fragment is cleaved from the N-

introns in the DNA. Amino acids 1-57 serve as the transit sequence, which is removed after entry of the polypeptide into the chloroplast. Although a number of base changes occur in the two nucleotide sequences, the polypeptides from the two genes differ only at one position *in the transit sequence*. (Figure courtesy of A. R. Cashmore.)

terminus after synthesis of the polypeptide remains to be established. Because many polypeptides recovered from the chloroplast have their N-terminal amino group blocked, by addition of acetate or another acid, determination of the sequence of amino acids in chloroplast proteins has been difficult. One case in which processing of a chloroplast product has been established involves the "photogene" product (see p. 177). The initially translated polypeptide has a molecular weight of about 35,000, but the functional polypeptide in thylakoid membranes has a molecular weight of 32,000.

B. Regulation of Subunit Production

The plant nuclear genome contains a small, multigene family of about 4–6 different copies of the small subunit gene. How closely these separate genes are linked is not known. On the other hand, from nearly 100 to several thousand genes for the large subunit may exist in each cell in chloroplast DNA. An interesting question is how production of the carboxylase subunits is coordinated.

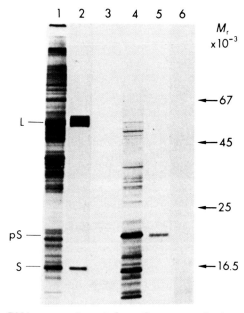

FIG. 8.15. Synthesis *in vitro* of the precursor to the small subunit of ribulose 1,5-bisphosphate carboyxlase. Poly(A)-rich RNA was purified from *Chlamydomonas reinhardtii* and translated in the presence of [³⁵S]methionine in a cell-free wheat germ system. Antibodies, prepared against the small subunit purified from the mature enzyme, were added to the translation mixture upon completion of protein synthesis. The samples were analyzed by electrophoresis and radioautography of dried gels to detect radioactive polypeptides. Lane 1 shows the pattern of soluble proteins synthesized *in vivo*. The positions of the large (L) and small (S) subunits of the mature carboxylase are marked. Lane 2 shows that addition of antibodies to the sample of proteins synthesized *in vivo* made it possible to recover both subunits, since the complete, mature holoenzyme was recovered in the immunoprecipitate. In lane 3, a control sample of the wheat germ system with no added RNA was run. Lane 4 shows the pattern of polypeptides translated by the wheat germ system from poly(A)-rich RNA from *Chlamydomonas*. Lane 5 shows the single polypeptide that was immunoprecipitated when antibodies against the small subunit were added to the *in vitro* translation mixture. This polypeptide (pS) was about 5000 daltons larger than the small subunit in the mature enzyme, but the immunoreactivity established its relatedness to the small subunit. This precursor polypeptide cannot be detected after *in vivo* protein synthesis because of the rapid rate of entry of cytoplasmically made precursors into the chloroplast and concomitant processing. If intact chloroplasts are added to the *in vitro* translation mixture, the precursor to the small subunit is taken up by the chloroplasts, processed, and incorporated along with the large subunit into the holoenzyme. The large subunit was not present in lanes 4 and 5, because its mRNA lacks a 3′ poly(A) tail. Lane 6 shows a control sample with preimmune immunoglobulins. (From Dobberstein *et al.*, 1977. Reprinted with permission.)

In higher plants, such as wheat and alfalfa, synthesis of the small subunit appears to be rate limiting. In nuclear polyploid strains of these plants, the carboxylase activity per cell increased in direct proportion to the nuclear gene dosage (Table 8.1). This observation suggests that the cells' capacity to synthesize the large subunit is considerably greater than that for the small subunit. The question then becomes whether expression of the large subunit gene is controlled at the transcriptional level, or whether accumulation of the polypeptide is regulated at the translational or posttranslational levels.

In experiments in which synthesis of large subunit was blocked by inhibitors of chloroplast ribosomes, the small subunit did not accumulate. Although the inhibitors did not directly affect *synthesis* of small subunit on cytoplasmic ribosomes, under these conditions the small subunit was rapidly degraded (Mishkind and Schmidt, 1983). This degradative activity apparently also occurs in the chloroplast and destroys small subunits that enter the chloroplast in excess of the amount of large subunits available for assembly of the enzyme. The reverse also seems to apply in cells treated with inhibitors of cytoplasmic ribosomes. Synthesis of large subunits proceeds under these conditions, but without small subunits available for assembly, the large subunits are also degraded.

Therefore, synthesis of the subunits is not tightly regulated at the level of transcription or translation. Although in etiolated cells, light stimulates transcription of these genes, this response may be coincidental to the regulation of production of the subunits at subsequent steps.

C. Thylakoid Membrane Polypeptides

Although both the chloroplast and cytoplasm contribute polypeptides to developing thylakoid membranes (see p. 186), most of the polypeptides are made in the cytoplasm by translation of polyadenylated mRNA. It also has been established that these polypeptides are made as precursors. Apel (1979) and Tobin (1981) showed that in barley and duckweed, respectively, the major chlorophyll *a/b*-binding polypeptide (see p. 70) was synthesized *in vitro* as a com-

TABLE 8.1. Correlation between Nuclear Ploidy and the Amount of Ribulose 1,5-Bisphosphate Carboxylase in the First Leaves of Three Wheat *(Triticum)* Species[a]

	T. monococcum (2×)	T. dicoccum (4×)	T. aestivum (6×)
Nuclear DNA/cell (pg)	12.4	24.2	34.6
RuBPCase/cell (pg)	764	1517	2242
RuBPCase/cell:nDNA/cell	62	63	65
Chloroplast DNA/cell (pg)	3.46	6.9	5.05
Plastids/cell	54±3	103±5	133±6
RuBPCase/plastid (pg)	14	15	17

[a]RuBPCase, ribulose 1,5-bisphosphate carboxylase. The constant ratio between nuclear DNA and the amount of the carboxylase per cell suggests that the small subunit is limiting and thus the controlling factor in the synthesis of the enzyme. (Data from Dean and Leech, 1982. Similar data were obtained with alfalfa leaves by Meyers *et al.*, 1982.)

ponent about 4000 daltons larger than the major 25,000-dalton membrane poly-peptide. Cumming and Bennett (1981) and Schmidt *et al.* (1981) found that in pea leaves the primary product of translation of the major 28,000-dalton membrane polypeptide is a 33,000-dalton precursor (Fig. 8.16). A less prominent polypeptide, which is immunochemically related and about the same size as this major polypeptide in pea, is made as a precursor, 32,000 in molecular weight. In the alga *Chlamydomonas*, two primary translation products were found, 31,500 and 30,000 in molecular weight, that are 2000 and 4000 larger than the mature polypeptides (Fig. 8.17). The precursor forms of these membrane polypeptides are soluble in water. But after transport into the plastid and conversion to the mature forms, they are inserted into thylakoid membranes as integral, water-insoluble components. Isolated intact chloroplasts will take up these *in vitro*-made products and integrate them into their proper environment. The final disposition of the polypeptides *in vitro* is indistinguishable from that achieved *in vivo*. Thus, the information that determines where a polypeptide eventually resides is contained within the polypeptide itself.

Fig. 8.18 shows the predicted amino acid sequence of the major chlorophyll *a/b*-binding polypeptide from pea. The sequence initially was determined in an indirect fashion. mRNA that codes for the polypeptide was used as a template for the synthesis of a complementary strand of DNA (**complementary DNA** or **cDNA**) by the enzyme **reverse transcriptase.** After hydrolysis of the RNA by alkaline conditions, the single strand of DNA was used as template for synthesis of the second strand. The double-stranded cDNA fragment was then integrated into a plasmid, which was inserted into cells of the bacterium *Escherichia coli.* As the bacterial cells multiplied, the cDNA-containing plasmid was

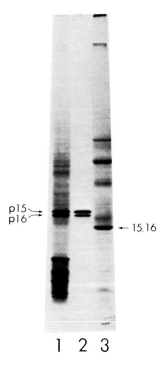

p15
p16

← 15,16

1 2 3

FIG. 8.16. *In vitro* synthesis of the precursors to the major polypeptides, 15 and 16, of the light-harvesting chlorophyll *a/b*-protein complex. Poly(A)-rich RNA was purified from pea seedlings and translated in the presence of [³⁵S]methionine in the wheat-germ cell-free system. The translation mixture was then immunoprecipitated with antibodies against purified polypeptide 15. The samples were analyzed by electrophoresis and radioautography to detect radioactive polypeptides. Lane 1 shows the pattern of total products of *in vitro* translation. The positions of the two precursor polypeptides are marked as p15 and p16. These polypeptides were recovered in the immunoprecipitate from the translation mixture, as shown in lane 2. Lane 3 shows the pattern of thylakoid membrane polypeptides labeled *in vivo*, in which the positions of the mature 15 and 16 are marked. The precursor polypeptides, which are synthesized in the cytoplasm of the cell, are taken up by the chloroplast, processed by removal of the transit sequence, and the smaller, mature forms are incorporated into the membrane. The transit sequence accounts for 4000–5000 of the total molecular weight of 33,000 for the precursors. (From Schmidt *et al.*, 1981. Reprinted with permission of the Rockefeller University Press.)

also replicated. Reisolation of the cDNA from plasmids, prepared from a batch of the bacterial cells, provided sufficient material for determination of the nucleotide sequence (Coruzzi et al., 1983).

The cDNA was then used to select the nuclear gene. Fragments of nuclear DNA were obtained with a restriction endonuclease and inserted into plas-

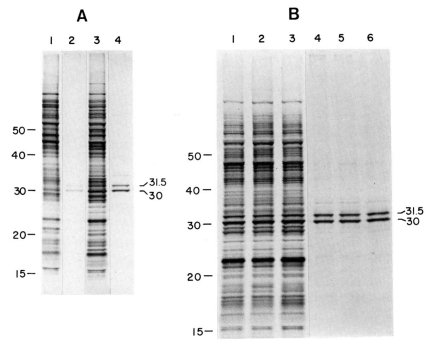

FIG. 8.17. Analysis of the amount of translatable mRNA for the chlorophyll a/b-binding polypeptides in Chlamydomonas reinhardtii y-1. Poly(A)-rich RNA was purified from the algal cells and translated in the presence of [^{35}S]methionine in a rabbit reticulocyte lysate. The translation mixture then was treated with antibodies against polypeptide 11 (see p. 68). Since polypeptides 16 and 17 are immunochemically related to polypeptide 11, precursors to all three were immunoprecipitated. The polypeptides were separated by electrophoresis, and the amounts synthesized were determined by radioautography of the gel. A: RNA was prepared from etiolated cells after growth in the dark for three days at 25°C (the normal growth temperature). Lane 1 shows the initial pattern of total translated polypeptides. Lane 2 shows that very little mRNA for the membrane polypeptides was present in the etiolated cells. However, when these etiolated cells were exposed to white light and allowed to green for 3 hr, the amount of mRNA increased markedly. Lane 3 shows the pattern of total translated polypeptides for RNA obtained from the greening cells, and lane 4 shows the polypeptides that were immunoprecipitated from the sample in lane 3. The numbers ($\times$ 10^3) on the left indicate molecular weight. The number ($\times$ 10^3) on the right indicate the molecular weight of the precursor polypeptides. The prominent product of translation, about 21,000 in molecular weight, was the precursor of the small subunit of ribulose 1,5-bisphosphate carboxylase. B: Etiolated C. reinhardtii y-1 cells were incubated 2 hr at 38°C in the dark before isolation of poly(A)-rich RNA. Lane 1 shows the pattern of total and lane 4 the immunoprecipitated polypeptides synthesized from this RNA. The 31,500- and 30,000-molecular-weight polypeptides were quite prominent in this sample. If cells were exposed to light at 38°C for 15 min (lanes 2 and 5) or 60 min (lanes 3 and 6) before preparation of the RNA, no further change was observed in the total or immunoprecipitated translation products. Thus, whereas light was required for the increase in translatable RNA at 25°C (A), the higher temperature (38°C) caused this effect without light (B). (From Hoober et al., 1982. Reprinted with permission of the Rockefeller University Press.)

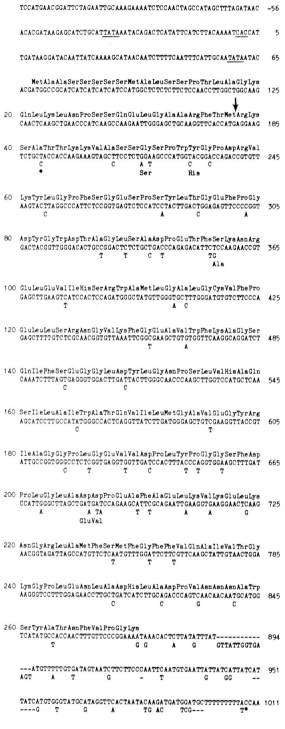

```
TCCATGAACGGATTCTAGAATTGCAAAGAAAATCTCCAACTAGCCATAGCTTTAGATAAC   -56

ACACGATAAGAGCATCTGCATTATAAATACAGACTCATATTCATCTTACAAAATCACCAT     5

TGATAAGGATACAATTATCAAAAGCATAACAATCTTTTCAATTTCATTGCAATATAATAC    65

        MetAlaAlaSerSerSerSerMetAlaLeuSerSerProThrLeuAlaGlyLys
        ACGATGGCCGCATCATCATCATCATCCATGGCTCTCTCTTCTCCAACCTTGGCTGGCAAG   125

   20 GlnLeuLysLeuAsnProSerSerGlnGluLeuGlyAlaAlaArgPheThrMetArgLys
        CAACTCAAGCTGAACCCATCAAGCCAAGAATTGGGAGCTGCAAGGTTCACCATGAGGAAG   185

   40 SerAlaThrThrLysLysValAlaSerSerGlySerProTrpTyrGlyProAspArgVal
        TCTGCTACCACCAAGAAAGTAGCTTCCTCTGGAAGCCCATGGTACGGACCAGACCGTGTT   245
                  C           C      A T      C C
                  *                  Ser      His

   60 LysTyrLeuGlyProPheSerGlyGluSerProSerTyrLeuThrGlyGluPheProGly
        AAGTACTTAGGCCCATTCTCCGGTGAGTCTCCATCCTACTTGACTGGAGAGTTCCCCGGT   305
              C                   A       C            A

   80 AspTyrGlyTrpAspThrAlaGlyLeuSerAlaAspProGluThrPheSerLysAsnArg
        GACTACGGTTGGGACACTGCCGGACTCTCTGCTGACCCAGAGACATTCTCCAAGAACCGT   365
                        T  T   C T          TG
                                            Ala

  100 GluLeuGluValIleHisSerArgTrpAlaMetLeuGlyAlaLeuGlyCysValPhePro
        GAGCTTGAAGTCATCCACTCCAGATGGGCTATGTTGGGTGCTTTGGGATGTGTCTTCCCA   425
              T                      A C

  120 GluLeuLeuSerArgAsnGlyValLysPheGlyGluAlaValTrpPheLysAlaGlySer
        GAGCTTTTGTCTCGCAACGGTGTTAAATTCGGCGAAGCTGTGTGGTTCAAGGCAGGATCT   485
                        T              A

  140 GlnIlePheSerGluGlyGlyLeuAspTyrLeuGlyAsnProSerLeuValHisAlaGln
        CAAATCTTTAGTGAGGGTGGACTTGATTACTTGGGCAACCCAAGCTTGGTCCATGCTCAA   545
              C           C

  160 SerIleLeuAlaIleTrpAlaThrGlnValIleLeuMetGlyAlaValGluGlyTyrArg
        AGCATCCTTGCCATATGGGCCACTCAGGTTATCTTGATGGGAGCTGTCGAAGGTTACCGT   605
                    C                              T

  180 IleAlaGlyGlyProLeuGlyGluValValAspProLeuTyrProGlyGlySerPheAsp
        ATTGCCGGTGGGCCTCTCGGTGAGGTGGTTGATCCACTTTACCCAGGTGGAAGCTTTGAT   665
                    C  T      T   C       T T     T

  200 ProLeuGlyLeuAlaAspAspProGluAlaPheAlaGluLeuLysValLysGluLeuLys
        CCATTGGGCTTAGCTGATGATCCAGAAGCATTCGCAGAATTGAAGGTGAAGGAACTCAAG   725
                 A     A TA      T T     A     A      G
                 GluVal

  220 AsnGlyArgLeuAlaMetPheSerMetPheGlyPhePheValGlnAlaIleValThrGly
        AACGGTAGATTAGCCATGTTCTCAATGTTTGGATTCTTCGTTCAAGCTATTGTAACTGGA   785
                        T   T          T

  240 LysGlyProLeuGluAsnLeuAlaAspHisLeuAlaAspProValAsnAsnAsnAlaTrp
        AAGGGTCCTTTGGAGAACCTTGCTGATCATCTTGCAGACCCAGTCAACAACAATGCATGG   845
                      C          C        G          C

  260 SerTyrAlaThrAsnPheValProGlyLys
        TCATATGCCACCAACTTTGTTCCCGGAAAATAAACACTCTTATATTTAT-----------   894
                 T                 G G    A    G    GTTATTGGTGA

        ---ATGTTTTGTGATAGTAATCTTCTTCCCAATTCAATGTGAATTATTATCATTATCAT   951
        AGT    A    T      G     -   T        G  GG          --

        TATCATGTGGGTATGCATAGGTTCACTAATACAAGATGATGGATGCTTTTTTTTTACCAA  1011
        ----G    T    G    A      TG AC    TCG---           T*

        ATTTTAAATTTTATGTTTCATGCTTTCCATTGCTAGACAT
```

FIG. 8.18. The nucleotide sequences of genes from pea for the chlorophyll *a/b*-binding polypeptides. The complete sequence is for one of the family of these genes that was cloned from nuclear

mids, which were again inserted into bacterial cells, as described in the last paragraph for the cDNA. Those bacterial cells that had taken up a plasmid bearing the gene for the chlorophyll *a/b*-binding polypeptide were detected by hybridization of cellular DNA with the cDNA. Subsequently, after growth of these clones, the DNA fragment containing the gene was recovered in sufficient quantity and sequenced. The nucleotide sequences of the gene and the cDNA are shown in Fig. 8.18, along with the amino acid sequences deduced from the DNA sequence with the genetic code.

The two nucleotide sequences shown in Fig. 8.18 are similar but not entirely homologous. The differences in the sequences suggest that a multigene family also exists in pea for this polypeptide, as occurs with the carboxylase small subunit. The cDNA apparently was synthesized from mRNA transcribed from a different gene than that selected from nuclear DNA. In contrast to the gene for the carboxylase small subunit, the gene shown in Fig. 8.18 does not contain an intron. The transit sequence at the N-terminus of the precursor polypeptide is no more than 39 amino acids in length, which is consistent with the difference in size between the precursor and the mature polypeptide.

The identity of the N-terminal amino acid has not been established for the chlorophyll *a/b*-binding polypeptide. However, a segment containing the sequence -Ser-Ala-Thr-Thr-Lys-Lys- is known to exist within the molecule (Mullet, 1983). Inspection of the amino acid sequence showed that this peptide occurs near the N-terminus of the precursor, starting at position 40 (see Fig. 8.18). The mature polypeptide contains only a few more amino acids, if any, at that end. Establishing the position of this peptide within the protein molecule, however, has answered an important question regarding the orientation of the polypeptide within the membrane. Mullet (1983) found that treatment of thylakoid membranes with trypsin, a proteolytic enzyme, released the six-amino acid peptide from the chlorophyll-binding polypeptide. Since this peptide exists near the N-terminus, and previous data suggested that the polypeptide spans the membrane, it is the N-terminal end that is exposed on the stromal side.

D. Regulation of Assembly of the Chlorophyll *a/b*-Protein Complex

Formation of the major light-harvesting chlorophyll–protein complex has several particularly interesting features. When greening cells are transferred to the dark, to stop synthesis of chlorophyll, further accumulation of the polypep-

---

DNA. A second sequence was obtained by the synthesis of a cDNA (complementary DNA) copy with the enzyme reverse transcriptase (RNA-dependent DNA polymerase) and pea mRNA as the template. The sequence of the cDNA is shown on the lower line only where it differs from the gene. Deleted sequences are indicated by ----. A predicted complete amino acid sequence for the precursor polypeptide is shown above the nucleotide sequence. Where the cDNA predicts amino acid changes in the second polypeptide, these are indicated on the lower line. The polypeptides encoded by the two sequences differ by five amino acids. No introns are present in this gene. The vertical arrow indicates a possible cleavage site for removal of a 37-amino-acid transit sequence, which would be consistent with a difference of 4000 to 5000 in molecular weights of the precursor and mature polypeptides. (Figure courtesy of A. R. Cashmore.)

tides also soon comes to a halt. Conversely, if synthesis of the polypeptides on cytoplasmic ribosomes is inhibited with cycloheximide during greening in the light, synthesis of chlorophyll immediately stops. These observations suggest that accumulation of the components is closely linked to assembly of the complex.

Some information is beginning to emerge regarding the relationship between synthesis of chlorophyll and accumulation of the polypeptides. Greening of most plant cells exhibits a lag phase of several hours before a significant rate of chlorophyll synthesis is achieved. During this time the amounts of the mRNA for the polypeptides gradually increase, from a barely detectable level (Fig. 8.17). The lag, therefore, reflects the time required to achieve sufficient transcription in the nucleus of the genes for the chlorophyll-binding polypeptides. The overall rate of greening, in fact, seems to be limited by the rate at which the polypeptides of the chlorophyll a/b-protein complex can be made.

As these results indicate, transcription of the genes for the chlorophyll-binding polypeptides is strongly stimulated by light. But the illumination does not need to be continuous nor is white light required. Exposure of etiolated seedlings to intermittent light (2 min of white light separated by 98 min of darkness) promotes transcription of the genes. Under this regime, only a small amount of chlorophyll a and no chlorophyll b are made. Alternatively, accumulation of the mRNA will also occur in the dark in etiolated higher plants following a single 15-sec pulse of red light, which activates phytochrome, an important regulatory factor (see p. 252).

Although under normal growth conditions illumination of cells markedly increases the amount of the mRNA for the thylakoid polypeptides, light apparently is not an absolute requirement for this effect. In etiolated cells of *Chlamydomonas reinhardtii* y-1, these specific mRNA species also increase without any exposure to light following an increase in the temperature of the culture from 25°C to 38°C. Maximal levels are achieved within 1–2 hrs at the higher temperature (Fig. 8.17). Neither the mechanism of this induction by heat nor that brought about by light is understood.

In each of these three cases, i.e., a single pulse of red light, a regime of intermittent light, or an elevation in temperature, subsequent exposure of the cells to continuous light results in immediate and rapid greening, as expected if synthesis of the polypeptides is the rate-limiting step. However, translation of the mRNA into the polypeptides is not linked to the synthesis of chlorophyll and occurs efficiently in the dark. The question then becomes why the chlorophyll-binding polypeptides do not accumulate in the dark even when significant amounts of the mRNA are present and translated. The evidence suggests that if chlorophyll synthesis cannot occur and thus the polypeptides are unable to assemble into a complex with chlorophyll in the membrane, the polypeptides are rapidly degraded (Bennett, 1981).

Synthesis of chlorophyll b, in particular, is required for stabilization of the chlorophyll a/b-binding polypeptide. A mutant strain of barley exists that lacks chlorophyll b. Chlorophyll b similarly is very low in plants exposed only to intermittent light. In both situations, mRNA for the chlorophyll a/b-binding polypeptide is transcribed and translated into the precursor form of the polypeptide. The precursor subsequently is taken up by the chloroplast and processed to the mature form in the normal manner. But in the absence of chlorophyll b, the polypeptide is then degraded (Bellemare et al., 1982).

So an interesting state of affairs exists in that the cellular levels of the major thylakoid polypeptides are controlled at two sites. Normally, light is required for transcription of the genes for these polypeptides in the nucleus. The mechanism of this regulatory process and the identity of the functional inducing substance are not known. In Chlamydomonas, this transcriptional control process apparently is overridden by an increase in temperature, and, as a result, the mRNA molecules are made even in the dark (Hoober et al., 1982). However the availability of the mRNA is not sufficient for accumulation of the polypeptide. After translation of the mRNA in the cytoplasm, the precursor polypeptides enter the chloroplast. The second regulatory process then operates to coordinate the amount of incoming polypeptides with the amount of chlorophyll synthesized. Those polypeptides present in excess over what can be accommodated by the expanding membrane are degraded. Binding of both chlorophyll a and b seems to be required to rescue the polypeptides from degradation.

The complementary situation also occurs. Inhibition of protein synthesis in the cytoplasm causes an abrupt halt in chlorophyll synthesis. There is abundant evidence that synthesis of the major thylakoid polypeptides in the cytoplasm is necessary for chlorophyll synthesis and thylakoid membrane assembly. It is not known where the chlorophyll–protein complexes form or whether the polypeptides bind completed chlorophyll molecules or the precursor chlorophyllides. Nevertheless, by binding to the end products of the pathway, the polypeptides mitigate feedback inhibition of chlorophyll synthesis. Thus, each of the components of the major light-harvesting chlorophyll a/b-protein complex in some manner regulates synthesis of the other components. The conclusion is that thylakoid membrane assembly, using this complex as an example, is an exceptionally highly regulated process.

V. How Are Polypeptides Taken up by Chloroplasts?

Suggestions are beginning to emerge regarding the mechanism of association between cytoplasmically made precursor polypeptides and the chloroplast envelope. Chloroplasts from spinach, and probably other plants as well, carry a high negative charge, with an average of one negative electrostatic charge for every 6.5 nm^2. The pH at which no net charge occurs on the surface (the isoelectric point or pI) is 4.5, which is near the pI for the carboxyl groups on the side chains of glutamic and aspartic acids. Thus acidic proteins may contribute to the surface charge of the chloroplasts.

The transit sequences in the precursor forms of the small subunit of ribulose 1,5-bisphosphate carboxylase and the chlorophyll-binding polypeptide are rich in the basic amino acids lysine and arginine. Acidic amino acids are rare. Therefore, it seems probable that the initial attraction of the polypeptides to the chloroplast surface occurs by electrostatic interaction. Specific receptors also seem to exist in the envelope, and binding of precursor polypeptides to these sites can occur without transport or subsequent processing (Pfisterer et al., 1982). Moreover, the carboxylase small subunit precursor from Chlamydo-

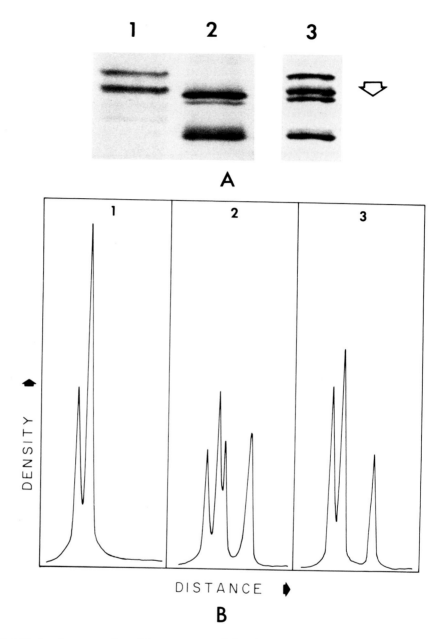

FIG. 8.19. *In vitro* processing of the precursors of the chlorophyll *a/b*-binding polypeptide from *Chlamydomonas reinhardtii*. Poly(A)-rich RNA was translated *in vitro* as described in Fig. 8.17. Then a cell extract, prepared by centrifugation of broken *Chlamydomonas* cells at 120,000g for 2 hr, was added to the translation mixture. After 1 hr of incubation at 30°C, antibodies against polypeptide 11 were added, as in Fig. 8.17. The immunoprecipitated polypeptides were analyzed by electrophoresis and radioautography. Only the middle portion of lanes on the radioautogram is shown.

A: Lanes 1 and 2 show precursors made *in vitro* and mature membrane polypeptides made *in vivo*, respectively, that were precipitated with the antibody preparation. Lane 3 shows the pattern obtained after adding the cell extract to the translation mixture. Partial conversion to the mature-sized polypeptides is evident. The arrow shows the direction of migration. B: To establish

monas is not taken up by isolated pea chloroplasts, which possibly is a demonstration of the specificity of the receptor proteins. Although the transit peptides on both the *Chlamydomonas* and pea small subunit precursors are rich in basic amino acids, the sequences show little homology.

Transport of precursor polypeptides into the chloroplast requires energy in the form of ATP, which can be added exogenously or produced *in situ* by photophosphorylation. The mechanism of this transport process is not known. Whether the polypeptides must pass completely into the chloroplast stroma, or only partially cross the inner envelope membrane, before the transit peptide is cleaved also is not known.

VI. Processing of Incoming Chloroplast Proteins

Since the transit sequence is required for entry of a polypeptide into the chloroplast, it seems logical that cleavage of the precursor molecule should occur within the organelle. Schmidt *et al.*, (1979) observed that the precursor of the carboxylase small subunit, made by *in vitro* translation, was cleaved to the mature polypeptide by a protease ("transit peptidase") in an extract of *Chlamydomonas* cells. In pea leaves this activity was recovered after cell fractionation with chloroplasts. The protease occurs as a soluble protein and thus remains in the supernatant fraction after sedimentation of membranes. These results suggest that the processing of precursors occur in the chloroplast stroma (Smith and Ellis, 1979).

The protease appears to be remarkably specific for cleavage of the transit sequence. Uptake of precursors into the chloroplast is accompanied by precise processing to the mature polypeptides (see Figs. 8.15 and 8.16). Also, the addition of the extracted protease to precursors made in a translation system results in cleavage only to the mature-sized polypeptides, with no further degradation. Fig. 8.19 shows *in vitro* processing of the precursors of the chlorophyll *a/b*-binding polypeptides in *Chlamydomonas*. After incubation, the decrease in the amounts of the precursors can be accounted for by the increase in the mature-size polypeptides.

The extracted "transit peptidase" has permitted identification of the precursor–product relationships between the two precursor polypeptides and the mature forms in *Chlamydomonas*. Although it was expected that the 31,500-

the pattern of processing to mature polypeptides, radioautograms were scanned with a densitometer to obtain quantitative data. Panel 1 shows the scan of the immunoprecipitated precursors in the untreated sample. When the translation mixture was incubated with the cell extract, and then polypeptides were immunoprecipitated, partial cleavage of both precursors to 29,000- and 26,000-molecular-weight products was obtained (panel 2). Panel 3 shows the result when the polypeptides were first immunoprecipitated and then treated with the cell extract. The decrease in amount of the 30,000-molecular-weight precursor was accompanied only by an increase in the 26,000-molecular-weight component. (Data of D. B. Marks, B. J. Keller, and J. K. Hoober.)

and the 30,000-molecular-weight precursors become the 29,500- and 26,000-molecular-weight membrane polypeptides, respectively, this relationship could not be established from the results shown in Fig. 8.19A. However, digestion of an immunoprecipitate of the precursor polypeptides with the protease resulted in cleavage of the 30,000-molecular-weight precursor to the 26,000-molecular-weight polypeptide, with no decrease in the larger precursor.

After entry of the precursors into the chloroplast, and subsequent cleavage of the transit sequence, the polypeptides associate with membranes. If the chloroplast is capable of making chlorophylls, the polypeptides become incorporated into stable complexes within thylakoid membranes. In the dark, without chlorophyll the polypeptides rapidly are degraded by proteases apparently associated with the membrane.

VII. Source of Chloroplast Proteins—A Summary

Fig. 8.20 summarizes the current view on the production of structures within the chloroplast. RNA molecules are transcribed from nuclear DNA, processed and modified by polymerization of additional AMP residues (from ATP) at the 3′ end. These RNAs are translated in the plant cell cytoplasm into precursors of the plastid proteins. The N-terminal extensions on these precursors facilitate entry into the chloroplast. After proteolytic processing to the mature polypeptide, the cytoplasmically made products join polypeptides synthesized within the chloroplast to form functional assemblies.

Among the *soluble proteins* made by the cytoplasm are the enzymes of the

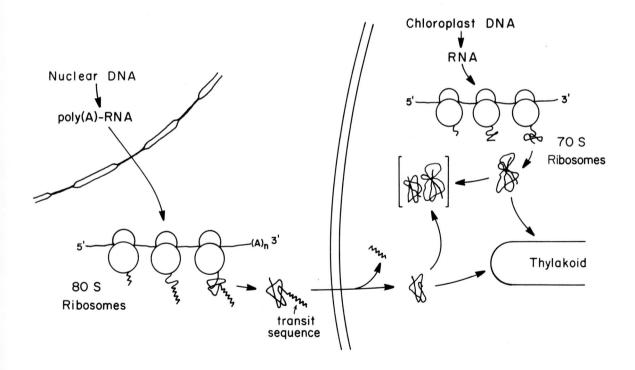

C_3 cycle (except for the large subunit of ribulose 1,5-bisphosphate carboxylase), the amino acyl–tRNA synthetases, RNA polymerase, ferredoxin, and ferredoxin–NADP$^+$ reductase. The enzymes involved in chlorophyll synthesis, at least those in the pathway between δ-aminolevulinic acid and chlorophyllides, are apparently made in the cytoplasm. But in addition to the large subunit of the carboxylase, identified stromal proteins that are made in the chloroplast are the elongation factors Tu (EF-Tu$_{Chl}$) and G (EF-G$_{Chl}$) involved in protein synthesis.

The majority of *thylakoid membrane proteins*, including the polypeptides of the light-harvesting complexes, are made in the cytoplasm. On the other hand, membrane polypeptides made in the plastid are the α, β, and ε subunits of CF$_1$ (the chloroplast coupling factor), a 32,000-dalton polypeptide associated with photosystem 2, cytochromes f and b-559, the apoprotein of the chlorophyll a–protein complex of photosystem 1, and the photosystem reaction center polypeptides.

Assembly of ribosomes also involves cooperation of the nuclear–cytoplasmic and chloroplast systems. The ribosomal RNA, which is transcribed from chloroplast DNA, associates with proteins made in both cellular compartments. Only five or six of the 32 or 33 proteins of the large ribosomal subunit are made in the chloroplast. Of the 24 or 25 integral proteins of the small ribosomal subunit, 12 appear to be made in the chloroplast (Schmidt et al., 1983).

Phycobilisomes are major light-harvesting complexes associated with thylakoid membranes in red algae (see Chapter 3). The major pigmented polypeptides, the α and β subunits of the biliproteins, and a protein that anchors the complexes to the membrane are synthesized on chloroplast ribosomes. However, other nonpigmented polypeptide components of the phycobilisomes are made on cytoplasmic ribosomes (Egelhoff and Grossman, 1983).

From studies with isolated chloroplasts, the total list of proteins made within the organelle is considerably longer. Proteins synthesized by preparations from pea, spinach, and *Euglena* were labeled by the incorporation of high-specific-activity amino acids (e.g., [^{35}S]methionine). When the stromal fraction of these labeled chloroplasts were separated by a high-resolution, two-dimensional gel electrophoresis system, about 80 labeled polypeptides were detected. However, the identities of only a few are known.

The concept that emerges from these data is that, although the chloroplasts

←

FIG. 8.20. Summary diagram illustrating the origins and fluxes of polypeptides involved in assembly of chloroplast structures. Most membrane polypeptides, including the chlorophyll a/b-binding polypeptides, are synthesized on 80 S cytoplasmic ribosomes with mRNA transcribed from nuclear DNA. The mRNA molecules contain a poly(A) sequence at their 3′ end and carry information for synthesis of transit sequences at the N-termini of the polypeptides. Entry of these precursor molecules into the plastid is accompanied by removal of the transit sequence. Other polypeptides, including the reaction center polypeptides in photosystems 1 and 2, are synthesized on 70 S chloroplast ribosomes with mRNA transcribed from chloroplast DNA. The actual site of assembly of the thylakoid membrane is not known but probably is within the membrane itself. Soluble stromal enzymes, including the small subunit of ribulose 1,5-bisphosphate carboxylase, are made on cytoplasmic ribosomes from poly(A)-RNA. After entry into the chloroplast and loss of the transit sequence, the small subunit polypeptide combines with the large subunit, which is made on chloroplast ribosomes. Relatively few plastid polypeptides are made on chloroplast ribosomes.

are capable of synthesizing a number of proteins, many of the proteins essential for the structure, function, and reproduction of the organelle are produced by the nuclear–cytoplasmic system. Assembly of the chloroplast structures is, therefore, a coordinated process, and integration of the synthetic capability of both cellular compartments is required for chloroplast development.

VIII. Light and Chloroplast Development

Photoreduction of protochlorophyllide is an important and necessary step in chloroplast development, but this is not the only photoreaction that promotes development of the chloroplast. Although the effects of these other reactions have been recognized for many years, the mechanisms by which they promote the developmental process is still not known.

A. The Regulatory Factor Phytochrome

One of the most enigmatic regulatory factors in plant cells is **phytochrome.** A large number of responses of plants to red (650–670 nm) light (see Table 8.2) have been described that are inhibited or reversed if the plants subsequently are irradiated with far-red (730–750 nm) light. Borthwick and his colleagues (1952) found that this photoreversibility persisted over nearly 100 cycles of red and far-red light, which suggested to them that the effects may be caused by two forms of the same pigment. Over 20 years passed, since these observations in the early 1950s, before the factor was successfully purified. During purification it became clear that the factor is a chromoprotein, which exists in two forms. Interconversion between these forms requires no cofactor or activator other than light. Since the two forms have different absorption spectra (Fig. 8.21), phytochrome was assayed during purification by the change in the absorbance ratio $A_{665 \, nm}:A_{730 \, nm}$ induced by irradiation of a preparation with red light. Vierstra and Quail (1983) and Litts et al. (1983) have succeeded in isolating undegraded phytochrome. The protein moeity is a single polypeptide 124,000 daltons in mass, which may exist in vivo as a dimer. The chromphore is an open-chain tetrapyrrole similar to the biliprotein chromophores (p. 64). In phytochrome, the tetrapyrrole structure (Fig. 8.22) is covalently attached to a cys-

TABLE 8.2. Representative Photoreversible,
Phytochrome-Mediated Responses

Stimulation of seed germination
Promotion of hypocotyl elongation
Enhancement of rate of chloroplast development
Chloroplast division rate in Polytrichum
Chloroplast movement in Mougeotia
Leaflet movement of Mimosa pudica
Modulation of bioelectric potentials
Induction of mRNA synthesis in nucleus and chloroplast
Induction of anthocyanin biosynthesis
Stimulation of chlorophyll b synthesis

teine residue in the protein through a thioether linkage formed by addition of the sulfhydryl group to the ethylidene group on ring A.

Absorption of red (665-nm) light results in a conformational change in the chromophore structure. The initial form, designated P_r, consequently is converted to the presumed active form, P_{fr}. The chemistry of the reorientation of the chromphore in the P_{fr} form is still uncertain but seems to involve electronic rearrangements induced by additional covalent interactions with the protein. Irradiation with far-red (730-nm) light causes reversal to the initial, inactive P_r form. In a simplified manner, without showing intermediate stages, the processes can be represented by the following scheme:

$$P_r \xrightleftharpoons[\text{Far-red light (fast)}]{\text{Red light (fast)}} P_{fr} \rightarrow \rightarrow \rightarrow \text{Morphogenic response} \qquad [8.7]$$

$$\text{Dark (slow)}$$

Fig. 8.21 shows the calculated absorption spectra of pure P_r and P_{fr}. Since the spectra overlap, in reality photoconversion to the P_{fr} form does not go to completion. Rather, a photoequilibrium is reached by irradiation with red light in which 80–85% of the phytochrome is in the P_{fr} form. However, most responses in the plant cell can be observed with a much lower ratio of P_{fr} to P_r. Photoconversion is rapid, with a half-life at 8°C of less than 0.5 sec. The dark, thermal reversal of P_{fr} to P_r is slow, with a half-life on the order of 30 min at 25°C.

Phytochrome in unirradiated oat tissues behaves as a soluble protein, dispersed throughout the cytoplasm. After photoconversion to the P_{fr} form, a significant amount of the protein becomes associated with cellular structures. The cause of this association is not known, but work by Hahn and Song (1982) indicates that a hydrophobic area on the protein surface is more exposed in the P_{fr}

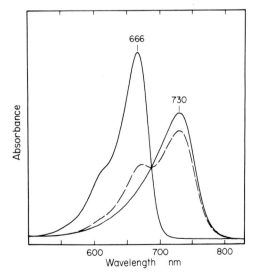

FIG. 8.21. The calculated absolute absorption spectra for the P_r (λ_{max} = 666 nm) and the P_{fr} (λ_{max} = 730nm) forms of oat (*Avena sativa*) phytochrome. Because the spectra of the two forms overlap, it is not possible to convert the inactive form (P_r) completely to the active form (P_{fr}). Irradiation with 665-nm light, which is absorbed maximally by the P_r form, achieves a photoequilibrium between the two, with about 85% of the phytochrome in the P_{fr} form (----). (From Vierstra and Quail, 1983. Reprinted with permission).

form. Exposure of this area may favor interactions of the phytochrome with other cellular components through hydrophobic interactions. As with the morphogenic responses, if photoconversion to the P_{fr} form is immediately followed by return to P_r with far-red light, the phytochrome remains soluble. Because of this association with cellular structures, and the changes in permeability properties that accompany generation of P_{fr}, membranes have been suggested as a site of action for phytochrome.

Expression of a large number of genes also seems to be controlled by phytochrome. A particularly elegant demonstration of such control was performed by Apel (1979). In etiolated barley seedlings, a 15-sec pulse of red (655-nm) light alone caused a significant accumulation of the mRNA for the chlorophyll a/b-protein of thylakoid membranes (Fig. 8.23). Up to 4 hr, the single pulse of red light was as effective as continuous white light in stimulating transcription of this gene. However, when the red light pulse was followed immediately with far-red (759-nm) light, the increase in the RNA was much less. Therefore, phytochrome plays a role in controlling synthesis of thylakoid membrane polypeptides in higher plants.

Although chlorophyll synthesis does not depend on phytochrome, the rate of total chlorophyll synthesis is enhanced by P_{fr}. Work in Mohr's laboratory has shown that synthesis of chlorophyll(ide) b in etiolated mustard seedlings in particular is stimulated by P_{fr} during the initial stages of greening (Oelze-Karow and Mohr, 1978, 1982). Chlorophyll(ide) b was detected during the first 30 min of irradiation with white light only in plants pretreated with the red light. Furthermore, if the plants were exposed to 756-nm light following the 668-nm-light exposure, no chlorophyll(ide) b was detected. They also observed that in etiolated cells, in which protochlorophyll(ide) was fully converted to chlorophyll(ide) a by providing a saturating dose of white light, chlorophyll(ide) b appeared during a subsequent dark period only in plants pretreated with 668-nm light. In this experiment, chlorophyll(ide) b seemed to appear at the expense of chlorophyll(ide) a.

It is evident, therefore, that two light reactions are involved in chloroplast development and thylakoid membrane formation in higher plants. The two photoreactions have been identified as the light-mediated reduction of proto-

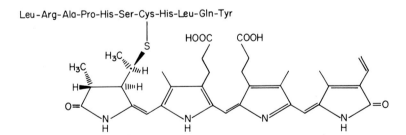

Leu-Arg-Ala-Pro-His-Ser-Cys-His-Leu-Gln-Tyr

FIG. 8.22. The structure of the chromophore in oat phytochrome and the linkage of the chromophore to the polypeptide. The polypeptide was digested with the proteases pepsin and thermolysin. A peptide, 11 amino acids long, was purified that contained the bound chromophore. The linear tetrapyrrole structure of the chromophore, and its linkage to the polypeptide, is the same as that in C-phycocyanin. (Redrawn from Lagarias and Rapoport, 1980.)

chlorophyll(ide) and the photoconversion of P_r to P_{fr}. Because the action spectra of these two reactions overlap, it is often difficult to separate their effects. Moreover, the mechanisms by which these reactions are involved in regulatory processes are elusive.

B. Effects of Blue Light

Although effects of phytochrome have been studied extensively in higher plants, no unequivocal evidence exists for the presence of phytochrome in algae. A response that does occur in both types of plant cells, but which is particularly evident in algae, is that to blue light (380–450 nm). The receptor for blue-light responses has not been conclusively identified, but the data support the suggestion that it is usually a flavin or flavoprotein.

Two types of responses are most common. As exemplified by nitrate reductase, an enzyme complex involved in reducing nitrate to metabolically useful ammonia, irradiation with blue light converts an inactive to an active form of the enzyme. The enzyme complex contains a flavin (FAD), which probably is the light receptor, a cytochrome (cyt b-557), and molybdenum as cofactors.

A second type of response is induction of synthesis of several enzymes in blue-light-irradiated cells. Klein and Senger (1978) extensively studied the dramatic effect of blue light on chlorophyll biosynthesis in the alga *Scenedesmus*. This alga contains two pathways for synthesis of δ-aminolevulinic acid, the pathway from glutamate and that from glycine and succinyl-coenzyme A catalyzed by δ-aminolevulinic acid synthetase (see pp. 223–225). In mutant strains

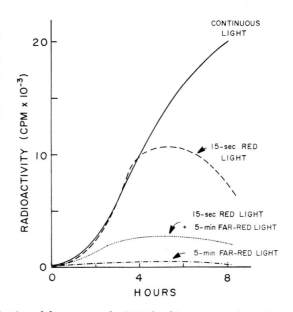

FIG. 8.23. Time course of the increase in the amount of translatable mRNA for the light-harvesting chlorophyll a/b-protein in barley seedlings after treatment of etiolated seedlings with continuous illumination (———), a 15-sec pulse of red light followed by darkness (— — —), a 15-sec pulse of red light followed by 5 min of far-red light and then darkness (· · · · · ·), or a 5-min treatment of far-red light alone followed by darkness (—·—·—). After the various treatments, the mRNA was assayed by *in vitro* translation of poly(A)-rich RNA extracted from plants at each time point. The light-harvesting chlorophyll a/b-protein, labeled during translation with [^{35}S]methionine, was immunoprecipitated from the translation mixture, and the amount of radioactivity recovered in the precipitate was an indication of the amount of mRNA for this protein in the cells. The large and prolonged increase in the mRNA promoted by only 15 sec of red light, and the extensive reversal of this effect by far-red light, are characteristic of responses to activation of phytochrome. (Adapted from Apel, 1979.)

of *Scenedesmus*, which require light for chlorophyll biosynthesis, the synthesis of enzymes in both pathways is dependent on blue light. Consequently, initiation of chlorophyll biosynthesis and chloroplast development require both red light, for photoreduction of protochlorophyllide, and blue light, for synthesis of the enzymes needed for δ-aminolevulinate synthesis.

C. Are There Additional Light Receptors?

Another light-mediated reaction seems to be involved in promoting chlorophyll synthesis, at least in algae. An orange light (600-nm) receptor, which strongly promotes chloroplast development, has been detected by Kaufman *et al.* (1982) in *Euglena*. This chromophore is different from protochlorophyllide, which absorbs blue and red light. In *Chlamydomonas*, monochromatic light near 650 nm, the absorption maximum in red light for phototransformable protochlorophyllide (see p. 228), does not support normal greening at 25°C. This result is puzzling, considering that 650-nm light is sufficient to mediate synthesis of chlorophyll. If, in addition, wavelengths of light near 600 nm are provided, greening occurs as rapidly as in white light. This situation is in contrast to that at 38°C, in which greening of this organism occurs with 650-nm light alone (Fig. 8.24). This apparent contradiction can be resolved by taking into account the fact that at 38°C the mRNA for the chlorophyll-binding polypeptides is transcribed in the dark. Consequently, the polypeptides are available to complex with chlorophyll synthesized in response to 650-nm light. It is possible, therefore, that in algae the light receptor involved in inducing synthesis of the membrane polypeptides has a maximal absorption near 600 nm. The mechanism of light-induced synthesis of proteins is an interesting question, but very little information exists that can provide an explanation.

IX. Summary

Chloroplasts in most higher plants and in a few algae require light for chloroplast development and thus are etiolated when grown in the dark. In dark-

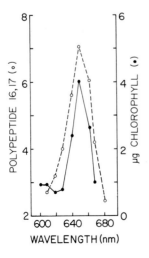

FIG. 8.24. Effects of light of different wavelengths on the synthesis of chlorophyll and the accumulation of chlorophyll *a/b*-binding polypeptides in *Chlamydomonas reinhardtii* y-1. Cells were incubated in the dark at 38°C for 2 hr and then exposed to light transmitted by interference filters (11-nm half-band width) for an additional hour. Chlorophyll (●——●) was synthesized most actively in cells exposed to light of 650 nm. Although the polypeptides were synthesized in the dark at 38°C (see Fig. 8.17), they were degraded when chlorophyll was unavailable for formation of chlorophyll–protein complexes. Thus, accumulation of the polypeptides (○- - -○) paralleled the increase in chlorophyll. (Adapted from Hoober and Stegman, 1976.)

ness, higher-plant chloroplasts develop from a proplastid stage into etioplasts, which contain an elaborate prolamellar body. In the light, a proplastid and an etioplast each will develop into a mature chloroplast over a period of about two days in higher plants. The primary morphological event during chloroplast development is formation of thylakoid membranes.

Synthesis of chlorophyll is necessary for chloroplast development. The synthesis of chlorophyll triggers the synthesis of many other thylakoid membrane components. Also, chlorophyll is necessary for assembly of the membrane. The enzyme protochlorophyllide reductase forms chlorophyllide *a* in a light-facilitated reaction, and the primary involvement of light in chloroplast development, as opposed to function, is in this step of the pathway. The first compound in the pathway committed to chlorophyll synthesis is δ-aminolevulinic acid, whose synthesis is also stimulated in the light. The release of the enzymes from feedback inhibition seems to be the major mechanism for stimulation of δ-aminolevulinate synthesis. Glutamate and α-ketoglutarate are the major precursors of δ-aminolevulinate used in chlorophyll synthesis. The pathway to chlorophyll from δ-aminolevulinate resides entirely within the chloroplast.

Synthesis of fatty acids for thylakoid membrane lipids occurs in the stroma of the chloroplast. However, the major site for synthesis of the thylakoid lipids is the chloroplast envelope. The process of transfer of lipids between the membranes is not known.

Proteins of the thylakoid membrane are made on both chloroplasts and cytoplasmic ribosomes. The major membrane proteins, along with most stroma proteins, are made on cytoplasmic ribosomes. These proteins from the cytoplasm are initially made as precursors that are 2000 to 5000 in molecular weight larger than the mature form found in the chloroplast. During transport into the plastid, the additional peptide segment (the "transit sequence") is cleaved off. Formation of ribulose 1,5-bisphosphate carboxylase is a particularly interesting process, since the large subunits are made as the mature polypeptide in the chloroplast whereas the small subunits are synthesized as a precursor polypeptide on cytoplasmic ribosomes.

The plastid structures are generated as complexes formed through interaction of cytoplasmic and organellar products. When chlorophyll is not made, and membranes cannot form, unincorporated polypeptides are rapidly degraded by a chloroplast proteolytic system. Consequently, excess proteins or subunits cannot be detected easily. Assembly of thylakoid membranes seems to occur by the insertion of individual components. However, the process is highly coordinated and results in the progressive accumulation of fully functional membranes. Little difference can be detected between membranes formed during the early stages of greening and those in mature chloroplasts. Thus, photosynthetic function develops in parallel with assembly of the membranes.

Although photoreduction of protochlorophyllide is the primary promotive as well as regulatory step involving light, another important regulatory factor affected by light is the chromoprotein phytochrome. Red light activates phytochrome, whereas far-red light reverses the activation. These wavelengths of light cause conformational changes in the tetrapyrrole chromophore. Activated phytochrome promotes transcription of the genes for a number of proteins and also promotes the rate of developmental reactions. In addition, absorption of blue light, probably by flavins or flavoproteins, promotes developmental processes in plants and particularly algae. The mechanisms of these effects are unclear.

Literature Cited

Akoyunoglou, G., and Argyroudi-Akoyunoglou, J. H. (1978) Control of thylakoid growth in *Phaseolus vulgaris*, *Plant Physiol.* **61**: 834–877.

Akoyunoglou, G., and Argyroudi-Akoyunoglou, J. H. (1979) The chlorophyll-protein complexes of the thylakoid in greening plastids of *Phaseolus vulgaris*, *FEBS Lett.* **104**:78–84.

Alberte, R. S., Thornber, J. P., and Naylor, A. W. (1972) Time of appearance of photosystems I and II in chloroplasts of greening jack bean leaves, *J. Exptl. Bot.* **23**:1060–1069.

Apel, K. (1979) Phytochrome-induced appearance of mRNA activity for the apoprotein of the light-harvesting chlorophyll a/b protein of barley (*Hordeum vulgare*), *Eur. J. Biochem.* **97**: 183–188.

Bellemare, G., Bartlett, S. G., and Chua, N.-H. (1982) Biosynthesis of chlorophyll a/b-binding polypeptides in wild type and the chlorina f2 mutant of barley, *J. Biol. Chem.* **257**: 7762–7767.

Bennett, J. (1981) Biosynthesis of the light-harvesting chlorophyll a/b protein. Polypeptide turnover in darkness, *Eur. J. Biochem.* **118**:61–70.

Block. M. A., Dorne, A.-J. Joyard, J., and Douce, R. (1983a) Preparation and characterization of membrane fractions enriched in outer and inner envelope membranes from spinach chloroplasts, I. Electrophoretic and immunochemical analyses, *J. Biol. Chem.* **258**:13273–13280.

Block. M. A., Dorne, A. -J., Joyard, J., and Douce, R. (1983b) Preparation and characterization of membrane fractions enriched in outer and inner envelope membranes from spinach chloroplasts, II. Biochemical characterization, *J. Biol. Chem.* **258**: 13281–13286.

Boffey, S. A., Selldén, G., and Leech. R. M. (1980) Influence of cell age on chlorophyll formation in light-grown and etiolated wheat seedlings, *Plant Physiol.* **65**:680–684.

Borthwick, H. A., Hendricks, S. B., Parker, M. W., Toole, E. H., and Toole, V. K. (1952) A reversible photoreaction controlling seed germination, *Proc. Natl. Acad. Sci. USA* **38**:662–666.

Cahen, D., Malkin, S., Shochat, S., and Ohad, I. (1976) Development of photosystem II complex during greening of *Chlamydomonas reinhardi* y-1, *Plant Physiol.* **58**:257–267.

Coruzzi, G., Broglie, R., Cashmore, A., and Chua, N.-H. (1983) Nucleotide sequences of two pea cDNA clones encoding the small subunit of ribulose 1,5-bisphosphate carboxylase and the major chlorophyll a/b-binding thylakoid polypeptide, *J. Biol. Chem.* **258**:1399–1402.

Cumming, A. C., and Bennett, J. (1981) Biosynthesis of the light-harvesting chlorophyll a/b protein. Control of messenger RNA activity by light, *Eur. J. Biochem.* **118**:71–80.

Dean, C., and Leech, R. M. (1982) Genome expression during normal leaf development, 2. Direct correlation between ribulose bisphosphate carboxylase content and nuclear ploidy in a polyploid series of wheat, *Plant Physiol.* **70**:1605–1608.

Dobberstein, B., Blobel, G., and Chua, N.-H. (1977) *In vitro* synthesis and processing of a putative precursor for the small subunit of ribulose-1,5-bisphosphate carboxylase of *Chlamydomonas reinhardtii*, *Proc. Natl. Acad. Sci. USA* **74**:1082–1085.

Dujardin, E. (1982) Extinction of the *in-vivo* low-temperature fluorescence of chlorophyll a by long-wavelength-absorbing quenchers formed from protochlorophyllide, in *Cell Function and Differentiation*, Part B (G. Akoyunoglou, A. E. Evangelopoulos, J. Georgatsos, G. Palaiologos, A. Trakatellis, and C. P. Tsiganos, eds.), Alan R. Liss, New York, pp. 43–52.

Egelhoff, T., and Grossman, A. (1983) Cytoplasmic and chloroplast synthesis of phycobilisome polypeptides, *Proc. Natl. Acad. Sci. USA* **80**:3339–3343.

Emerson, R., and Arnold, W. (1932) The photochemical reaction in photosynthesis, *J. Gen. Physiol.* **16**:191–205.

Gershoni, J. M., and Ohad, I. (1980) Chloroplast-cytoplasmic interrelations involved in chloroplast development in *Chlamydomonas reinhardi* y-1: Effect of selective depletion of chloroplast translates, *J. Cell. Biol.* **86**:124–131.

Gershoni, J. M., Shochat, S., Malkin, S., and Ohad, I. (1982) Functional organization of the chlorophyll-containing complexes of *Chlamydomonas reinhardi*, *Plant Physiol.* **70**:637–644.

Gibbs, S. P. (1979) The route of entry of cytoplasmically synthesized proteins into chloroplasts of algae possessing chloroplast ER, *J. Cell Sci.* **35**:253–266.

Griffiths, T. W. (1978) Reconstitution of chlorophyllide formation by isolated etioplast membranes, *Biochem. J.* **174**:681–692.

Hahn, T.-R., and Song, P.-S. (1982) Molecular topography of phytochrome as deduced from the tritium-exchange method, *Biochemistry* **21**:1394–1399.

Harel, E., and Klein, S. (1972) Light dependent formation of δ-aminolevulinic acid in etiolated leaves of higher plants, *Biochem. Biophys. Res. Commun.* **49**:364–370.

Harel, E., and Ne'eman, E. (1983) Alternate routes for synthesis of 5-aminolevulinic acid in maize leaves, II. Formation from glutamate, *Plant Physiol.* **72**:1062–1067.

Highfield, P. E., and Ellis, R. J. (1978) Synthesis and transport of the small subunit of chloroplast ribulose bisphosphate carboxylase, *Nature* (London) **271**:420–424.

Hill, R. (1951) Oxidoreduction in chloroplasts, *Adv. Enzymol. Rel. Sub. Biochem.* **12**:1–39.

Hoober, J. K., and Stegeman, W. J. (1976) Kinetics and regulation of synthesis of the major polypeptides of thylakoid membranes in *Chlamydomonas reinhardtii* y-1 at elevated temperatures, *J. Cell Biol.* **70**:326–337.

Hoober, J. K., Bednarik, D., Keller, B. J., and Marks, D. B. (1982) Regulatory aspects of thylakoid membrane formation in *Chlamydomonas reinhardtii* y-1, in *Cell Function and Differentiation,* Part B (G. Akoyunoglou, A. E. Evangelopoulos, J. Georgatsos, G. Palaiologos, A. Trakatellis, and C. P. Tsiganos, eds.), Alan R. Liss, New York, pp. 127–137.

Hoober, J. K., Marks, D. B., Keller, B. J., and Margulies, M. M. (1982) Regulation of accumulation of the major thylakoid polypeptides in *Chlamydomonas reinhardtii* y-1 at 25°C and 38°C, *J. Cell Biol.* **95**:552–558.

Janero, D. R., and Barrnett, R. (1981) Thylakoid membrane biogenesis in *Chlamydomonas reinhardtii* 137+: Cell cycle variations in the synthesis and assembly of polar glycerolipid, *J. Cell Biol.* **91**:126–134.

Joliot, P., and Kok, B. (1975) Oxygen evolution in photosynthesis, in *Bioenergetics of Photosynthesis* (Govindjee, ed.) Academic Press, New York, pp. 387–412.

Joyard, J., Grossman, A., Bartlett, S. G., Douce, R., and Chua, N.-H. (1982) Characterization of envelope membrane polypeptides from spinach chloroplasts, *J. Biol. Chem.* **257**:1095–1101.

Kannangara, C. G., Gough, S. P., and von Wettstein, D. (1978) The biosynthesis of δ-aminolevulinate and chlorophyll and its genetic regulation, in *Chloroplast Development* (G. Akoyunoglou and J. H. Argroudi-Akoyunoglou, eds.), Elsevier/North Holland, Amsterdam, pp. 147–160.

Kaufman, L. S., Lyman, H., and Grzesick, R. (1982) Regulation of chloroplast development by a 600 nm receptor. *J. Cell Biol.* **95**:275a.

Klein, O., and Senger, H. (1978) Biosynthetic pathways to δ-aminolevulinic acid induced by blue light in the pigment mutant C-2A of *Scenedesmus obliquus, Photochem. Photobiol.* **27**: 203–208.

Lagarias, J. C., and Rapoport, H. (1980) Chromopeptides from phytochrome. The structure and linkage of the Pr form of the phytochrome chromophore, *J. Am. Chem. Soc.* **102**:4821–4828.

Litts, J. C., Kelly, J. M., and Lagarias, J. C. (1983) Structure–function studies on phytochrome: Preliminary characterization of highly purified phytochrome from *Avena sativa* enriched in the 124-kilodalton species, *J. Biol. Chem.* **258**:11025–11031.

Lütz, C., and Nordmann, U. (1983) The localization of saponins in prolamellar bodies mainly depends on the isolation of etioplasts, *Z. Pflanzenphysiol.* **110**:201–210.

Lütz, C., Röper, U., Beer, N. S., and Griffiths, T. (1981) Sub-etioplast localization of the enzyme NADPH:protochlorophyllide oxidoreductase, *Eur. J. Biochem.* **118**:347–353.

Malkin, S., and Fork, D. C. (1981) Photosynthetic units of sun and shade plants, *Plant Physiol.* **67**:580–583.

Meyers, S. P., Nichols, S. L., Baer, G. R., Molin, W. T., and Schrader, L. E. (1982) Ploidy effects in isogenic populations of alfalfa, I. Ribulose-1,5-bisphosphate carboxylase, soluble protein, chlorophyll, and DNA in leaves, *Plant Physiol.* **70**:1704–1709.

Mullet, J. E. (1983) The amino acid sequence of the polypeptide segment which regulates membrane adhesion (grana stacking) in chloroplasts, *J. Biol. Chem.* **258**:9941–9948.

Mishkind, M. L., and Schmidt, B. W. (1983) Posttranscriptional regulation of ribulose 1,5-bisphosphate carboxylase small subunit accumulation in *Chlamydomonas reinhardtii, Plant Physiol.* **72**:847–854.

Murakami, S., and Ikeuchi, M. (1982) Biochemical characterization and localization of the 36,000-dalton NADPH:protochlorophyllide oxidoreductase in squash etioplasts, in *Cell Function and Differentiation,* Part B (G. Akoyunoglou, A. E. Evangelopoulos, J. Georgatsos, G. Palaiologos, A. Trakatellis, and C. P. Tsiganos, eds.), Alan R. Liss, New York, pp. 13–23.

Oelze-Karow, H., and Mohr, H. (1978) Control of chlorophyll *b* biosynthesis by phytochrome, *Photochem. Photobiol.* **27:** 189–193.

Oelze-Karow, H., and Mohr, H. (1982) Phytochrome action on chlorophyll synthesis—A study of the escape from photoreversibility, *Plant Physiol.* **70:**863–866.

Ohad, I., Siekevitz, P., and Palade, G. E. (1967) Biogenesis of chloroplast membranes, II. Plastid differentiation during greening of a dark-grown algal mutant (*Chlamydomonas reinhardi*), *J. Cell Biol.* **35:**553–584.

Pfisterer, J., Lachmann, P., and Kloppstech, K. (1982) Transport of proteins into chloroplasts. Binding of nuclear-coded chloroplast proteins to the chloroplast envelope, *Eur. J. Biochem.* **126:**143–148.

Robertson, D., and Laetsch, W. M. (1974) Structure and function of developing barley plastids, *Plant Physiol.* **54:**148–159.

Schmidt, G. W., Bartlett, S. G., Grossman, A. R., Cashmore, A. R., and Chua, N.-H. (1981) Biosynthetic pathways of two polypeptide subunits of the light-harvesting chlorophyll a/b protein complex, *J. Cell Biol.* **91:**468–478.

Schmidt, G. W., Devillers-Thiery, A., Desruisseaux, H., Blobel, G., and Chua, N.-H. (1979) NH$_2$-Terminal amino acid sequences of precursor and mature forms of the ribulose-1,5-bishphosphate carboxylase small subunit from *Chlamydomonas reinhardtii*, *J. Cell Biol.* **83:**615–622.

Schmidt, R. J., Richardson, C. B., Gillham, N. W., and Boynton, J. E. (1983) Sites of synthesis of chloroplast ribosomal proteins in *Chlamydomonas, J. Cell Biol.* **96:**1451–1463.

Shimakata, T., and Stumpf, P. K. (1982) Fatty acid synthetase of *Spinacia oleracea* leaves, *Plant Physiol.* **69:**1257–1262.

Smith, S. M., and Ellis, R. J. (1979) Processing of small subunit precursor of ribulose bisphosphate carboxylase and its assembly into whole enzyme are stromal events, *Nature* (London) **278:** 662–664.

Tobin, E. M. (1981) White light effects on the mRNA for the light-harvesting chlorophyll a/b-protein in *Lemna gibba* L. G-3, *Plant Physiol.* **67:**1078–1083.

Vierstra, R. D., and Quail, P. H. (1983) Photochemistry of 124 kilodalton *Avena* phytochrome *in vitro, Plant Physiol.* **72:** 264–267.

Wang, W. Y., Boynton, J. E., and Gillham, N. W. (1977) Genetic control of chlorophyll biosynthesis: Effect of increased δ-aminolevulinic acid synthesis on the phenotype of the y-1 mutant of *Chlamydomonas, Mol. Gen. Genet.* **152:**7–12.

Additional Reading

Battersby, A. R., Fookes, C. J. R., Matcham, G. W. J., and McDonald. E. (1980) Biosynthesis of the pigments of life: Formation of the macrocycle, *Nature* (London) **285:**17–21.

Bedbrook, J. R., Smith, S. M., and Ellis, R. J. (1980) Molecular cloning and sequencing of cDNA encoding the precursor to the small subunit of chloroplast ribulose-1,5-bisphosphate carboxylase, *Nature* (London) **287:**692–697.

Bogorad, L. (1975) Evolution of organelles and eukaryotic genomes, *Science* **188:**891–898.

Bogorad, L. (1976) Chlorophyll biosynthesis, in *Chemistry and Biochemistry of Plant Pigments* (T. W. Goodwin, ed.), Academic Press, New York, pp. 64–148.

Buetow, D. E., Wurtz, E. A., and Gallagher, T. (1980) Chloroplast biogenesis during the cell cycle, in *Nuclear-Cytoplasmic Interactions in the Cell Cycle* (G. L. Whitson, ed.), Academic Press, New York, pp. 9–34.

Castelfranco, P. A., and Beale, S. I. (1983) Chlorophyll biosynthesis: Recent advances and areas of current interest, *Annu. Rev. Plant Physiol.* **34:**241–278.

Ellis, R. J. (1981) Chloroplast proteins: Synthesis, transport, and assembly, *Annu. Rev. Plant Physiol.* **32:**111–137.

Gatenby, A. A., Castelton, J. A., and Saul, M. W. (1981) Expression in *E. coli* of maize and wheat chloroplast genes for large subunit ribulose bisphosphate carboxylase. *Nature* (London) **291:** 117–121.

Gibbs, S. P. (1981) The chloroplast endoplasmic reticulum: Structure, function and evolutionary significance. *Internatl. Rev. Cytol.* **72:**49–99.

Gilbert, C. W., and Buetow, D. E. (1981) Gel electrophoresis of chloroplast polypeptides: Comparison of one-dimensional and two-dimensional gel analysis of chloroplast polypeptides from *Euglena gracilis, Plant Physiol.* **67**:623–628.

Granick, S., and Beale, S. I. (1978) Hemes, chlorophylls and related compounds: Biosynthesis and metabolic regulation, *Adv. Biochem. Rel. Areas Mol. Biol.* **46**:33–203.

Kasemir, H. (1983) Action of light on chlorophyll(ide) appearance, *Photochem. Photobiol.* **37**:701–708.

McCarthy, S. A., Mattheis, J. R., and Rebeiz, C. A. (1982) Chloroplast biogenesis: Biosynthesis of protochlorophyll(ide) via acid and fully esterified biosynthetic branches in higher plants. *Biochemistry* **21**:242–247.

Mullet, J. E., and Chua. N.-H. (1983) *In vitro* reconstitution synthesis, uptake and assembly of cytoplasmically synthesized chloroplast proteins, *Meth. Enzymol.* **97**:502–509.

Nelson, N. (1983) Structure and synthesis of chloroplast ATPase, *Meth. Enzymol.* **97**:510–523.

Ohad, I., and Drews, G. (1982) Biogenesis of the photosynthetic apparatus in prokaryotes and eukaryotes, in *Photosynthesis: Development, Carbon Metabolism, and Plant Productivity,* Vol. II (Govindjee, ed.), Academic Press, New York, pp. 89–140.

Pratt, L. H. (1982) Phytochrome: The protein moiety, *Annu. Rev. Plant Physiol.* **33**:557–582.

Rebeiz, C. A., and Lascelles, J. (1982) Biosynthesis of pigments in plants and bacteria, in *Photosynthesis: Energy Conversion by Plants and Bacteria.* Vol. I (Govindjee, ed.), Academic Press, New York, pp. 699–780.

Schiff, J. A. (ed.) (1982) *On the Origins of Chloroplasts,* Elsevier, New York.

Senger, H. (1982) The effect of blue light on plants and microorganisms, *Photochem. Photobiol.* **35**:911–920.

Smith, W. O., Jr. (1981) Characterization of the photoreceptor protein, phytochrome, *Photochem. Photobiol.* **33**:961–964.

Stiekema, W. J., Wimpee, C. F., Silverthorne, J., and Tobin, E. M. (1983) Phytochrome control of the expression of two nuclear genes encoding chloroplast proteins in *Lemna gibba* L. G-3, *Plant Physiol.* **72**:717–724.

Treffry, T. (1978) Biogenesis of the photochemical apparatus, *Internatl. Rev. Cytol.* **52**:159–196.

Wettern, M., Owens, J. C., and Ohad, I. (1983) Role of thylakoid polypeptide phosphorylation and turnover in the assembly and function of photosystem II, *Meth. Enzymol.* **97**:554–567.

Williams, R. S., and Bennett, J. (1983) Synthesis and assembly of thylakoid membrane proteins in isolated pea chloroplasts, *Meth. Enzymol.* **97**:487–502.

Wang, W.-Y. (1978) Genetic control of chlorophyll biosynthesis in *Chlamydomonas reinhardtii, Internatl. Rev. Cytol.* **(Suppl. 8)**:335–354.

Werner-Washburne, M., Cline, K., and Keegstra, K. (1983) Analysis of pea chloroplast inner and outer envelope membrane proteins by two-dimensional gel electrophoresis and their comparison with stromal proteins, *Plant Physiol.* **73**:569–575.

9

Evolutionary Aspects of Chloroplast Development

I. Evolutionary History of Chloroplasts

Life began nearly 3.8 billion years ago, within a remarkably short time after formation of the earth. The rapidity in the appearance of living structures raises the possibility that this event could have happened more than once. Nevertheless, the near universality of the genetic code can be explained most directly by assuming that all current genomes are derived from a single, original genome.

Until recently, biological organisms were divided into two groups, the **eukaryotes** and the **prokaryotes.** However, two fundamentally different kingdoms are now known to exist among the prokaryotes, the **eubacteria** (or true bacteria) and the **archaebacteria** (Fox et al., 1980). The newly recognized kingdom of archaebacteria contains methane-producing, extremely halophilic, and thermoacidophilic organisms. Among their different characteristics, these organisms contain cell walls of different composition than that of eubacteria, different ribosomal structure and components, and an elongation factor that is sensitive to inactivation by diphtheria toxin. They also do not initiate protein synthesis with N-formylmethionyl-tRNA. The latter two characteristics have been considered as distinctly eukaryotic properties (Gray and Doolittle, 1982).

Evolutionary divergence into the eubacteria and archaebacteria apparently occurred very early. The origin of the nuclear-cytoplasmic system in eukaryotic cells is unclear. This line may also have diverged early, but as knowledge of the archaebacteria has grown, it has become apparent that the eukaryotes may have emerged from within this group. But it is also evident that eukaryotic cells have a chimeric phenotype, with one lineage for the nucleus, one for mitochondria and still another for chloroplasts.

The characteristics of the earliest forms of life are not known. The suggestion frequently has been made that the first cells were heterotrophic, in that they required absorption of complex, preformed precursors from their environment for biosynthetic reactions rather than synthesize these compounds *de novo* from simpler chemicals. However, from some indications, photosynthetic organisms also were among the earliest forms of life. In the early anaerobic atmosphere, energy for biosynthetic reactions had to come from either fermentation, a low yield process requiring a large supply of exogenous substrates, or from photosynthesis within the cell. Acquisition of the latter process, of course, would have been a decided advantage. It is interesting that purple photosynthetic bacteria produce bacteriochlorophyll and chromatophores only under anaerobic conditions. With time, as the environmental supply of nutrients became sufficiently high through photosynthesis to make this process universally unnecessary, some organisms may then have lost this capability and became heterotrophic. There are suggestions that the eubacteria are heterotrophic descendents of these early photosynthetic organisms.

About 2 billion years ago, with the advent of photosystem 2 and photosynthetic oxygen evolution, the accumulation of oxygen in the atmosphere permitted the process of respiration. This development of the use of oxygen as a terminal acceptor of electrons in metabolism seems to have been widespread. Because carbohydrates could be oxidized completely to CO_2 and H_2O, heterotrophic energy metabolism became much more efficient. Out of this progression came the progenitors of mitochondria.

Development of the first eukaryotic cell was a momentous event in the history of life. The fossil record does not tell us when this event occurred, but some estimates, in particular one based on the sequence of 5 S ribosomal RNA, place it near 1.8 billion years ago (Hori and Osawa, 1979). If the ancestral organism contained a cell wall similar to that of current eubacterial cells, a change in the characteristics of the wall was necessary, along with the development of eukaryotic cell structure, to enable this cell to capture the progenitors of mitochondria and chloroplasts. However, if instead their source was among the archaebacteria, with their variability in cell wall structure, the cell wall may not have been as much of a barrier.

Whether mitochondria or chloroplasts developed first is debatable. Mitochondria are most closely related to purple nonsulfur photosynthetic bacteria. After being captured by a eukaryotic cell, the ancestral symbiont lost photosynthetic activity but retained respiration. Chloroplasts, however, possess a dual ancestry. Those in the red algae and cryptomonads are derived from the cyanobacteria (formerly called blue-green algae), whereas the chloroplasts in chlorophyte (green) algae and the higher points stem from a noncyanobacterial, oxygenic-photosynthetic prokaryote. A modern example of this putative progenitor of chloroplasts in green plants in *Prochloron*, a prokaryote that contains chlorophyll *a* and *b* and both photosystems. The hypothesis best supported by current evidence suggests that chloroplasts originated through endocytosis by a eukaryotic cell of these photosynthetic prokaryotes. The subsequent evolution of this arrangement provided the organelle we now observe, with its variety of forms.

The evolutionary history of those classes of algae that contain four membranes surrounding the plastid, two of the chloroplast envelope and two of the chloroplast endoplasmic reticulum, is particularly interesting. The interest has been spurred by the recent suggestion that these eukaryotic organisms developed photosynthesis by capture of another *complete* eukaryotic algal cell (Whatley, 1981). The nucleomorph in the chloroplast of these organisms (see p. 41) may be the residual nucleus of the endosymbiont. The evolutionary relationships between these different algal species were illustrated on p. 43.

265

EVOLUTIONARY
ASPECTS OF
CHLOROPLAST
DEVELOPMENT

II. Chloroplasts Resemble Prokaryotic Cells

The chloroplasts in green algae and higher plants have evident prokaryotic characteristics (Gray and Doolittle, 1982; Wallace, 1982). The organelle genome is circular, as in prokaryotic cells, and chloroplast genes are expressed correctly when introduced into bacterial cells. Protein synthesis in chloroplasts has prokaryotic features, including initiation with N-formylmethionyl-tRNA, the size of the ribosomes, and the interchangeability of tRNA species, amino acyl-tRNA synthetases, and elongation factors between chloroplast and prokaryotic systems. The fatty acid synthesizing enzymes in chloroplasts also are similar to those in prokaryotic cells. In all these features, the chloroplast components are, in structure, markedly different from their counterparts in the cytoplasm of the eukaryotic host cell.

The sequences of chloroplast ribosomal and transfer RNA genes show extensive homology with the analogous genes in bacterial cells. The location of the gene for elongation factor Tu_{chl} on the chloroplast genome has been identified by homologous base-pairing between fragments of chloroplast DNA and a DNA fragment from *Escherichia coli* that contains the bacterial gene. Remarkable similarity also exists in the sequences of the gene for the large subunit of ribulose 1,5-bisphosphate carboxylase from prokaryotic cyanobacteria and the eukaryotic green algae and higher plants. Fig. 9.1 shows a comparison of the amino acid sequences in this polypeptide from several of these organisms. In greater than 80% of the 476 positions, from the spectrum of organisms, the resident amino acid is the same. At the nucleotide level, there is 70-75% homology between cyanobacteria and higher plants. Because of the constraint of maintaining function, most base changes are tolerated only at the third position of the code words, which usually results in no amino acid change. Three lysine residues, at positions 175, 202, and 335, are particularly important for the activity of the enzyme. Although the nucleotide sequences in the gene vary slightly, the amino acid sequences around these lysine residues in the polypeptides are identical between cyanobacteria and higher plants. Such a high degree of homology strongly implies an evolutionary heritage.

III. Development of Phylogenetic Trees

Mutations, i.e., base substitutions, deletions or insertions in DNA, occur with a low but definite frequency. Most changes that occur in the sequences coding for a protein result in loss of function. Loss of a required activity leads to heterotrophy, or a dependence upon the environment. If the environment is not supportive, the cell cannot survive. Consequently, natural selection will tend to "conserve" the prototype gene sequence among surviving progeny. In contrast, except for sites that control transcription, the nucleotide sequences in regions that flank the genes are conserved to a much lesser extent.

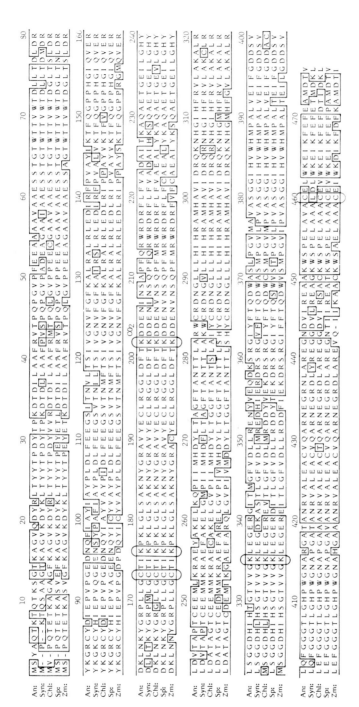

FIG. 9.1. A comparison of the amino acid sequences, deduced from nucleotide sequences in DNA, of the large subunit of ribulose 1,5-bisphosphate carboxylase from a filamentous cyanobacterium (*Anabena* = An), a unicellular cyanobacterium (*Anacystis* = Syn = Syn), unicellular green alga (*Chlamydomonas* = Chl), and two higher plants (*Spinacea* = Sp and *Zea mays* = Zm). The degree of homology between the cyanobacteria (formerly called blue–green algae) and the higher plants is greater than 80%, a remarkably high number. Boxes surround positions at which at least three of the polypeptides have the same amino acid. Circled lysine residues (K) are participants in the catalytic activity of the enzyme and reside within sequences that have been highly conserved. The numbering refers to positions in the *Anabena* sequence, which is one amino acid longer than the chloroplast polypeptide and five longer than the *Synechococcus* polypeptide. Amino acids are designated by the single letter abbreviations: A = alanine, R = arginine, D = aspartic acid, N = asparagine, C = cysteine, E = glutamic acid, Q = glutamine, G = glycine, H = histidine, I = isoleucine, L = leucine, K = lysine, M = methionine, F = phenylalanine, P = proline, S = serine, T = threonine, W = tryptophan, Y = tyrosine, and V = valine. (From Curtis and Haselkorn, 1983. Reprinted with permission.)

A number of base substitutions can be tolerated within the gene for a protein. Since the genetic code is degenerate (i.e., more than one codon exists for most amino acids), a number of changes can take place in DNA without changing the amino acid sequence of the protein. In addition, conservative amino acid changes will not necessarily affect the function of a protein, particularly if the new amino acid is similar in properties to the one it replaces.

The probability that defines the permitted rate of "drift" in nucleotide and amino acid sequences can be used to build evolutionary phylogenies. Many nucleotide changes do not show up as changes in amino acid sequences, as described above. Thus, nucleotide sequences provide the more accurate representation of phylogenetic relationships. The large number of DNA sequences that have been determined over the last few years has given this analysis an enormously valuable data base.

A phylogenetic tree assumes that related genes evolved from a common ancestor, and thus began divergence from the same genome. The corollary of this assumption is that all genes in such a family are subject to the same rate of mutation, and that the greater the differences the further apart the species have evolved. Conversely, species with fewer differences are assumed to be more closely related. These assumptions hold only if parallel changes and reversals are minimal. Since some genes apparently can tolerate more changes than others without losing function, accurate evolutionary trees can only be built with a single gene family. However, regardless of individual rates of drift, trees for individual families should fit on a composite tree. Schwartz and Dayhoff (1978) have used this type of information to build the phylogenetic tree shown in Fig. 9.2.

Calculations based on sequence data for the large subunit of ribulose 1,5-bisphosphate carboxylase, as in Fig. 9.1, have shown that the rate of amino acid substitution in this polypeptide is 0.25–0.5×10^{-9}/year/site (Shinozaki et al., 1983). This value is near that for cytochrome c, one of the most highly conserved proteins. For most proteins the rate of amino acid substitution is 3–4 times greater. Thus, the sequence of the carboxylase large subunit has also been highly conserved.

Evolutionary distances based on the rate of amino acid substitution within the carboxylase sequence suggest that the divergence time between cyanobacteria and the higher plant chloroplast was about 0.5 billion years ago. Since eukaryotic cells already had been around for nearly a billion years, this time may mark the endocytic event that began chloroplast development. Subsequent divergence of monocotyledons, such as maize, and dictotyledons, such as spinach, occurred 0.1–0.2 billion years ago.

Fig. 9.3 shows the interrelationships between chlorophyll-bearing organisms as developed by Olson (1981). The most ancient photosynthetic organism possibly contained only protochlorophyll a. Its descendant became capable of synthesizing chlorophyll a. Since additional modifications are required for synthesis of bacteriochlorophyll (see p. 54 for structures), the ability to perform these reactions came later in offspring that evolved into the photosynthetic bacteria. Aerobic bacteria and mitochondria developed from this family line.

Photosynthetic prokaryotes (cyanobacteria and *Prochloron*) and chloroplasts, which contain chlorophyll a and are capable of oxygen evolution, are possibly derived from a common ancestor. This ancestor developed the ability to oxidize water to oxygen with the advent of photosystem 2 probably a little over 2 billion years ago. From this line has developed the variety of plastid forms. The cyanobacteria have continued their line of descent to the present. Red algae are closely related to cyanobacteria in containing phycobiliproteins and only chlorophyll *a*. Another branch on this tree led to *Prochloron*, a prokaryotic cell that contains chlorophyll *a and b* and has thylakoid membranes similar to those in eukaryotic green algae. Thus, *Prochloron* is considered a likely modern representative of the ancestor of chloroplasts in green algae and higher plants.

IV. Was DNA Transferred from Chloroplast to Nucleus?

Although chloroplasts probably originated through endocytosis of a cyanobacterium by a eukaryotic cell, a puzzling aspect of this situation is that many

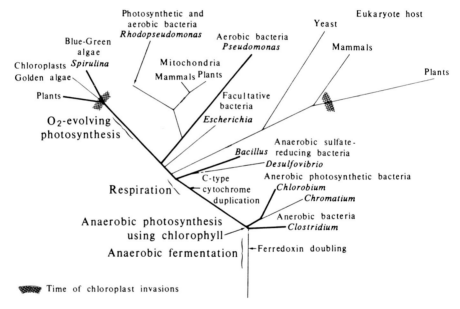

FIG. 9.2. A composite phylogenetic tree based on sequences of ferredoxin, c-type cytochromes and 5 S ribosomal RNA. Evolutionary relationships are shown as branches on the tree. Relative time advances along each branch from the point of departure; the tree does not indicate actual time. Anaerobic, fermentative (heterotrophic) organisms are shown as most ancient, with divergence of photosynthetic bacteria occurring early. The origin of the nuclear lineage of the eukaryotic hosts is shown to be near that of the eubacteria *Bacillus* and *Escherichia*. Mitochondria are shown branching from aerobic bacteria. The chloroplast lineage originates within cyanobacteria (bluegreen algae). Plants are shown on three branches, representing the nuclear, mitochondrial and chloroplast genomes. The point at which chloroplast symbiosis occurred is shaded. As knowledge of evolutionary relationships develop, the phylogenetic tree will become more refined. Thus, details of this tree may need to be modified (see Hori and Osawa, 1979; Fox *et al.*, 1980). In particular, this tree does not show the position of archaebacteria and also suggests that respiratory ability appeared before oxygen production, a point of controversy. (From Schwartz and Dayhoff, 1978. Reprinted with permission).

of the protein components of the chloroplast are now synthesized on cytoplasmic ribosomes and coded by nuclear genes. If the plastid did originate as a prokaryotic endosymbiont, then a transfer of genetic material possibly occurred from the endosymbiont to the nucleus of the recipient cell. How the transfer of such a large number of genes was accomplished is not known. Or did the early eukaryotic cell still carry genes for some photosynthetic functions, the residue of a photosynthetic ancestor? When an endosymbiont gene reiterated the function of one already in the nucleus, which one was discarded? Probably a significant number of genes from the endosymbiont simply were lost by genomic reduction as the result of evolutionary pressures.

Recently, evidence for transfer of DNA between organelles has been found. Homologous segments of DNA are present in both mitochondria and nuclear DNA in fungi. In plants, several common sequences have been found in both chloroplasts and mitochondria. In one case, a homologous copy of a segment of maize chloroplast DNA, which lies within the inverted repeat region and includes a gene for 16 S ribosomal RNA, also occurs within mitochondrial DNA (Stern and Lonsdale, 1982). And, of particular interest, Timmis and Scott (1983) observed that nuclear DNA in spinach contains sequences that are homologous to a significant portion of the chloroplast genome. As the techniques become more refined, and more searching is done, additional common sequences are being found among the organelles. However, the nuclear homologues of mitochondrial or chloroplast DNA represent usually only fragments of genes and often are rearranged from their functional structure. Perhaps

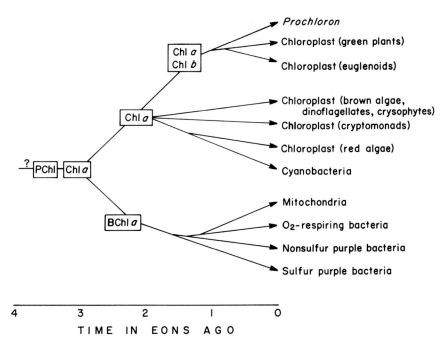

FIG. 9.3. Evolutionary relationships between photosynthetic and respiratory organelles, PChl = protochlorophyll, Chl = chlorophyll, BChl = bacteriochlorophyll. See text for further explanations. (Adapted from Olson, 1981.)

numerous DNA segments have been duplicated and a copy moved to other sites in the cell. DNA exchanges within plant cells could have occurred by reactions similar to the transposition events in bacterial cells. We can only guess at how the shuffling of DNA was performed in the evolving cell. However, considering the remarkable similarities in the chloroplast DNA of most green plant cells, this genetic restructuring must have occurred early.

To achieve the outcome we now observe, with the remaining genetic information gathered into a single, small, circular molecule of chloroplast DNA, any transfer of DNA to the nucleus must have been remarkably specific. But along with the transfer another extraordinary modification may have occurred. As each of those genes that code for a chloroplast protein was transferred to the nucleus, it had to include, at a precise position, a segment of DNA that carried information for returning the protein to the chloroplast. It seems unlikely that this additional segment of DNA, which codes for the transit sequence (see p. 235), would have been part of the original gene, since it is not retained in the functional protein. This segment may have contained an -A-T-G- sequence in frame to provide the initiating codon or was inserted immediately after the initiating codon in the original gene without destroying its reading frame. The result was introduction of an extension at the N-terminus of the protein, which is removed after synthesis as the protein enters the chloroplast. Whether a type of specific vector was involved in an organized transfer of DNA, which included this modification, is not known. The variation in length and sequence of the transit sequence on precursors of chloroplast proteins does not suggest an obvious origin.

There is no doubt that, although the chloroplast contains a number of genes that are expressed for the organelle's function, most of the chloroplast proteins are coded by nuclear DNA. What is the purpose of this division? There surely are advantages to this arrangement that at present we are unable to understand. The carboxylase poses a particularly interesting situation. This enzyme is by far the major protein made in photosynthetic organisms. In the prokaryotic progenitor, genes for both the large and small subunits presumeably were within the same genome. To achieve a true symbionic relationship, synthesis of this enzyme had to be brought under subjection of the overall economy of the more complex eukaryotic cell. Transfer of the small subunit gene to the nucleus apparently was the mechanism by which this necessary control was accomplished. But the situation with membrane proteins may include an aspect in addition to control of synthesis. Perhaps to achieve a functional orientation within the asymmetric membrane a protein must approach the structure from a specific side. Although uncertainty persists, the tacit assumption is made, nevertheless, that there is purpose to this division of synthetic labor.

V. Development of Photosynthetic Systems

The earliest organisms that converted light energy into chemical energy apparently contained cyclic systems similar to those in photosynthetic bacteria (see Fig. 4.19). Photosystem 1 in chloroplasts, which also drives a cyclic electron flow, is a descendant of these early systems. The major forms of chemical energy produced by such systems are NADPH (or NADH), ferredoxin and ATP. The availability of these compounds to drive biosynthetic reactions permitted a faster rate of growth of the host organism and thus greatly enhanced survival. (The uses of these compounds in biosynthetic reactions were described in Chapter 5.)

To maintain the activity of the cyclic photosystems, an electron donor was needed to "prime the pump." The oxidation–reduction potential of the electron donor had to be sufficiently low to reduce the reaction centers of these systems. This role in the early, highly reduced, oxygen-deficient biosphere possibly was provided by reduced inorganic sulfur compounds, a feature retained by the photosynthetic bacteria (see Fig. 4.19). Later, about 2 billion years ago, the additional activity of photosystem 2 developed, which through intermediate carriers provided an endogenous supply of electrons to photosystem 1 (see Fig. 4.7). As a result of excitation by light, and subsequent reduction of the primary donor, the reaction centers of photosystem 2 also became strong enough oxidants to abstract electrons from water. This remarkable development transferred the organism's dependence from strong reductants, present at low concentrations in its environment, to an electron donor present in limitless quantity, water.

The consequence of water oxidation, however, was the production of molecular oxygen. The appearance of oxygen was a mixed blessing. The positive aspect was that oxygen, as its concentration increased in the atmosphere, became a convenient electron acceptor for respiration. The efficiency of heterotrophic metabolism thus greatly improved, since now organic compounds could be oxidized completely to CO_2 and H_2O. But conversely, the reactivity of oxygen, and its propensity to enter radical-mediated reactions and to form excited states, required the development of protective mechanisms against its toxicity. In addition, the organism's supply of nitrogen and sulfur no longer existed in readily assimilated forms, such as ammonia and sulfide ions, but as the oxides nitrate and sulfate. Incorporation of these elements into cellular material became energetically expensive processes.

The enzymatic "dark" systems for synthesizing carbohydrate from CO_2 are probably as ancient as the photosystems. Ribulose 1,5-bisphosphate carboxylases from all photosynthetic organisms show significant homology, as described in Section II (see Fig. 9.1), although the enzyme in photosynthetic bacteria structurally is less complex than that from chloroplasts. The C_3 cycle originated first, more than 2.5 billion years ago, and provided the opportunity to use the high-energy compounds generated by the photosystems for the synthesis of glucose from CO_2. Whereas the high energy compounds (e.g., NADPH and ATP) produced by the photosystems are present in limited amounts and used within minutes by the cell itself, the C_3 cycle, by permitting carbohydrate synthesis, allowed trapping of the energy in a long-term storage form. These storage forms, as starch or another polymer of glucose, then were available for metabolic processes in the dark. Consequently, not only was the cell gaining independence over environmental reducing agents (except for the abundant water) and complex organic compounds (which it could now make from CO_2), it also was no longer continuously dependent upon light. With only a supply of H_2O, CO_2, the oxides of nitrogen, sulfur and phosphorus, and trace minerals (such as Fe^{3+}, Ca^{2+}, Mg^{2+}, Cu^{2+}, Mn^{2+}, Mo^{6+}) which were available in soil and at low but sufficient concentrations in the waters of the earth, the cell was highly developed to support a relatively independent lifestyle.

The stage of development described in the last paragraph apparently was reached during the evolution of cyanobacteria and *Prochloron*. After endocytosis of such an organism, the entire photosynthetic apparatus entered the eukaryotic cell and was retained by the developing chloroplast. The advantages that the photosynthetic mechanisms previously had given to the evolving prokaryotic cell were now presented to eukaryotic organisms. An additional characteristic of eukaryotic organisms, moreover, was the greater content and organization of genetic information in the nucleus, which provided the capacity to develop multicellular structures and to achieve specialized functions through differentiation of certain segments of the population. Out of these fundamental processes grew the diversity of terrestrial plants.

The currently ubiquitous flowering plants (Angiosperms), with their different cell types within leaf tissues, appeared only about 0.1–0.2 billion years ago. All contain the C_3 cycle. An outgrowth of the capability for differentiation was the development of metabolic appendages to the C_3 cycle. On the evolutionary time scale, the C_4 pathway and the crassulacean acid metabolism (CAM) (see p. 133) originated relatively recently. Although these additional pathways occur in only a minority of species, they are broadly distributed among the plant kingdom, which is evidence that each originated frequently and independently during evolution of plants. The presence of these pathways enhance survival of the host plant under stressful conditions by providing a mechanism for concentrating CO_2 at the site of operation of the C_3 cycle.

Not only did terrestrial plants proliferate abundantly, they also, by storing photosynthetic carbohydrate, provided the developing animal kingdom with a readily available supply of food. Without a doubt, the photosynthetic reactions within the chloroplast have played an indispensible role in the development, maintenance and health of the animal kingdom. Without plants, the animal kingdom would shortly eat itself into oblivion. And to think that after all these years, we at this point in time can understand the function of the chloroplast, and reflect upon its role in our lives, is astonishing.

VI. Summary

Traces of evolutionary heritage remain within chloroplast structure, at both the molecular and morphological levels. The German botanist Schimper first proposed an endosymbionic relationship for chloroplasts over a hundred years ago, an idea more fully developed by Mereshkovsky in Russia in 1905. Current research in DNA sequences has finally provided a means to study the lineage of these processes in a serious and objective manner. The potential of this work is enormous.

The evolution of chloroplast function has been established in broad outline form. It is recognized that without the development of the chloroplast, with its ability to capture light energy and transform it into a benign storage form, all other, nonphotosynthetic forms of life would be unable to survive. The animal kingdom is entirely dependent on the plant kingdom, a fact that must be considered in our treatment of the world's resources.

Literature Cited

Curtis, S. E., and Hasselkorn, R. (1983) Isolation and sequence of the gene for the large subunit of ribulose-1, 5-bisphosphate carboxylase from the cyanobacterium *Anabaena* 7120, *Proc. Natl. Acad. Sci. USA* **80:**1835–1839.

Fox, G. E., Stackebrandt, E., Hespell, R. B., Gibson, J., Maniloff, J., Dyer, T. A., Wolfe, R. S., Balch, W. E., Tanner, R. S., Magrum, L. J., Zablen, L. B., Blakemore, R., Gupta, R., Bonen, L., Lewis, B. J., Stahl, D. A., Luehrsen, K. R., Chen, K. N., and Woese, C. R. (1980) The phylogeny of prokaryotes, *Science* **209:**457–463.

Gray, M. W., and Doolittle, W. F. (1982) Has the endosymbiont hypothesis been proven? *Microbiol. Rev.* **46:**1–42.

Hori, H., and Osawa, S. (1979) Evolutionary change in 5S RNA secondary structure and a phylogenic tree of 54 5S species, *Proc. Natl. Acad. Sci. USA* **76:**381–385.

Mereschkowsky, C. (1910) Theorie der zwei Plasmaarten als Grundlage der Symbiogenesis, einer neuen Lehre von der Entstehung der Organismen, *Biol. Centralbl.* **30:**353–367. (German reprint of original paper published 1905 in Russian.)

Olson, J. M. (1981) Evolution of photosynthetic and respiratory prokaryotes and organelles, *Ann. N. Y. Acad. Sci.* **361:**8–17.

Schimper, A. F. W. (1883) Uber die Entwickelung der Chlorophyllkorner und Farbkorper, *Botanische Zeitung* **41:**105–114.

Schwartz, R. M., and Dayhoff, M. O. (1978) Origins of prokaryotes, eukaryotes, mitochondria, and chloroplasts, *Science* **199:** 395–403.

Shinozaki, K., Yamada, C., Takahata, N., and Sugiura, M. (1983) Molecular cloning and sequence analysis of the cyanobacterial gene for the large subunit of ribulose-1, 5-bisphosphate carboxylase/oxygenase, *Proc. Natl. Acad. Sci. USA* **80:**4050–4054.

Stern, D. B., and Lonsdale, D. M. (1982) Mitochondrial and chloroplast genomes of maize have a 12-kilobase DNA sequence in common, *Nature* (London) **299:**698–702.

Timmis, J. N., and Scott, N. S. (1983) Sequence homology between spinach nuclear and chloroplast genomes, *Nature* (London) **305:**65–67.

Wallace, D. C. (1982) Structure and evolution of organelle genomes, *Microbiol. Rev.* **46:**208–240.

Whatley, J. M. (1981) Chloroplast evolution—Ancient and modern, *Ann. N. Y. Acad. Sci.* **361:**154–164.

Additional Reading

Cavalier-Smith, T. (1981) The origin and early evolution of the eukaryotic cell, *Symp. Soc. Gen. Microbiol.* **32:**33–84.

Doolittle, W. F. (1980) Revolutionary concepts in evolutionary cell biology, *Trends Biochem. Sci.* **5:**146–149.

Fredrick, J. F. (ed.) Origins and evolution of eukaryotic intracellular organelles, *Ann. N. Y. Acad. Sci.* **361.**

Jukes, T. H. (1980) Silent nucleotide substitutions and the molecular evolutionary clock, *Science* **210:**973–978.

Moore, P. D. (1982) Evolution of photosynthetic pathways in flowering plants, *Nature* (London) **295:**647–648.

Schiff, J. A. (ed.) (1982) *On the Origins of Chloroplasts*, Elsevier, Amsterdam.

Stackebrandt, E., and Woese, C. R. (1981) The evolution of prokaryotes, *Symp. Soc. Gen. Microbiol.* **32:**1–31.

Van Valen, L. M. (1982) Phylogenies in molecular evolution: *Prochloron, Nature* (London) **298:**493–494.

Index